Springer Polar Sciences

Springer Polar Sciences

Springer Polar Sciences is an interdisciplinary book series that is dedicated to research on the Arctic and sub-Arctic regions and Antarctic. The series aims to present a broad platform that will include both the sciences and humanities and to facilitate exchange of knowledge between the various polar science communities.

Topics and perspectives will be broad and will include but not be limited to climate change impacts, environmental change, polar ecology, governance, health, economics, indigenous populations, tourism and resource extraction activities.

Books published in the series will have ready appeal to scientists, students and policy makers.

More information about this series at http://www.springer.com/series/15180

Gail Fondahl • Gary N. Wilson

Editors

Northern Sustainabilities: Understanding and Addressing Change in the Circumpolar World

Springer

Editors
Gail Fondahl
University of Northern British Columbia
Prince George, BC, Canada

Gary N. Wilson
University of Northern British Columbia
Prince George, BC, Canada

ISSN 2510-0475
Springer Polar Sciences
ISBN 978-3-319-46148-9
DOI 10.1007/978-3-319-46150-2

ISSN 2510-0483 (electronic)

ISBN 978-3-319-46150-2 (eBook)

Library of Congress Control Number: 2016961943

Printed on acid-free paper

This Springer imprint is published by Springer Nature
The registered company is Springer International Publishing AG
The registered company address is: Gewerbestrasse 11, 6330 Cham, Switzerland

Foreword

The present book – *Northern Sustainabilities: Understanding and Addressing Change in the Circumpolar World* – highlights some of the most important challenges of the present and future Arctic. The 23 chapters illustrate the complexity and diversity of Arctic social sciences. They are also very representative of the dynamic ICASS VIII Conference in Prince George 2014. This event is, and has been since the early 1990s, one of the most prominent features of the International Arctic Social Sciences Association (IASSA). More than 450 researchers gathered for meetings, presentations, and discussions. The general theme of the conference was sustainability, something an increasing awareness of present and future challenges tells us is of uttermost importance. This book is an important contribution to knowledge production and offers an improved understanding of this very complex thing called sustainability.

I do not think there is any keyword that is more prominent and frequently mentioned in the whole discussion about the Arctic than sustainability. It has turned into a guiding star and a pronounced ambition of everyone concerned with the Arctic, whoever and wherever they are. The Kiruna Declaration of the ministerial Arctic Council meeting recognizes that the environment needs to be protected as a basis for sustainable development and emphasizes economy and business innovation as crucial for the same purpose. The Arctic Council even has a working group for sustainable development that constitutes a dynamic arena for various research projects that all strive to contribute to our understanding of what sustainability is and how we can give it the best conditions to develop. The US chairmanship of the Arctic Council has pronounced efforts to take action for a sustainable development in the Arctic. Moreover, the integrated European Union policy for the Arctic that was presented in late April 2016 has in its very first sentence the statement that a safe, stable, sustainable, and prosperous Arctic is important not just for the region itself but for the European Union (EU) and for the world.

During the Arctic Frontiers conference in Tromsø in 2015, the report *Growth from the North: How Can Norway, Sweden and Finland Achieve Sustainable Growth in the Scandinavian Arctic?* was launched. It locates main drivers of growth and suggests instruments like a shared regulatory framework, human capacity, infrastructure, and unity. This underlines the fact that sustainability is a core concept for the planning processes in the Arctic. Nevertheless, we have to conclude that there are so many different understandings and uses of sustainability. Researchers, politicians, and locals often do not mean the same thing when they talk about sustainability, and its implementation varies greatly between different levels of governance. It is, however, indisputable that sustainability will be an important field for the forthcoming arctic research. The International Conference on Arctic Research Planning III (ICARP III) has as one of three goals to improve the understanding of vulnerability and resilience of arctic environments and societies and to support sustainable development.

There is a strong correlation between knowledge and improved capacity. The scope of sustainability is extensive and its meaning disputed and sometimes blurred. I am happy to see the broad spectrum of fields into which the authors of this book lead us. Perspectives on environment, economy, politics, resources, stakeholders, indigenous peoples, institutions, gender, health, urbanization, education, and human capital are opened. We can see that there is a lot of important research that already carries out the tasks addressed by ICARP III. Some of them are present in this book. The changes that are about to come involve fundamentals, where infrastructures and legislations are included together with production of tolerance, knowledge, and trust.

The International Arctic Science Committee (IASC) has prioritized crosscutting initiatives among its working groups, and it has turned out that sustainability is a perfect match to these efforts. When natural science, technology, medicine, and social sciences and humanities meet, the shared ambition to produce research results that are relevant and valuable for a sustainable development is a starting point for the discussions.

The Arctic is in motion, and perpetual change is what we can count on for the future too. The Arctic is also very hot, in terms of attention and in terms of degrees. Many share an ambition to make the Arctic a more secure, vibrant, and resilient place, and they see a distinct relation to global warming and climate change in this process leading out to the global community. Research has a very important role to play in this development, and the chapters in this book are excellent contributions to an increased knowledge base and give important advice on how to understand sustainability.

ARCUM Arctic Research Centre, Umeå University   Peter Sköld
Umeå, Sweden

Acknowledgments

Assembling an edited volume on such a diverse and complex topic certainly has its challenges and also its rewards. We would like to acknowledge Margaret Deignan at Springer Press, who suggested the idea of a volume and who patiently encouraged us to pursue the project. Molly Fredeen provided invaluable help with reviewing the manuscripts, editing those chapters written by our colleagues whose first language is not English, and providing consistent formatting throughout. Her editorial work was largely funded by a publication grant from the University of Northern British Columbia's (UNBC) Office of Research.

We owe a huge debt of gratitude to a number of individuals and organizations for their hard work in supporting the Eighth International Congress of Arctic Social Sciences (ICASS VIII), the conference at which these chapters were originally presented: the International Arctic Social Sciences Association (IASSA), Cher Mazo (IASSA Secretary, 2011–2014), the organizing committee of ICASS VIII and the conference sponsors. The generosity of the following sponsors enabled several of the contributors to this volume to attend ICASS VIII: the US National Science Foundation (NSF grants PLR #1360365 and #1338850), UNBC, the International Arctic Science Committee, the Canadian Polar Commission, the Social Sciences and Humanities Research Council of Canada, the Association of Canadian Universities for Northern Studies, the Korea Maritime Institute, the Oak Foundation, the Arctic Institute of North America, and the Arctic Research Consortium of the United States. Both the conference and this volume were greatly enriched by their ability to participate.

Each chapter was separately peer-reviewed; once complete, the entire manuscript was also peer-reviewed. We extend our sincere appreciation to all those individuals who performed this important but time-consuming and often under-acknowledged part of the scholarly process.

As scholars, both of us have been profoundly influenced by our longtime association with the University of Northern British Columbia. UNBC's mandate to serve northern communities is at the heart of our research agendas, and we are indebted to our colleagues and to the university for the support that they provided throughout this whole process.

Last, but certainly not least, we thank our respective spouses and families for all the support that they provided to us over the life of this project. We are extremely grateful to have such loving and understanding families.

Contents

Contributors

Ingrid Bay-Larsen Nordland Research Institute, Bodø, Nordland, Norway

Trevor Bell Department of Geography, Memorial University of Newfoundland, St. John's, NL, Canada

Matthew Berman Institute of Social and Economic Research, University of Alaska Anchorage, Anchorage, AK, USA

Lill Rastad Bjørst Centre for Innovation and Research in Culture and Living in the Arctic (CIRCLA), Aalborg University, Aalborg, Denmark

Lauren A. Brooks-Cleator University of Ottawa, Ottawa, ON, Canada

Shauna BurnSilver School of Human Evolution and Social Change, Arizona State University, Phoenix, AZ, USA

Susan A. Crate Department of Environmental Science and Policy, George Mason University, Fairfax, VA, USA

Ashlee Cunsolo Labrador Institute of Memorial University, Happy Valley-Goose Bay, NL, Canada

Herminia Din Department of Art, University of Alaska Anchorage, Anchorage, AK, USA

Jenanne Ferguson Department of Anthropology, University of Nevada-Reno, Reno, NV, USA

Gail Fondahl Geography Program, University of Northern British Columbia, Prince George, BC, Canada

Ulrik Pram Gad Department of Political Science, University of Copenhagen, Copenhagen, Denmark

Audrey R. Giles University of Ottawa, Ottawa, ON, Canada

Gunhild Hoogensen Gjørv Department of Sociology, Political Science, and Community Planning, UiT The Arctic University of Norway, Tromsø, Norway

Catherine T.R. Glass University of Ottawa, Ottawa, ON, Canada

Christina Goldhar Nunatsiavut Secretariat, Nunatsiavut Government, Nunatsiavut, NL, Canada

Heather Sauyaq Jean Gordon Indigenous Studies, University of Alaska Fairbanks, Fairbanks, AK, USA

Caroline Hervé Interuniversity Centre for Aboriginal Studies and Research, Québec, QC, Canada

Grete K. Hovelsrud Nord University and Nordland Research Institute, Bodø, Nordland, Norway

Uffe Jakobsen Department of Political Science, University of Copenhagen, Copenhagen, Denmark

Sámal T.F. Johansen Søvn Landsins (The Faroese National Archive), Hoyvík, Faroe Islands

Anna Karlsdottir Department of Geography and Tourism Studies, University of Iceland, Reykjavik, Iceland

Gary Kofinas Institute of Arctic Biology and School of Natural Resources and Extension, University of Alaska Fairbanks, Fairbanks, AK, USA

Michał Łuszczuk Maria Curie-Skłodowska University, Lublin, Poland

Kjell Nilsson NORDREGIO – Nordic Centre for Spatial Development, Stockholm, Sweden

Umut Riza Ozkan University of Montreal, Montreal, QC, Canada

Andrey N. Petrov University of Northern Iowa, Cedar Falls, IA, USA

Stacy Rasmus Center for Alaska Native Health Research, University of Alaska Fairbanks, Fairbanks, AK, USA

Rasmus Ole Rasmussen NORDREGIO – Nordic Centre for Spatial Development, Stockholm, Sweden

Rudolf Riedlsperger Department of Geography, Memorial University of Newfoundland, St. John's, NL, Canada

Toril Ringholm Lillehammer University College, Lillehammer, Norway

Rémy Rouillard School of Psychoeducation, Université de Montréal, Montréal, QC, Canada

Stephan Schott Carleton University, Ottawa, ON, Canada

Tom Sheldon Department of Lands and Natural Resources, Environment, Nunatsiavut Government, Nunatsiavut, NL, Canada

Inez Shiwak 'My Word': Storytelling & Digital Media Lab, Rigolet Inuit Community Government, Rigolet, NL, Canada

Lena Sidorova Department of Culturology, North Eastern Federal University, Yakutsk, Sakha Republic (Yakutia), Russian Federation

Jeppe Strandsbjerg Department of Business and Politics, Copenhagen Business School, Frederiksberg, Denmark

Jukka Terräs NORDREGIO – Nordic Centre for Spatial Development, Stockholm, Sweden

Olga Ulturgasheva Department of Social Anthropology, University of Manchester, Manchester, UK

Laur Vallikivi Department of Ethnology, University of Tartu, Tartu, Estonia

Liliia Vinokurova Institute for Humanities Research and Indigenous Studies of the North, Siberian Branch, Russian Academy of Sciences, Yakutsk, Sakha Republic (Yakutia), Russian Federation

Ryan Weber NORDREGIO – Nordic Centre for Spatial Development, Stockholm, Sweden

Emma Wilson Scott Polar Research Institute, Cambridge, UK

Gary N. Wilson Department of Political Science, University of Northern British Columbia, Prince George, BC, Canada

Michele Wood Nunatsiavut Department of Health and Social Development, Happy Valley-Goose Bay, NL, Canada

Lyudmila Zalkind Department of Urban Socio-Economic Development, Kola Science Centre, Apatity, Murmansk Oblast, Russian Federation

Chapter 1
Exploring Sustainabilities in the Circumpolar North

Gail Fondahl and Gary N. Wilson

Abstract 'Sustainability' is a major concern in the North, given the rapid environmental and social (including political, economic and cultural) changes the region is undergoing. Yet the definition of what sustainability is, and how it might be achieved, are still much debated. Where one is located, both geographically and socially, influences how one perceives 'sustainability'. This volume addresses various facets of northern sustainability in a variety of places across the Circumpolar North and from a variety of perspectives, thus contributing to our understandings of the multiple dimensions of sustainability in the arctic and sub-arctic regions of the world.

Keywords Sustainability • Circumpolar • Arctic • Social sciences • North

1.1 Introduction

At a "Business Dialogue" meeting convened by the Arctic Council's Sustainable Development Working group (SDWG) in the fall of 2012, the terms "sustainability" and "sustainable development" came up frequently. Participants using the term included representatives from the oil and gas, shipping, mining, fisheries, and tourism sectors, as well as from environmental non-governmental organizations. At one point Helena Omma, then Sami Council representative to and co-chair of the SDWG, asked: "What do we mean when we talk about sustainability?" She suggested that the term appeared to hold rather different meanings for the various

G. Fondahl (✉)
Geography Program, University of Northern British Columbia,
3333 University Way, Prince George, BC, Canada, V2N 4Z9
e-mail: gail.fondahl@unbc.ca

G.N. Wilson
Department of Political Science, University of Northern British Columbia,
3333 University Way, Prince George, BC, Canada, V2N 4Z9
e-mail: gary.wilson@unbc.ca

© Springer International Publishing Switzerland 2017
G. Fondahl, G.N. Wilson (eds.), *Northern Sustainabilities: Understanding and Addressing Change in the Circumpolar World*, Springer Polar Sciences,
DOI 10.1007/978-3-319-46150-2_1

participants. "Sustainability" is indeed an elusive term, deployed frequently and with appreciably different connotations. As Ulrik Pram Gad, Uffe Jakobsen and Jeppe Strandsberg write in the first paper in Part I of this volume, "there is little agreement on what 'sustainable' means. For different actors (governments, indigenous peoples, NGOs, etc.) the concept implies different sets of opportunities and precautions."

Yet if "sustainability" has become a buzzword deployed by various actors for various ends, it has also become a major concern in the North over the past several decades (Petrov et al. 2016). The North – the arctic and subarctic regions of the Earth – is experiencing change, both environmental and social, at a rate that is exceeding that of more southerly latitudes and accelerating at an unprecedented pace. Concomitantly, the North is ever more intimately interlinked with more southerly areas. Our understandings of those linkages and their significance to a whole range of socio-ecological changes are embryonic and constantly evolving. Moreover, changes in ice and snow regimes, the depopulation of small villages, urbanization, immigration, language loss, and cultural transformations have both local and global dimensions. To what extent do these and other transformations encourage or erode sustainability? Indeed, what does sustainability in the North look like? What are its drivers and barriers? How might it be most effectively pursued and achieved?

Where one is located – both geographically and socially – matters in terms of some of the most pressing issues of sustainability in the North today, be they social, economic, or environmental. This book tackles various facets of northern sustainability in a variety of places (Fig. 1.1) and from a variety of perspectives. The authors include social scientists – anthropologists, economists, geographers, and political scientists, and humanities scholars – artists, historians and linguists. Many reside in the North. Attending to place, the authors explore local conceptualizations and understandings of sustainability, local challenges to sustainability, and local pathways to pursuing more sustainable development. Thus, rather than attempting to answer the question 'what is sustainability in the North', this book contributes to our understandings of the multiple dimensions of sustainability in the arctic and sub-arctic regions of our world.

1.2   Genesis of This Volume

It was in part Helena Omma's comment that inspired our choice of "Northern Sustainabilities" for the theme of the Eighth International Congress of Arctic Social Sciences (ICASS VIII), from which the contributions to this volume originate. ICASS is the triennial congress of the International Arctic Social Sciences Association (IASSA), the premier organization dedicated to social sciences and humanities research in the Arctic and Subarctic. The first Congress was held in 1992 in Montreal, Canada; since then it has been convened in Rovaniemi, Copenhagen, Quebec City, Fairbanks, Nuuk and Akureryi. In 2014, ICASS VIII was held on the main campus of the University of Northern British Columbia, in Prince George, Canada. Some 470 attendees presented 411 papers as well as 38

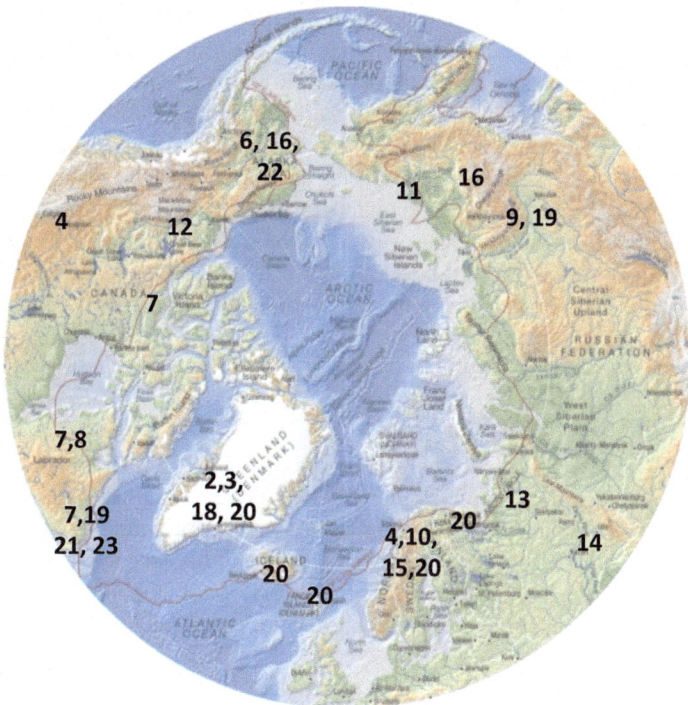

Fig. 1.1 Approximate location of places examined in the chapters of this volume (Basemap produced by Philippe Rekacewicz, UNEP/GRID-Arendal. Used with permission)

posters, in 110 different sessions. While the Congress entertained a diversity of sessions and papers that reached far beyond the conference theme, many scholars offered papers on sustainability in the North. Collectively, these papers spanned a wide variety of topics, places, and approaches.

From these rich and diverse conference presentations, we solicited but a small sample to publish in this volume. We intentionally asked the authors to produce manuscripts that were somewhat shorter than the average journal article or chapter, in order to include a greater number and variety of papers. We strove to include papers that would offer wide topical and geographic representation, from a diversity of authors in terms of nationality, place of employment, disciplinary background and career stage. This volume includes articles on topics ranging from sustainable security to sustainable art, from gendered responses to climate change to barriers to sustainable health promotion, and from sustainability and corporate social responsibility to sustainably conducting research in the North. The chapters touch on most countries and regions within the Circumpolar North (Fig. 1.1). In addition to representing many disciplines, as noted above, the authors also come from across the Circumpolar North, as well as from non-arctic states. Among them are academics (ranging from PhD students and post-doctoral fellows to Full Professors), representatives of local and regional governments, and community researchers. Several of the authors are indigenous northerners. As is common with edited volumes, a few of the authors who we had asked to participate found that they could not complete their contributions. This resulted in gaps in coverage, both in terms of authors' nationalities and the geographic scope of the book. That said, we feel that this volume is broadly representative of the peoples and places of the circumpolar North.

In exploring the experiences of a diverse set of regions and communities around the circumpolar North, this volume underscores that northern regions and communities have their own unique histories and problems, a reality that has set them on different trajectories in terms of development. Indeed, as the title suggests, there are numerous sustainabilities – and numerous norths.

1.3 Chapter Overview

The chapters that comprise this volume address issues of cultural, economic, environmental, political and social sustainability in the Circumpolar North. Indeed, most examine more than one dimension of sustainability, underscoring the interrelated and interdependent nature of different types of sustainability. The authors do so by exploring the experiences of a diverse set of communities and regions around the Circumpolar North.

Given the overlapping nature of this topic, we have organized the chapters of this collection into three sections:

- Conceptualizing and Measuring Northern Sustainability;
- Challenges to Northern Sustainability; and
- Advancing Northern Sustainability.

Each section emphasizes a different dimension of sustainability and, in doing so, makes an important contribution to our conceptual, analytical and applied understandings of this term. Several of the chapters could have been assigned to more than one of the sections, a fact that underscores the interconnectedness of this volume's contributions.

1.3.1 Conceptualizing and Measuring Northern Sustainability

A common theme in this first set of chapters is the question of how sustainability is understood, conceptualized, defined, packaged and repackaged, and measured. The section starts with a discussion of how the Arctic, as a discrete world region, shapes discourses of sustainability, as well as how the concept shifts in meaning both among the regions and communities that constitute the Arctic and across scales. Ulrik Pram Gad, Uffe Jakobsen and Jeppe Strandsbjerg describe understandings of 'sustainability'; then, to illustrate their assertions, they examine the shifting politics of sustainability in Greenland as it moves towards increasing autonomy and sovereignty. In doing so, they propose a research agenda for examining the concept of sustainability in the Arctic.

Lill Rastad Bjørst continues the discussion by examining the specific question of how the concept of sustainability threads through the political debate regarding the future of uranium mining in Greenland. Assertions regarding the necessity of mining uranium for Greenland's economic sustainability collide with fears regarding the threats that mining poses to its environmental sustainability. The particular re-conceptualization of sustainability and associated policy shift seen in Greenland is representative of developments played out around the North that require trade-offs among cultural, economic, environmental and social sustainability. Bjørst also underscores the agency of Greenlanders in co-producing, together with industry, a receptive environment for mining uranium, in part by reframing sustainability discourses to emphasize economic sustainability as the most critical policy goal.

Gunhild Hoogensen Gjørv continues the discussion of resource development by exploring the tensions between conceptions of economic, environmental and energy security in northern Canada and Norway. Focusing specifically on development in the Alberta oil sands and offshore oil and gas projects off the coast of Norway, Hoogensen Gjørv shows how hydrocarbon development in these sub-arctic and arctic regions has been pitched as an "ethical" alternative to other less-environmentally and politically acceptable forms of resource development.

While critical discourses of security have elevated 'soft' securities to the limelight in terms of sustainable development (e.g., Huebert et al. 2012; Hoogensen et al. 2014), approaching traditional security studies through the lens of sustainability has yet to receive much attention by scholars of the Circumpolar North. Michal Łuszczuk engages with such an approach, exploring how it might inform our understandings of the arctic security environment and military cooperation among arctic states.

How do we actually measure sustainability? In a rapidly changing North, Matthew Berman, Gary Kofinas and Shauna BurnSilver argue that achieving sustainability requires adaptive capacity and, in some cases, transformative capacity. These capacities can be seen as forms of capital that communities can draw upon to tackle vulnerability to external drivers. Can their role in reducing vulnerability be measured? Using a case study of rising fuel prices in two Alaskan villages, the authors review the challenges to performing such measures in the context of the rural North.

Fate control – the ability to determine one's future – and its relationship to local sustainable development have recently received increased attention (Dahl et al. 2010; Ozkan and Schott 2013; Kimmel 2014). Umut Riza Ozkan and Stephan Schott review political and fiscal indicators as measures of self-determination for the three Inuit dominated areas of the Canadian North – Nunavut, Nunavik and Nunatsiavut – noting that financial dependence limits the ability of these regional governments to truly self-govern. Caroline Hervé looks specifically at the political organizations of Nunavik, and how Inuit in that region conceptualize and develop their relations with these organizations as a means of interacting productively and equitably with the Quebec provincial and Canadian federal governments. Like

Bjørst, she underscores the agency of local residents – in this case, describing how the Nunavimmiut have established a productive level of political autonomy within the non-Indigenous political structures in which Nunavik is embedded.

1.3.2 Challenges to Sustainability in the North

Globalization and climate change are powerful external forces that present significant challenges to the sustainable development of the world's northern regions (AHDR-II 2014). The challenges associated with globalization include increasing migrations of peoples, the spread of technologies and ideas to and from the North, demands from the South for a multitude of arctic resources, cultural erosion, competing political processes such as devolution and centralization, and many other social forces. Climate change compounds many of these challenges by modifying access to resources, and threatening the viability of traditional activities and the stability of infrastructure. The chapters in the second section of this volume illustrate specific challenges to achieving sustainability in the face of both climate change and globalization.

Climate change is the focus of two chapters, by Liliia Vinokurova and by Ingrid Bay-Larsen and Grete Hovelsrud. Vinokurova considers the gendered dimensions of climate change in northern Russia by examining evaluations of, and responses to, flooding in rural parts of Sakha Republic (Yakutia). Men and women are anxious about different outcomes of climate change: men cite the loss of traditional occupations as the greatest concern, while women focus mainly on issues of health and safety. Differences according to age are also noted. Such differences will likely affect the adaptive capacities of particular communities and groups of people.

Bay-Larsen and Hovelsrud seek to address the question of *when* adaptive capacity is activated in the face of change. Thus far, research has focused mainly on the factors that determine adaptive capacity, with relatively little attention paid to the questions of whether and under what circumstances such adaptive capacity is deployed. In examining fishing communities in northern Norway, they found that fishers do not perceive anthropogenic climate change as a threat: narratives of self-reliance, extensive knowledge about the marine environment, professional competency and independence diminish their concerns about climate change. Policy-makers attempting to address climate change need to be aware of such differing perceptions, and recognize that failure to acknowledge them could negatively affect the adaptive capacity of northern regions and communities. Likewise, as Vinokurova reminds us, they need to be aware of gendered and generational divergences that may affect individuals' engagement with adaptive actions.

Lena Sidorova, Jenanne Ferguson and Laur Vallikivi consider cultural sustainability through an examination of the linguistic landscape in a settlement in the northeastern corner of Sakha Republic (Yakutia). While the area is home to a variety of indigenous peoples (Chukchi, Yukaghir and Eveny), the authors note the absence of indigenous languages in the linguistic landscape of the town of Chersky.

Signs include Russian, Sakha and even English, but those portraying indigenous peoples are 'mute'. The absence of indigenous languages in the urban landscape mirrors a hierarchy of language valuation that contributes to their continued erosion.

Sustainable health is the focus of two chapters in this section, albeit in markedly different ways. Audrey Giles, Lauren Brooks-Cleator and Catherine Glass consider the obstacles to sustainable health promotion and injury prevention in the Northwest Territories. In addition to financial issues and staffing, program content and delivery are identified as barriers that need to be considered in future health promotion planning. Rémy Rouillard interrogates ideas about adaptations to the arctic environment within both popular and scientific discourses, through an examination of two groups of ethnic Russian oil-workers in Nenets Autonomous *Okrug* (district) in northwestern Russia: those who came north a generation ago and those who have arrived in the past 15 years. Rouillard's work also brings to the fore ideas regarding issues of economic sustainability for extractive industries, especially in relation to labour availability and mobility.

Looking at extractive industries from another angle, Emma Wilson documents the local and corporate understandings of social license and social responsibility, through an examination of the oil industry in Komi Republic in northern Russia. Local understandings of corporate social responsibility diverge from those of the corporations, even when these corporations claim, and are seen, to be socially responsible, according global sustainability standards. Wilson argues for the need to involve local people in (re)defining what social license entails in northern development projects. Toril Ringholm also considers the interactions between a multinational resource corporation and local communities, through a study of the development of the Goliat oilfield off the coast of northern Norway. She explores the efforts of municipal governments to ensure local benefits from the development through evocations of social license. From her interviews Ringholm deduces a process of mutual learning and the building of trust between local officials and company representatives, a trust lacking in the Komi case that Wilson describes.

This section ends with a consideration of the role of human capital as a factor in sustainable development in the North. Andrey Petrov connects the development of human capacity to the evolution of creative capital and the knowledge economy. He argues that these elements provide an important means of sustainable economic diversification that will allow northern regions and communities to move away from an over-reliance on non-renewable resources.

1.3.3   Advancing Sustainability in the North

The chapters in the third section of this volume focus on manifold ways of advancing the sustainability of the North. The first two chapters address the performance of sustainable research in the North. Stacy Rasmus and Olga Ulturgasheva advocate a new methodological approach to research, that of peer observation. As indigenous scholars from the USA and the Russian Federation working together in northern indigenous communities in each of these countries, they consider how identities inscribed on them by community members affected their research experiences and relations. Observing each other's reactions and reflecting together on their responses, they argue, encourages them to consider how their research positions are negotiated within the communities where they conduct research and how their identities, both performed *and* ascribed, affect their research outcomes. In her chapter on "Building Relationships in the Arctic", Heather Gordon, another indigenous scholar, continues the discussion about sustainable research relations by relating her experiences working with Greenlandic Inuit. She identifies trust as paramount, and describes the actions scientists need to take to build and sustain trust when working with indigenous communities.

Susan Crate offers a framework for pursuing research with rural communities to explore rural sustainability, based on her extensive work in Sakha Republic (Yakutia) and in Nunatsiavut, in Labrador, Canada. Research in the two regions demonstrated the efficacy of two innovative tools for building community sustainability: a knowledge exchange process and an on-line atlas project. Crate also underscores that the forces of economic globalization that threaten local subsistence activities and cause the outmigration of youth from rural communities in both regions are considered by locals to be more serious threats than those posed by climate change.

Distinctly less attention has been paid to urban environments and urban sustainability in the circumpolar North, than to rural areas, although the majority of the region's population is urban (AHDR 2014). Turning to arctic urban environments, Ryan Weber, Rasmus Ole Rasmussen, Lyudmila Zalkind, Anna Karlsdottir, Sámal Johansen, Jukka Terräs and Kjell Nilsson discuss land use planning as a tool for sustainable development. Examining six urban areas in Faroe Islands, Finland, Greenland, Iceland, Russia and Sweden, they note key common issues affecting the desire of northern urban residents to have easy access to nature, and the role that complex governance structures play in enabling such access.

Cultural sustainability is the focus of the chapter by Ashlee Cunsolo, Inez Shiwak, Michele Wood and the *IlikKuset-Ilingannet Team* from Nunatsiavut. They describe a youth mentorship program in the region, developed to promote cultural resilience to a changing physical and social environment. In her chapter, Herminia Din discusses a different sort of cultural program, one focused on sustainable art. Her work, which involves students and faculty at the University of Alaska–Anchorage and the University of the Arctic's Thematic Network on Arctic Sustainable Art and Design, encourages students to think about environmental conservation through artistic activities.

ICASS VIII hosted three plenary panels on Northern Sustainabilities, chosen through a competitive, peer-reviewed process. One of these showcased the *SakKijânginnatuk Nunalik* (the Sustainabilities Communities Initiative), an ongoing project in Nunatsiavut. We end this volume on northern sustainabilities with an account of this project. Rudolf Riedlsperger, Christina Goldhar, Tom Sheldon and Trevor Bell describe the challenges the project has faced as well as the success it has experienced, and the lessons learned regarding how to overcome such challenges.

They argue that a focus on local understandings of, and approaches to, sustainability may offer a much needed counterpoint to sustainability initiatives imported from the South and informed by comprehensions distant from local values, philosophies and practices.

The range and scope of chapters in this book speak to the significance of sustainability as an overarching research theme in arctic social sciences and humanities. By organizing and compiling this edited volume, it was our intention to highlight some of the important and ground-breaking research that is influencing our understandings of the multiple facets of sustainability in the Circumpolar North. In its examination of sustainabilities, theoretically and empirically, in various places and at various scales, we hope that this book makes a contribution to both the academic and public discourse about this dynamic and diverse region, and the challenges that it faces at the start of the twenty-first century.

References

AHDR-II. (2014). *Arctic human development report. Regional processes and global linkages* (J. N Larsen & G. Fondahl, Eds.). Copenhagen: Nordic Council of Ministers.

Dahl, J., Fondahl, G., Petrov, A., & Fjelheim, R. (2010). Fate control. In J. N. Larsen, P. Schweitzer, & G. Fondahl (Eds.), *Arctic social indicators* (pp. 129–146). Copenhagen: Nordic Council of Ministers.

Hoogensen Gjørv, G., Bazely, D. R., Goloviznina, M., & Tanentzap, A. J. (Eds.). (2014). *Environmental and human security in the Arctic*. London: Routledge.

Huebert, R., Exner-Pirot, H., Lajeunesse, A., & Gulledge, J. (2012). *Climate change & international security: The Arctic as a bellwether*. Arlington: The Center for Climate and Energy Solutions.

Kimmel, M. (2014). Fate control and human rights: The policies and practices of local governance in America's Arctic. *Alaska Law Review, 31*(2), 179–210.

Ozkan, U. R., & Schott, S. (2013). Sustainable development and capabilities for the polar region. *Social Indicators Research, 114*, 1259–1283.

Petrov, A., BurnSilver, S., Chapin, T., Fondahl, G., Graybill, J., Keil, K., Nilsson, A., Riedlspieger, R., & Schweitzer, P. (2016). *Arctic sustainability research: A white paper for the International Conference on Arctic Research Planning III (ICARP III). Summary and findings*. Available from http://icarp.iasc.info/images/articles/Themes/WP_Summary_Sustainability_science_ICARP3_draft1.pdf

Part I
Conceptualizing and Measuring Arctic Sustainability

Chapter 2
Politics of Sustainability in the Arctic: A Research Agenda

Ulrik Pram Gad, Uffe Jakobsen, and Jeppe Strandsbjerg

Abstract The concept of sustainability has become central in arctic politics. However, there is little agreement on what 'sustainable' means. For different actors (governments, indigenous people, NGOs, etc.) the concept implies different sets of opportunities and precautions. Sustainability, therefore, is a much more fundamental idea to be further elaborated depending on contexts than a definable term with a specific meaning. This paper suggests a set of theoretical questions, which can provide the first steps toward a research agenda on the politics of sustainability. The approach aims to map and analyze the role of sustainability in political and economic strategies in the Arctic. Sustainability has become a fundamental concept that orders the relationship between the environment (nature) and development (economy), however, in the process rearticulating other concepts such as identity (society). Hence, we discuss, first, how, when meeting the Arctic, sustainability changes its meaning and application from the global ecosphere to a regional environment, and, second, how sustainability is again conceptually transformed when meeting Greenlandic ambitions for postcoloniality. This discussion leads us to outline an agenda for how to study the way in which sustainability works as a political concept.

Keywords Concept of sustainability • Political theory • Discourse theory • Postcolonial identity • Greenland

U.P. Gad (✉) • U. Jakobsen
Department of Political Science, University of Copenhagen,
Øster Farimagsgade 5, 1353 Copenhagen, Denmark
e-mail: upg@ifs.ku.dk; uj@ifs.ku.dk

J. Strandsbjerg
Department of Business and Politics, Copenhagen Business School,
Solbjerg Plads 3, 2000 Frederiksberg, Denmark
e-mail: js.dbp@cbs.dk

© Springer International Publishing Switzerland 2017
G. Fondahl, G.N. Wilson (eds.), *Northern Sustainabilities: Understanding and Addressing Change in the Circumpolar World*, Springer Polar Sciences, DOI 10.1007/978-3-319-46150-2_2

2.1   Introduction: Sustainability as a Political Concept in the Arctic

Changes to the climate, global power balances, demands for natural resources, and aspirations for self-determination set the stage for new political struggles in the Arctic. Central to the struggles is the notion of the Arctic as a special place characterized by a nature at once hostile and fragile. In this clash between fragility and the drive towards development, the concept of sustainability has become pivotal. Yet there is neither consensus on what sustainability should refer to, or on how it should be achieved. Despite, or rather because of, its salience for policymaking, there is no consensus about the precise contents of the concept. And this is exactly what makes sustainability such an interesting and politically potent concept. With this chapter, we want to present and advocate for a particular take on sustainability that posits sustainability as a political concept rather than a technical concept.

As the social sciences have been invited to contribute both to perfecting our understanding of sustainability and to implementing it, much scholarship has embraced the concept. In contrast, some critics have advocated a wholesale rejection of the concept on accounts of neo-colonialism. Rather than joining one of these two camps, we suggest an approach that steps back and investigates the diverse political consequences of sustainability becoming a buzzword in the Arctic. For different actors (governments, indigenous people, non-governmental organizations (NGOs), etc.) the concept implies different sets of opportunities and precautions. There are significant differences between businesses and state governments that tend to see sustainability as a precautionary note in the pursuit of wealth in a fragile setting, indigenous communities that often note that it is their particular lifestyle that should be sustained, and global NGOs such as Greenpeace and the WWF which tend to act as spokespersons for a fragile nature in the context of global environmental balance and biodiversity.

As a concept, sustainability has entered an arctic political reality that may be characterized as postcolonial: Indigenous peoples hold a prominent place and have relatively strong organizations in the Arctic. Their relations to their respective states involve a variety of autonomy arrangements designed to remedy histories of colonialism, paternalism and exploitation. As an extreme case, Greenland, once a colony but now a self-governing territory within the realm of Denmark, regularly declares its ambitions to be independent. Greenland explores new strategies for economic development while negotiating a tension between a postcolonial and an indigenous political identity. Political debates in Greenland play out as a negotiation of how to prioritize and combine, in a sustainable way, political self-government with cultural self-sufficiency in terms of human resources, indigenous cultural practices (Inuit language, social norms, hunting and consumption of wild animals etc.) and imperative elements of Western modernity (Western judicial system, representative democracy, welfare state programs, market economy, etc.). These complexities are features of politics and living conditions generally in the Arctic. But in

Greenland these complexities take on a special character in the light of the unique ambitions of becoming a sovereign nation state – the first involving one of the Arctic's indigenous peoples.

It is a consequence, we suggest, of these complexities that sustainability requires further theorization as a *political* concept. That means that we should look at sustainability as a concept that does something to the way in which politics unfolds. We will elaborate this notion below. For the sake of argument, in this chapter we make the assumption that sustainability has become a concept that plays a central role in all arctic development discussions. The important question we should ask is how we should understand this idea. The main ambition of this chapter, then, is to present an approach and a set of questions that could be seen as the first steps toward a new research agenda on sustainability in the Arctic. Because Greenlandic politics embrace all the dilemmas invoked by sustainability, we use Greenland as a case study to show how the concept of sustainability operates politically. The argument is structured as follows: we commence with an outline of sustainability as a political concept followed by a discussion of sustainability in Greenland's postcolonial politics. This allows us to draw out the key analytical questions which we suggest should be asked when pursuing research on the politics of sustainability in the Arctic.

2.2 Sustainability as a Fundamental Concept

Since the Brundtland Report, sustainability has invoked – for lack of a better term – traditional, technical-rational authority to inform development policies. However, we suggest that sustainability has become a much more fundamental concept ordering the relationship between the environment (nature), development (economy), and identity (how can society develop while staying the same).

Whereas the concept of sustainability can be traced back centuries (Warde 2011), its rise to prominence as a political program rather than a tool for academic analysis only came about in the 1980s. As it became clear that the ecosystem of the planet was under threat from the production and development strategies of an ever more industrialized world, the reconciliation of society's developmental goals with the planet's environmental limits became the foundation of an idea that achieved political attention from the mid-1980s. The 1987 report "Our Common Future" by the so-called Brundtland Commission (also known as the World Commission on Environment and Development) was concerned with how to achieve sustainable development defined as "development that meets the needs of the present without compromising the ability of future generations to meet their own needs" (WCED 1987).

An overarching aim was to reinstate scientific and technological knowledge production in societies' efforts to achieve environmentally sustainable improvements in human well-being (Kates 1999). Four distinct research programs had developed: biological research relating humanity to its natural resource base; geophysical research relating human activities to the earth's climate; social research placing human institutions, economic systems and beliefs in nature as its environment; and finally technological research on the design of devices and systems to produce more social goods with less harm to the natural environment (Kates 2000). Current research on sustainability in the Arctic generally stays within one of these distinct research programs, committing normatively to turning unsustainabilities into sustainabilities. However, in committing to sustainability, much research blinds itself to the political effects of employing the concept of sustainability (cf. Sachs 1990; Banerjee 2003; Lélé 1991; Beckerman 1994).

We, therefore, suggest an approach that investigates what political role is played by the concept of sustainability and the practices (including knowledge production) induced by the concept. In this light, politics could be seen as a struggle between competing visions of the future (Palonen 2006) where concepts like sustainability, development, and identity are employed to implicitly or explicitly prognosticate and prescribe specific futures (Koselleck 1985). Since the arrival of Europeans in the Arctic, a discussion has been taking place about how to value and mediate between identity and development. From nineteenth century administrators to early twentieth century explorers and anthropologists, the question was: Can and should the Inuit stay true to their original culture – or must they develop according to a Western model, lest they die out (Høiris 1986)?

The key for analyzing sustainability is to identify its referent object – in other words, what needs to be sustained – and investigate how sustainability helps organize concepts in coherent narratives (Ricœur 1988). By entering established discourses structured around identity and development, the concept of sustainability changes them. Generations of Inuit leaders have submitted different reformulations of the problematique, trying to combine indigenous identity with modern development in various ways (Thomsen 1996). Particularly with the de-legitimization of authorities in the 1960s and 1970s, a new generation of indigenous leaders has presented colonialism and modernization as a threat to their identity (Gad 2005, 2013).

Relative to identity and development, sustainability is a newcomer to political struggles in the Arctic. At first, sustainability in the Arctic was all about protecting a fragile environment (Tennberg and Keskitalo 2002); later it branched out to encompass also the sustainability of human societies in the Arctic (Tennberg 2000). To talk about sustaining human societies diverts the meaning of sustainability from the technical character that came to the fore in the 1980s to one referring to a particular identity.

What is common, however, to the various discourses on sustainability and development in the Arctic is the emphasis of a unique regional environment which, in the more abstract sense, involves the particular characteristics of the materiality of arctic space. The cultural identities of peoples living in the Arctic are seen as shaped by the harshness and remoteness of arctic space (Lorentzen et al. 1999). Economic development has been seen as inhibited by the climate and distances of the Arctic, but also potentially facilitated and even necessitated by its natural resources (Howard

2009). So, arctic space constitutes both the natural environment as fragile, and sustainability as a particularly fragile balancing act between identity, state authority and economic development.

2.3 Sustainability in Greenlandic Politics of Postcoloniality

Greenland is a self-governing territory within the realm of Denmark. It was a Danish colony from 1721 to 1953. After the formal decolonization process in the wake of World War II, Greenland experienced some devolution of powers from Denmark but also, and somewhat paradoxically, a growing Danish presence and a "Danification" of private businesses and public services. One could say that Greenland was decolonized by being integrated (Beukel et al. 2010). This generated protests and gave birth to a national independence movement that resulted in the introduction of Home Rule in 1979. This process of "Greenlandification" developed further, and in 2009 an Act on Self-Government was adopted. In the present situation, Greenland enjoys a large degree of autonomy in domestic matters, but does not retain decision-making power on questions pertaining to citizenship, monetary, foreign, defense and security policy (Ackrén and Jakobsen 2015).

The 2009 Act, however, included a promise of full political independence from Denmark. The preamble of the Act on Self-Government stated that "the people of Greenland is a people pursuant to international law with the right of self-determination". In the Self-Government Act the conditions for independence are specified. On the one hand, a "Decision regarding Greenland's independence shall be taken by the people of Greenland" (21(1)). On the other hand, the procedure states, "An agreement between Naalakkersuisut [the Government of Greenland] and the [Danish] Government regarding the introduction of independence for Greenland shall be concluded … with the consent of the Folketing [Danish Parliament]" (21(3)). Before the Danish Parliament concludes, the agreement shall have "the consent of Inatsisartut and shall be endorsed by a referendum in Greenland" (21(3). Hence, this is the process through which Greenland can obtain political independence from Denmark (cf. Kleist 2010).

The economy remains a significant obstacle to this aim. It follows from the constitutional arrangement that increasing political autonomy from Denmark requires an economic surplus on Greenland's budget balance and thus, simply speaking, independence requires economic development (Strandsbjerg 2014). Obviously, the transfer of an annual grant of more than 3.5 billion Danish kroner (US$ 550 million) that Greenland receives from the Danish government budget, would stop once Greenland becomes independent from Denmark. Moreover, Greenland paid a crucial price for the formal recognition of its right to independence. In pursuant of the 2009 Act and in contrast to the provisions of the 1978 act, Greenland has to pay for further devolution. According to Article 5(1), the annual block grant is fixed at the 2009 level. Moreover, Article 6(1) states: "Fields of responsibility that are assumed by the Greenland Self-Government authorities … shall be financed by the Self-

Government authorities from the date of assumption". During the Home Rule years, every field of responsibility 'taken home' had a cheque attached to it in the form of an increased block grant.

So, in a speech on "Greenland's way forward" at the international conference 'Arctic Frontiers' in Tromsø, Norway in January 2014, then Greenlandic Premier Aleqa Hammond declared that Greenland's short term goal is a sustainable economy in order to obtain the long term goal of political independence: "I want Greenland to have a self-sustaining economy based on our own resources with a greater degree of integration into the world economy. Greenland's long-term political goal is independence" (Hammond 2014). Both the long-term goal of independence (however defined) and the immediate task of a self-sustaining economy outlasted Aleqa Hammond's brief period in power. Indeed, they are generally accepted across most of the political spectrum in Greenland, although differences pertain to the details of the roadmap for independence and the urgency of progress.

A further complication to the politics of sustainability in Greenland is the unsustainable nature of not only the financial side of the economy but also the human resources situation (Lang 2008). Greenland insists on proceeding as a technologically advanced welfare state, even if the level of education among the general population cannot sustain it. The result is a steady import of humanpower from the former colonizing power, Denmark and a continued reliance on the Danish language. This postcolonial re-enactment of colonial dependence forms the background of Aleqa Hammond's claim at the presentation of her government's working programme in April 2013 that "a special Greenlandic element should be to include culture in the concept of sustainable development. The process of reconciliation and forgiveness will be a central element in a sustainable development. Hence, the initiation of a series of activities, e.g. conferences, seminars and debates, aimed at uncovering the 'effects of colonial times'" (Aleqa Hammond in *Rigsombudsmanden* 2013: 6; our translation). With this, Hammond explicitly tied sustainability and potential sovereignty to a particular vision of Greenlandic culture conditioned by postcolonial ties to Denmark.

2.3.1   *Greenland in the Politics of Sustainability in the Arctic*

As one case among other arctic societies, Greenland has been approached by scholars as a struggle between indigenousness and modernization, both at the level of concrete societal practices and at the level of identity discourses. This has shown how Greenlandic politics is shaped as a negotiation of the specific combination of practices and aims promoted as indigenous with developments deemed necessary for prospering culturally, economically, and politically in a modern world (Thomsen 1996; Gad 2009). In this perspective, Greenland stands out as unique in the Arctic by aiming to become the first sovereign, indigenous nation state (Strandsbjerg 2014).

When the concept of sustainability is introduced to the Arctic, it changes its meaning and application from the global ecosphere to a regional environment. In this regard, sustainability seems to be conceptually transformed to allow rather than limit development in a fragile arctic environment. Scholarship often points to the Arctic as a special case; both nature and societies here are presented as particularly fragile (Lorentzen et al. 1999). Hence, the Arctic has become an arena for clashes between, on the one hand, institutions and NGOs promoting a global model for environmental management and, on the other hand, local knowledge and the cultural significance of the Inuit way of life (Caulfield 1997). These clashes illustrate the tension between sustainability as a universal concern and as a local concern.

In the Arctic, sustainability often means the sustainability a particular way of life (Berman et al. 2004; Buckler and Wright 2009), an understanding which might contradict universal attempts to regulate and manage the environment in a globally sustainable manner. This tension is but one example of what happens when a universal discourse on sustainability meets the discourses on arctic particularity and the regional interests of arctic politics. We argue, that the peculiarity of arctic space makes a difference – but we still need to see the full picture of what this peculiarity means in order to understand how the concept of sustainability works in the Arctic.

The point we want to make here is that we need to understand what difference the Arctic as a region with specific characteristics does to sustainability, and the different ways in which the concept of sustainability is employed in current struggles to define postcolonial statehood in Greenland and elsewhere in the Arctic. In Greenland, as discussed above, discourses on the particularity of arctic sustainability, arctic identity, arctic security, and arctic development are configured in a particular way due to the unique double role of the nation-state in Greenland. As a self-governing territory within the realm of Denmark, Greenland does not yet enjoy full sovereignty, but Self-Government is a promise of full sovereignty in the future. In this way a separate, future sovereign state is built into the constitutional arrangement of an existing post-imperial state.

We argue, that this arrangement makes a difference when the global struggles over the reconfiguration of arctic space are articulated in Greenland. It makes a difference whether one has in mind the sustainability, identity, security, and development of a future Greenlandic nation-state with its own independent national economy and human resource base, or whether Greenlandic identity is bound to be developed in a sustainable way within a Danish state ultimately in charge of citizenship, fiscal, foreign, defense and security policy. In sum, we propose that this makes a difference, but we still need to understand just what difference this peculiar version of post-coloniality means for how the concept of sustainability works in Greenland.

To recapture the argument, Greenland is in midst of a local struggle over how state authority is to be configured. This struggle is fueled by a developing climate change narrative that combines actual developments and political aspirations. It is said that arctic global warming means melting ice, both ice sheet and sea ice, and that melting ice means more accessibility to on shore and off shore natural resources, more possibilities for sailing in arctic waters and growing feasibilities for new shipping routes through the Arctic Ocean. Furthermore, more access to natural resources means more mining to meet growing demands on a global scale and more exploration for oil and gas in the Arctic, and more possibilities for new shipping routes mean more attractiveness for Asian interest in the Arctic.

Greenlandic political discourses combine the Asian interests in Greenland's natural resources with the possibility of economic sustainability as the pre-condition for political independence. These factors and this climate change narrative set the stage for a renegotiation not only of the materiality of arctic space but also of post-colonial sovereignty and statehood in Greenland. The climate change narrative, however, also implies that sustainability is conceptually transforming to allow rather than to limit development in this fragile arctic environment. Thus, in her opening speech of the Greenland Parliament in September 2013, then Premier Aleqa Hammond stated that "climate change and the receding ice mean that new business opportunities become available" and that the "mining industry can expand the exploration of raw materials", and that the more ice-free arctic waters in the future may play a role as "an alternative route for container traffic to and from Asia" (Hammond 2013).

So, one plausible scenario is that the goal of economic sustainability driven by exploitation of natural resources in order to obtain political independence marginalizes notions of cultural, social, and political sustainability. The consequence of such developments is that sustainability is transformed from a concept meant to limit development to a concept meant to allow development to take place in an otherwise fragile arctic environment.

2.4 Politics of Postcoloniality and Sustainability in the Arctic: Towards a Research Agenda

This chapter has been motived by the observation that sustainability has become an important and widely applied concept in arctic development discourses while, at the same time, there is little or no agreement between these discourses about the meaning of the concept. This has spurred us to pursue a theoretical approach – or research agenda – to capture the rise of sustainability discourses as a political process renegotiating the relationship between nature, society and development in the political struggles unfolding in the Arctic. This calls for a more nuanced analysis of how and where good and bad futures are envisioned when talking about sustainable development in the Arctic (Tennberg et al. 2014; Sejersen 2014, 2015). The intricacies should be systematically investigated in a research agenda involving both a mapping and a systematic analysis of the role of sustainability in various political and economic strategies in the Arctic.

To acquire a better understanding of arctic development, we need to capture sustainability as a political concept. Sustainability cannot be taken for granted – neither with regards to its substantial meaning nor to its political effects. We need to analyze the uses of the concept of sustainability, rather than assume that it works as a sign-post for problem solving and the rational balancing of interests.

The aim of such a research agenda is to theorize the changes that take place in the Arctic by investigating how the concept of sustainability is given radically different meanings and how these different meanings inform different political strategies. The agenda involves a series of consecutive steps:

- The first task will be to identify and map separate discourses of sustainability in the Arctic. Scholarly reports, political debates, regulatory texts as well as statements from all types of stakeholders in the Arctic should be analyzed to distill claims about *what* should be sustained, in relation to *what* environment or larger community or greater good, as well as *who* is responsible for getting us to sustainability.
- A second task will involve charting the genealogies of each discourse. From where do central ideas come? Did international governance bodies or national regulatory traditions provide the language in which each sustainability discourse is couched? Who promotes each discourse? How do the promoters work together or fight each other? How have the discourses clashed and merged? And what scenarios can be built to understand and predict future clashes or mergers?
- A final task will be to investigate how concepts of nature, identity and development are being reconfigured in these different discourses.

The research following this agenda should pay specific attention to the way in which discourses play out and order distinct scales. First, how is arctic space renegotiated in struggles over the meaning of sustainability and how is a global concept of sustainability given distinct meaning when articulated to arctic space? Second, how is postcolonial statehood and sovereignty renegotiated, especially when the struggles over the meaning of sustainability in the Arctic meet Greenlandic strategies for postcoloniality? We need to understand how the different ways in which the concept of sustainability is employed in current struggles to define postcolonial statehood in Greenland and independence from Denmark, and in parallel processes in other parts of the Arctic.

Hence, two research questions relating to specific changes in geographical scale are each in need of theoretical and empirical investigation. First, what happens when global discourses on sustainability meet the regional particularities of arctic material space? Second, what happens when the resulting discourses on arctic sustainability meet the prospects of Greenland as an indigenous nation state in the Arctic – and, in parallel, when they meet the way other distinct arctic communities envision each their futures? In both of these changes in scale, two core analytical questions are central: How is the concept of sustainability given radically different meanings? And how do these different meanings inform different political strategies?

By pursuing these questions we would get a better understanding of how sustainability works as a concept, but there is also the normative implication that by highlighting the political character – as opposed to its technical-rational appearance – of sustainability, the referent object, and hence what should be sustained, is opened for a political discussion proper.

Acknowledgement The authors would like to thank Marc Jacobsen and Nikoline Schriver for their valuable input to this chapter; in particular with their effort to study global and arctic discourses and the scholarly literatures on sustainability.

References

Ackrén, M., & Jakobsen, U. (2015). Greenland as a self-governing sub-national territory in international relations: Past, present and future perspectives. *Polar Record, 51*(4), 404–412.

Banerjee, S. B. (2003). Who sustain whose development? Sustainable development and the reinvention of nature. *Organization Studies, 24*(1), 143–180.

Beckerman, W. (2006 [1994]). Sustainable development': Is it a useful concept?. In M. Redclift (Ed.), *Sustainability, critical concepts in the social sciences, Volume II: Sustainable development* (pp. 236–255). London: Routledge (Original source: *Environmental Values* (1994), 3, 191–209).

Berman, M., et al. (2004). Adaptation and sustainability in a small Arctic community: Results of an agent-based simulation model. *Arctic, 57*(4), 401–414.

Beukel, E., Jensen, P. F., & Rytter, J. E. (2010). *Phasing out the colonial status of Greenland, 1945–54*. Monographs on Greenland, vol. 347. *Man and Society*, vol. 37. Copenhagen: Museum Tusculanum Press.

Buckler, C., & Wright, L. (2009). *Securing a sustainable future in the Arctic: Engaging and training the next generation of northern leaders*. Winnipeg: International Institute for Sustainable Development.

Caulfield, R. A. (1997). *Greenlanders, whales, and whaling: Sustainability and self-determination in the Arctic*. Hanover: University Press of New England.

Gad, U. P. (2005). Dansksprogede grønlænderes plads i et Grønland under grønlandisering og modernisering. En diskursanalyse af den grønlandske sprogdebat – læst som identitetspolitisk forhandling. *Eskimologis Skrifter, 19*.

Gad, U. P. (2009). Post-colonial identity in Greenland? When the empire dichotomizes back – Bring politics back in. *Journal of Language and Politics, 8*(1), 136–158.

Gad U. P. (2013). Greenland projecting sovereignty – Denmark protecting sovereignty away. In R. Adler-Nissen & U. P. Gad (Eds.), *European integration and postcolonial sovereignty games. The EU Overseas Countries and Territories* (pp. 217–234). London: Routledge ('New International Relations' series).

Hammond, A. (2013, September). *Åbningstale ved Formand for Naalakkersuisut*. Retrieved September 8, 2015, from http://naalakkersuisut.gl/~/media/Nanoq/Files/Attached%20Files/Taler/DK/Aabningstale_EM_2013_AH_DK.pdf

Hammond, A. (2014, June). *Health, wealth and independence* (Paper presented at the 'Arctic Frontiers' conference, Tromsø). Retrieved September 8, 2015, from http://arcticjournal.com/politics/362/health-wealth-and-independence

Høiris, O. (1986). *Antropologien i Danmark, Museal etnografi og etnologi 1860–1960*. København: Nationalmuseet.

Howard, R. (2009). *The Arctic gold rush: The new race for tomorrow's natural resources*. London: Continuum Publishing Corporation.

Kates, R. W. (1999). *Our common journey: A transition toward sustainability*. Washington, DC: National Academy Press.

Kates, R. W. (2000, May). *Sustainability science*. Paper presented at the World Academies conference transition to sustainability in 21st Century, Tokyo.

Kleist, M. (2010). Greenland's self-government. In N. Loukacheva (Ed.), *Polar law textbook* (pp. 171–198). Copenhagen: Nordic Council of Ministers.

Koselleck, R. (1985). *Futures past: On the semantics of historical time* (K. Tribe, Trans.). New York: Columbia University Press.

Lang, I. L. (2008). Barrierer for rekruttering af hjemmehørende grønlandsk arbejdskraft til Hjemmestyret – En undersøgelse af grønlandiseringen i forbindelse med rekruttering til det grønlandske hjemmestyre. Projekt- og Karrierevejledningens Rapportserie Nr. 234/2008, Københavns Universitet: Det samfundsvidenskabelige Fakultet.

Lélé, S. M. (2006 [1991]). Sustainable development: A critical review. In M. Redclift (Ed.), *Sustainability. Critical concepts in the social sciences, Volume II: Sustainable development* (pp. 165–190). London: Routledge (Original source: *World Development* (1991), 19 (6), 607–21).

Lorentzen, J., Jensen, E. L., & Gulløv, H. C. (Eds.). (1999). *Inuit, kultur og samfund. En grundbog i eskimologi*. Århus: Systime.

Palonen, K. (2006). Two concepts of politics: Conceptual history and present controversies. *Distinktion, 12*, 11–25.

Ricœur, P. (1988 [1985]). *Time and narrative* (Vol. 3) (K. Blamey & D. Pellauer, Trans.). Chicago: University of Chicago Press.

Rigsombudsmanden (2013): 'Indberetning fra Rigsombudsmanden i Grønland', 24 May, Retrieved October 28, 2013, from http://www.ft.dk/samling/20121/almdel/gru/bilag/45/1253674/index.htm.

Sachs, W. (2006 [1990]). On the archaeology of the development idea. In M. Redclift (Ed.), *Sustainability. Critical concepts in the social sciences, Volume II: Sustainable development* (pp. 328–353). London: Routledge (Original source: *The Ecologist*, (1990) 20(2), 42–43).

Sejersen, F. (2014). Klimatilpasning og skaleringspraksisser. In M. Sørensen & M. F. Eskjær (Eds.), *Klima og menneske: Humanistiske perspektiver på klimaforandringer* (pp.59–79). Copenhagen: Museum Tusculanums Forlag.

Sejersen, F. (2015). *Rethinking Greenland and the Arctic in the era of climate change*. London/New York: Routledge.

Strandsbjerg, J. (2014). Making sense of contemporary Greenland: Indigeneity, resources and sovereignty. In K. Dodds & C. Powell (Eds.), *Polar geopolitics? Knowledges, resources and legal regimes* (pp. 258–276). Cheltenham: Edward Elgar.

Tennberg, M. (2000). The politics of sustainability in the European Arctic. In L. Hedegaard & B. Lindström (Eds.), *The NEBI yearbook 2000: North European and Baltic Sea integration* (pp. 117–126). Berlin: Springer.

Tennberg, M., & Keskitalo, C. (2002). Global change in the Arctic and institutional responses – Discourse analytic approaches. In J. Käyhkö & L. Talve (Eds.), *Understanding the global system. The Finnish perspective* (pp. 225–228). FIGARE: Turku.

Tennberg, M., Vola, J., Espiritu, A. A., Fors, B. S., Ejdemo, T., Riabova, L., Korchak, E., Tonkova, E., & Nosova, T. (2014). Neoliberal governance, sustainable development and local communities in the Barents region. *Barents Studies: Peoples, Economics and Politics, 1*(1), 41–72.

Thomsen, H. (1996). Between traditionalism and modernity. In B. Jacobsen (Ed.), *Cultural and social research in Greenland 95/96: Essays in honour of Robert Petersen* (pp. 265–278). Nuuk: Ilisimatusarfik/Atuakkiorfik.

Warde, P. (2011). The invention of sustainability. *Modern Intellectual History, 8*(1), 153–170.

WCED. (1987). *Our common future: Report from the 'Brundtland' world commission on environment and development*. Oxford: Oxford University Press.

Chapter 3
Uranium: The Road to "Economic Self-Sustainability for Greenland"? Changing Uranium-Positions in Greenlandic Politics

Lill Rastad Bjørst

Abstract How did the government of Greenland in just a few weeks take on a clear pro-uranium position in the eyes of the industry? I introduce a case study of the production of tolerance towards the mining of Greenland's uranium as developed in the recent political debate about resource development, and particularly, uranium, and to the knowledge practices which help to legitimize varying arguments in the debate. The concept of sustainability is often mentioned in the debate but is given radically different meanings by different actors. In this study I question how these different meanings inform various political strategies, in the context of increased global attention to the possibility of the industrial development of one of the world's last underground treasures.

Keywords Greenland • Mining • Uranium • Sustainability • Development • Social license to operate • Arctic

3.1 Introduction

An old ambition of the former Danish colonial power of profiting from the mining of Greenland's uranium has reappeared. On October 24th, 2013, the Greenlandic Parliament, *Inatsisartut*, lifted a decade-long moratorium on mining radioactive elements. It had been previously following a zero-tolerance policy toward uranium (Sørensen 2013). This paved the way for Greenland (and the Kingdom of Denmark) – to become the newest western (and arctic) supplier of uranium (Vestergaard 2015:153). But as the debate around the acceptability of this move accelerated, it became clear that uranium was not just another mineral, but one capable of

L.R. Bjørst (✉)
Centre for Innovation and Research in Culture and Living in the Arctic (CIRCLA),
Aalborg University, Aalborg, Denmark
e-mail: rastad@cgs.aau.dk

© Springer International Publishing Switzerland 2017   25
G. Fondahl, G.N. Wilson (eds.), *Northern Sustainabilities: Understanding and Addressing Change in the Circumpolar World*, Springer Polar Sciences,
DOI 10.1007/978-3-319-46150-2_3

penetrating regional, local, and global energy, resource, environmental, power, and security agendas.

How did the government of Greenland, in just a few weeks, take on a clearly pro-uranium position in the eyes of the industry? This chapter provides an introduction to a case study of the growth of tolerance towards uranium mining, during the debate about resource development, and specifically uranium, that occurred in Greenland between 2013 and 2015. This chapter will also discuss the knowledge practices which help to legitimize varying arguments in the uranium debate. The concept of sustainability is often mentioned in the debate but it is assigned radically different meanings by different actors. In this study, I want to question how these different meanings inform various political strategies in the context of increased global attention to the possibility of the industrial development of one of the world's last underground treasures.

3.2　A New "Mining Friendly" Geopolitical Regime

Until 2009, mining in Greenland was under the institutional control of Denmark. A repatriation of the political and economic responsibility for mineral resources followed the introduction of the Greenland Self-Rule in 2009. Greenland has a long history of mining, including the extraction of coal, cryolite, gold, copper, and other minerals. Most mining activities took place in a period with limited or no environmental focus on the delicate arctic ecosystems. Some of the mining sites were later analysed and found to have traces of pollution with heavy metals from tailings (Sejersen 2014b). Mining legacies in the form of social and ecological impacts are known, but play a small part of the current political debate in Greenland – a debate framed by questions of agency, respect, and Greenland's right to development (Bjørst forthcoming).

With the introduction of the Self-Rule Act in 2009, Greenland had to look for new sources of income. The downturn of economic activities in 2012 and 2013, which continued in 2014, suggests that Greenland might face substantial economic problems in the years to come (Christensen and Jensen 2014). Scientists, politicians, and the Greenlandic business community have more-or-less accepted this 'inconvenient truth' and are looking for alternative ways to create growth and attract investors (CGMRBS 2014; Rambøll Rapport 2014; Fremtidssenarier for Grønland 2013).

The Self-Government Act stipulates that the subsidy the Greenland Self-Government receives from Denmark will be reduced as revenue from Greenland's mineral sector grows (Act nr. 473 2009). With this in mind, the introduction of a new "mining friendly" geopolitical regime in Greenland can be seen as an integrated part of the ongoing nation-building process and the road to build an independent, sustainable economy. From an economic point of view, it seems to be in the interest of both Denmark and Greenland that Greenland becomes a mining nation (Bjørst forthcoming). While the government of Greenland is preparing for what it characterizes as "sustainable mineral resource development" in the newest oil and mineral strategy (Government of Greenland 2014a:90), resistance in the urban centers, especially in the capital, Nuuk, is growing. Conflicting claims about 'what is sustainable for Greenland' is part of the debate. Whereas non-governmental organisations (NGOs) are mostly concerned about health and environmental problems related to uranium mining, industry advocates for mining as the road to job creation and local economic development.

3.3   Ambivalence Towards Mining of Greenland's Uranium

The mining company Greenland Minerals and Energy A/S (GME) began to operate in Greenland in 2007. Its activities in Greenland are concentrated in a licensed area in Kvanefjeld by the town of Narsaq in the south of Greenland. According to one source, Kvanefjeld contains 575 million pounds of uranium and 10.33 million tonnes total of rare earth oxide (Proactive Investors 2015). The company is a subsidiary of Greenland Minerals and Energy Ltd., which is listed in Australia and has its headquarters in Perth (GME 2015). GME seems confident that, in the years to come, it can finalize a cooperation agreement on the regulation of uranium production and exports from Greenland. Yet it still needs 'the social license to operate', which will not be achieved without some resistance. Studies from 2013 show that many local actors in Greenland are still undecided about uranium mining in Greenland (Bjørst forthcoming). Politicians have responded to these concerns by suggesting a local referendum about uranium. As of this chapter's writing, the question of whether referendum on uranium would be held was still undecided.

Another initiative was a public pre-hearing, which was held in Greenland in late-2014. The pre-hearing was complicated by an unexpected parliamentary election that was called because an expenses scandal prompted Prime Minister Aleqa Hammond to step down as leader of her party. The election gave new life to the public dispute on the mining of Greenland's uranium. A few weeks before the election, the leader of the opposition party Inuit Ataqatigiit (IA), Sara Olsvig, was quoted in a Danish newspaper saying that regardless of the outcome of the upcoming election and the promised local referendum on uranium mining, her (personal) suggestion was to vote against the mining of uranium in Narsaq (Klarskov 2014). In the Greenlandic newspapers she claimed afterwards that the quote was a misinterpretation of her statement (Duus 2014b). In any case, she had revealed her own ambivalence towards mining of Greenland's uranium, something the other parties would use against her during the campaign. In the Greenlandic newspaper *Sermitsiaq* she was accused of "speaking with a forked tongue" (Duus 2014a). Siumut-candidate Julie Rademacher, also running for a seat in parliament, called Olsvig's opinion "highly problematic," and the Greenland Minerals and Energy's managing director John Mair declared:

I am very surprised by the announcement – because then everybody has been ridiculed: the Greenlandic voters, who have to vote for no reason; the international investors, who have wasted their time; and GME, which so far has spent eight years to prepare a fantastic possibility for Greenland in close cooperation with the former Government and everyone else interested (Klarskov 2014).[1]

What is significant in this quote is a shift in paradigm, which has turned mining *in* Greenland into mining *for* Greenland, and legitimizes an argument about mining as "a fantastic possibility" and the only road to development (Bjørst 2016).

3.3.1   A State in Formation and the Pro-uranium Position

The positive discourse about mining is closely linked to the political project of Greenland as a state in formation (Gad 2014). As is true of most Greenlandic politics, this dispute can be read as a negotiation of how that national project is to be configured (Gad 2009). Recent local studies show that mining, and especially the Kvanefjeld project, is positioned as the solution to the problem of what Greenland will need to survive in the future. Therefore, the storyline of "saving the community" (promoted by GME, among others) and doing something for the benefit of Greenland is of high value among Greenlandic politicians (Bjørst 2016). In the election programs of the Greenlandic political parties prior to the parliamentary election in 2014, the question of how uranium penetrated political discussions was noteworthy. The issue ended up being a determining factor for the formation of the final political coalition (Østergaard 2014). In December 2014, Greenland got its new government. After long days of negotiations, what finally united the parties Siumut, Demokraterne and Atassut, and determined the formation of the coalition was their support for uranium exploration in Greenland. These parties promised to work to ensure "economic self-sustainability for Greenland" (Coalition Agreement 2014:3), which thrilled industry. Proactive Investors representative of GME John Mair proclaimed to the international news and media: "All coalition parties are of a pro-uranium position, and we anticipate that the government will be proactive in quickly moving to continue the work with Denmark on uranium regulation" (Proactive Investors 2015).

With this statement, Mair emphasized that the future extraction of rare mineral resources in Kuannersuit (near Narsaq) could not take place until the framework conditions were fulfilled, the required information was provided by the Danish state and Greenland and the requirements met. These included compliance with the International Atomic Energy Agency's safety guidelines and the requirements outlined in Euroatom's cooperation agreement (Coalition Agreement 2014). The Coalition Agreement states that its ambition is to "submit proposals to Inatsisartut to determine an upper limit to the uranium content required to be able to extract this mineral in the mineral resources sector" (Coalition Agreement 2014:7).

---

[1] All translations are those of the author unless otherwise stated.

On the question of the 'social license to operate', GME describe their version of the "conditions" for having an exploitation license approved as "largely dependent on establishing an economically robust, and environmental and socially acceptable development scenario" (GME 2014a:10). In other words, the "environmental and socially acceptable development scenario" was still being negotiated and the outcome of those negotiations would result in a 'social license to operate' and eventually, an exploitation license. Building trust with local communities is crucial for mining companies: it is no longer enough for mining companies to solely meet the formal obligations for a license to mine (Moffat and Zhang 2014:69). As part of this process GME has modified the Kvanefjeld Project numerous times. Having followed this process since 2012, I identified inconsistent information and a lack of transparency as being characteristics of the project. As the project is in its early stages, GME's own material seems to be the primary source of information. GME's communication shows that its public relations' staff is well trained and have continuously adapted the Kvanefjeld project to feed into Greenland's political agendas (Bjørst 2016). All the potential benefits mentioned by GME seem to be congruent with what the municipality (Kommune Kujalleq) feels are important needs for the community (Simonsen 2013:30). In June 2014, a raw materials strategy and action plan was adopted by the municipal council in Kommune Kujalleq (Kommune Kujalleq 2014). As part of its vision, this plan asserts that the "precondition for the sustainable use of the non-renewable resources is to ensure that the overall result of the activities leaves the local community in a positive position with continuous economic growth even after mining has ceased" (Kommune Kujalleq 2014:19). To be left in a "positive position" could mean many things, but at the moment the major issue locally seems to be securing jobs, and bringing Narsaq "back on track "in order to put the region back on the map (Bjørst 2016).

The quest for jobs and development takes into account the industrialisation of the Arctic and is framed as the primary benefit of attracting such a project to the town (Sejersen 2014a). Industry was very much aware of who it should mention as the primary stakeholders in the negotiation, and as a final remark to its *Quarterly Report* of December 2014, GME wrote about ambitions for future collaboration: "Greenland Mineral and Energy Ltd. will continue to advance the Kvanefjeld project in a manner that is in accord with both the Greenlandic Government's and local community expectations, and looks forward to being part of continued stakeholder discussions on the social and economic benefits associated with the development of the Kvanefjeld Project." (GME 2014a:19).

In this *Quarterly Report* (GME 2014a), as well as in the mining strategy of Kommune Kujalleq, 'sustainability' was primarily economically based. Any mention of the environment in outreach materials was mostly about creating a "stable investment environment" (GME 2014b). In her analysis of the economy of appearance, Anne Tsing has questioned the investor-driven process in which profit must be imagined before it can be extracted, and the potentials of companies, countries, regions, and towns are dramatized as places for investment. Based on her findings, she claims that: "Dramatic performance is the prerequisite of their economic performance" (Tsing 2000: 118). The predominately optimistic discourse of mining as a positive force of change in society and as something that brings growth has gained influence among politicians and the business community in Greenland. Take for example Greenland's Oil and Mining Strategy from 2014. In the preamble to the Strategy, the opening lines state:

> The Government of Greenland wishes to promote the prosperity and welfare of Greenland's society. One way of doing so is to create new income and employment opportunities in the area of mineral resources activities. The Government of Greenland's goal is to further the chances of making a commercially viable oil find. In addition, Greenland should always have five to ten active mines in the long term (Government of Greenland 2014a:7).

In the spring of 2014, the Strategy (in a short version with many illustrations) was also distributed along the coast of Greenland via the Greenlandic newspaper *Sermitsiaq* in both Greenlandic and Danish, with the title "Our natural resources [raw materials] must create growth (Government of Greenland 2014b). Judging from the political debate in Greenland, Denmark and elsewhere during the last 5 years, there seems to be no alternative to depending on mineral resources, oil and gas for economic growth (Bjørst 2016).

3.3.2   Urani? Naamik (Uranimium? No)

The decision to overturn the uranium ban has attracted widespread criticism from Greenlandic and international NGOs. In April 2013, 48 NGOs from around the world signed an appeal to the Greenlandic and Danish governments to uphold the uranium zero-tolerance policy in the Danish realm. Part of their argument was that rare earth elements can be extracted in Greenland without uranium (Avataq 2013). Avataq, which co-signed the appeal, is a Greenlandic environmental group that is not against mining in Greenland as such, but has expressed concerns that the repercussions from mining operations will have serious long-term consequences. A statement issued by the group clearly illustrates its argument against mining Greenland's uranium: "In the long term, the environmental impacts from uranium mining could constitute comprehensive radioactive contamination, which – because of the health risks – would make it dangerous to live, and necessary to ban fishing, hunting, agriculture and animal husbandry, in significant parts of Southern Greenland" (Avataq 2013:2).

GME, the Kommune Kujalleq, and the Government of Greenland see the mining of rare earth elements (REE) and uranium as the road to development and growth. Conversely, NGOs see a rather alarming future for the region from contamination, health risks, and – if damage is done – high cleanup costs of the residues of uranium mining, which would be covered by the tax payers in both Denmark and Greenland (Avataq 2013). The debate thus is characterized by the co-construction of different geopolitical imaginaries that range from boom to doom (Arbo et al. 2013) and by conflicting spatial storylines about "saving" or "destroying" the local community

(Bjørst 2016). As Law (2004:55) has observed: "Different realities are being created and mutually adjusted so they can be related – with greater or lesser difficulty."

When it comes to the mining of Greenland's uranium, a divided Greenland has significant implications for small arctic communities and populations on a daily basis. Studies of mining in other parts of the Arctic show that the social and cultural cost of mining operations cannot be ignored and need to properly addressed (Tester and Blangy 2013). Social and economic benefits are debatable. Impact and Benefit Agreements (IBAs) have been mentioned as a road to empowering local people and stakeholders. Based on her experience from Greenland, Anne Merrild Hansen (2014:15) asks: "Is IBA a tool to secure 'local' acceptance to achieve 'social license to operate' or the means to empower the locals to take part in the development processes?" There is a big difference between being a partner, a stakeholder or an ordinary citizen who is compensated via benefits. Hansen emphasizes that Greenlanders generally welcome development and that hopes for the future are high, but the public also feels a certain degree of anxiety concerning uncertainty about how life in Greenland will unfold when new projects are implemented (Hansen 2014). Demonstrators in Nuuk, Narsaq, and Copenhagen (from 2013 to 2014) who did not want to be a partner, stakeholder or beneficiary protested with the refrain, "Urani? naamik" (Uranium? No). Yet, while the resistance movement in Greenland has been growing, the majority of the population still supports the current government and its stand on the mining of Greenland's uranium.

3.4   Conclusion: Uranium and "Economic Self-Sustainability for Greenland"

How did the government of Greenland, in just a few weeks, adopt a "clear pro-uranium position" in the eyes of the industry? The answer is that the Greenlandic politicians see the mining of uranium as one of the important ways to strengthen the economy of Greenland and to ensure what they call "economic self-sustainability for Greenland" (Coalition Agreement 2014:3). The positive discourses and related storylines about the mining of Greenland's uranium, as developed in the political debate of 2013–2015, are mostly to be understood in that context. Similarly to other places in the Arctic, social issues are ignored under the rhetoric of "employment" (Tester and Blangy 2013). Discussions on the mining of Greenland's uranium illustrate how extractive industries affect local ideas of (sustainable) development in the Arctic. They also demonstrate that Greenlanders cannot be reduced to passive victims of mining capital. Rather, Greenlanders are co-producing the aspirations of the mining industry when they negotiate what kind of society they are willing to tolerate. 'Sustainability' is a political concept and part of what is required to be granted and maintain a 'social license to operate'. Parallel and conflicting storylines are constantly being produced and reproduced and have led to several splits between civil society and expert perceptions of risk and human impact. But that does not stop another spill-over of uranium ambivalence into the uranium debate. There could be another major shift in uranium-positions when more information about the technical, social, and environmental impacts is available and communicated to relevant stakeholders and decision-makers. The lift on the zero-tolerance policy towards uranium mining might be just the first of many changes following the election of the new Greenlandic government. Development scenarios are still being negotiated among actors inside and outside Greenland. It remains to be seen whether (or when) Greenland will become a supplier of uranium.

Acknowledgements This study was funded by Aalborg University in Denmark and will contribute to new research on how extractive industries affect local ideas of (sustainable) development in the Arctic.

References

Act on Greenland Self-Government. (2009). Act nr. 473 of 12 June 2009.

Arbo, P., Iversen, A., Knol, M., Ringholm, T., & Sander, G. (2013). Arctic futures: Conceptualizations and images of a changing arctic. *Polar Geography, 36*(3), 163–182.

Avataq. (2013). Appeal to the Greenlandic and Danish governments not to abolish the uranium zero tolerance policy in the Danish realm. Retrieved January 9, 2014, fromhttps://www.nirs. org/international/westerne/Statement%20on%20uranium%20mining%20in%20 Greenland%2026%20April.pdf

Bjørst, L. R. (2016). Saving or destroying the local community? Conflicting spatial storylines in the Greenlandic debate on uranium. *The Extractive Industries and Society, 3*(1), 34–40.

Bjørst, L. R. (forthcoming). Arctic resource dilemmas: Tolerance talk and the mining of Greenland's uranium. In R. Thomsen & L. Bjørst (Eds.), *Heritage and change in the Arctic.* Aalborg: Aalborg University Press.

CGMRBS (Committee for Greenlandic Mineral Resources to the Benefit of Society). (2014). To the Benefit of Greenland. A report written by The Committee for Greenlandic Mineral Resources to the Benefit of Society. University of Copenhagen and the University of Greenland, Ilisimatusarfik. http://nyheder.ku.dk/groenlands-naturressourcer/rapportogbaggrundspapir/ To_the_benefit_of_Greenland.pdf

Christensen, A. M., & Jensen, C. M. (2014). *Aktuelle tendenser i den grønlandske økonomi* [*Current trends in the economy of Greenland*]. Danmarks Nationalbank, Kvatalsoversigt 2. kvartal 2014 53. årgang nr. 2, pp. 71–76.

Coalition Agreement. (2014). *Coalition agreement, election term 2014–2018 "Fellowship – Security – Development"* (Please note: This is a translation–the Greenlandic version applies). Retrieved January 20, 2014, from http://naalakkersuisut.gl/~/media/Nanoq/Files/Attached%20 Files/Naalakkersuisut/DK/Koalitionsaftaler/Koalitionsaftale%202014-2018%20engelsk.pdf

Duus, S. D. (2014a). *Kritikere lugter blod efter Olsvig-forvirring om uran* [*Critics smell blood after Olsvig-confusion about uranium*]. Sermitisaq.ag, 23. Retrieved November 2014, from http://sermitsiaq.ag/kritikere-lugter-blod-olsvig-forvirring-uran

Duus, S. D. (2014b). *Olsvig forklarer sig efter Politiken-artikel om uran og folkeafstemning* [*Olsvig explains after Politiken article on uranium and referendum*]. Sermitsiaq.ag, Retrieved November 23, 2014, from http://sermitsiaq.ag/olsvig-forklarer-politiken-artikel-uran-folkeafstemning

Fremtidsscenarier for Grønland. (2013, September). *Instituttet for Fremtidsforsknings scenariebeskrivelser for Grønland* [*Future Scenarios for Greenland*]. CIFS future scenario descriptions for Greenland.

Gad, U. P. (2009). Post-colonial identity in Greenland?: When the empire dichotomizes back – Bring politics back in. *Journal of Language and Politics, 8*(1), 136–158.

Gad, U. P. (2014). Greenland: A post-Danish sovereign nation state in the making. *Cooperation and Conflict, 49*(1), 98–118.

GME (Greenland Mines and Energy). (2014a). December 2014 quarterly report. Thursday 29th January, 2015 (Highlights). Retrieved February 5, 2015, from http://www.ggg.gl/docs/quarterly-reports/Q4_2014_Quarterly_Activity_Report.pdf

GME. (2014b). Company announcement, December 5th, 2014: New coalition government formed in Greenland. Retrieved February 14, 2015, from http://www.ggg.gl/docs/ASX-announcements/New-Coalition-Government-December2014.pdf

GME. (2015). *About Greenland minerals and energy.* Retrieved June 1, 2015, from http://gme.gl/en/about-greenland-minerals-and-energy

Government of Greenland. (2014a). *Greenland's oil and mineral strategy 2014–2018.* Report available at: http://naalakkersuisut.gl/~/media/Nanoq/Files/Publications/Raastof/ENG/Greenland%20oil%20and%20mineral%20strategy%202014-2018_ENG.pdf

Government of Greenland. (2014b). Vores råstoffer skal skabe velstand [Our raw materials have to create prosperity] (2014). The Government of Greenland, Departementet for Erhverv, Arbejdsmarked og Handel. Published 05.06.2014: http://naalakkersuisut.gl/~/media/Nanoq/Files/Publications/Raastof/DK/Olie%20og%20Mineralstrategi%20DA.pdf

Hansen, A. M. (2014, February 1). Community impacts: Public participation, culture and democracy. Unpublished working paper, University of Copenhagen. Retrieved March 28, 2014, from http://vbn.aau.dk/ws/files/186256309/Community_Impacts.pdf; http://www.govmin.gl/images/stories/about_bmp/publications/Greenland_oil_and_mineral_strategy_2014-2018_ENG.pdf

Klarskov, K. (2014). *Grønlandsk toppolitiker vil blæse på resultatet af en folkeafstemning* [*Greenlandic politician are indifferent to outcome of a referendum*]. Politiken, Internationalt, 22. nov. 2014 KL. 22.30: http://politiken.dk/udland/ECE2462672/groenlandsk-toppolitiker-vil-blaese-paa-resultatet-af-en-folkeafstemning/

Kommune Kujalleq. (2014). *Råstofstrategi og handlingsplan* [*Mining strategy and action plan*]. Juni 2014: http://www.narsaq.gl/images/stories/pressemeddelelser/2014/06/Råstofstrategi Endelig20140606DK.pdf

Law, J. (2004). *After method: Mess in social science research.* New York: Routledge.

Moffat, K., & Zhang, A. (2014). The paths to social licence to operate: An integrative model explaining community acceptance of mining. *Resources Policy, 39*, 61–70.

Østergaard, C. (2014). *Med ny regering rykker Grønland tættere på uranmine* [*With the new government Greenland moves closer to uranium mining*]. Ingeniøren, 5 December 2014: http://ing.dk/artikel/med-ny-regering-rykker-groenland-taettere-paa-uranmine-172750

Proactive Investors. (2015). *Greenland minerals and energy MD John Mair talks with proactive investors.* Retrieved February 20, 2014, from http://www.proactiveinvestors.com.au/companies/news/60143/greenland-minerals-and-energy-md-john-mair-talks-with-proactive-investors-60143.html

Rambøll Rapport. (2014). *Hvor skal udviklingen komme fra? Potentialer og faldgrupper i den grønlandske erhvervssektor frem mod 2015* [*Where should development come from? Potentials and pitfalls in the Greenlandic sector until 2015*]. Rambøll Marts 2014.

Sejersen, F. (2014a). A job machine powered by water. In I. K. Hastrup & C. Rubow (Eds.), *Living with environmental change: Waterworlds* (pp. 102–105). Oxon: Routledge Falmer.

Sejersen, F. (2014b). *Efterforskning og udnyttelse af råstoffer i Grønland i historisk perspektiv* [*Exploration and exploitation of resources in Greenland in a historical perspective*]. Working paper, Open Access København. University of Copenhagen. http://nyheder.ku.dk/groenlands-naturressourcer/rapportogbaggrundspapir/Efterforskning_og_udnyttelse_af_r_stoffer_i_Gr_nland_i_historisk_perspektiv.pdf

Simonsen, S. (2013). *Borgmester Simon Simonsens nytårstale* [Mayor Simon Simonsen's New Year's speech]. *Atuagagdliutit*, AG nr. 03 Week 03, 16 January, 30.

Sørensen, S. P. (2013). *Qullissat: byen der ikke vil dø* [*Qullissat, a city that could not die*]. København, Frydenlund.

Tester, F. J., & Blangy, S. (2013). Introduction: Industrial development and mining impacts. *Études/Inuit/Studies, 37*(2), 11–14.

Tsing, A. L. (2000). Inside the economy of appearances. *Public Culture, 21*(1), 115–144.

Vestergaard, C. (2015). Greenland, Denmark and the pathway to uranium supplier status. *The Extractive Industries and Society, 2*(1), 153–161.

Chapter 4
Tensions Between Environmental, Economic and Energy Security in the Arctic

Gunhild Hoogensen Gjørv

Abstract The notion of security is being increasingly employed in debates regarding energy consumption, economies, and human relationships to the environment, not least the issue of climate change. This chapter looks at the tensions present across many arctic communities and states reliant upon or impacted by natural resource development, where environmental concerns collide with economic and energy vulnerabilities. The purpose of this chapter is to elucidate different understandings of security in relation to the extraction and consumption of non-renewable energy resources, and what is valued or prioritized within these different conceptions. The chapter then moves briefly to the "ethical oil" debate that focused on the Alberta oil sands in 2010, and the ways in which Norwegian oil and gas politics are also making ethical claims about continued extraction.

Keywords Security • Values • Ethics • Extractive industries • Oil sands • Norway • Canada

4.1 Introduction

In the wake of the United Nations Climate Change Conference of the Parties Twenty-first Session (COP 21) meetings in December 2015 in Paris, a brief moment of euphoria seemed to wash over participants and spectators to the event: finally an agreement was reached (BBC News 2015; CBC News 2015). The euphoria seems short-lived, however, when it has become increasingly clear that the agreement was still quite weak in committing states to adhering to the goal of a maximum temperature rise of 2 °C (with an ideal or hopeful target of 1.5 °C). It remains to be seen, therefore, how this target will be achieved.

G.H. Gjørv (✉)
Department of Sociology, Political Science, and Community Planning,
UiT The Arctic University of Norway, Tromsø, Norway
e-mail: gunhild.hoogensen.gjorv@uit.no

© Springer International Publishing Switzerland 2017
G. Fondahl, G.N. Wilson (eds.), *Northern Sustainabilities: Understanding and Addressing Change in the Circumpolar World*, Springer Polar Sciences,
DOI 10.1007/978-3-319-46150-2_4

35

The COP 21 Paris agreement was a globally negotiated process, with 195 countries signing the document to curb emissions contributing to climate change. The COP 21 agreement, following earlier international attempts since the first COP meeting in Berlin in 1995, reflects the increasing global acknowledgement that climate change is caused by significantly higher carbon (CO^2) emissions resulting from human activity. This conclusion has been documented by years of rigorous research upon which there is solid and reliable scientific consensus (Oreskes 2004; Cook et al. 2013).

The COP 21 agreement will have different impacts across the globe, demanding political willpower to alter lived realities for many who have relied upon non-renewable/carbon-based industries and energy resources which have contributed to increased CO^2 emissions. This means there will also be differing responses by signatory states. Carbon-based non-renewables have played a significant role as a source of income as well as a reliable and reasonably priced energy source for numerous states, not least those located across the arctic regions of the world where these resources are still estimated to be in very good supply (up to 30 % of undiscovered gas reserves, and 13 % of undiscovered oil reserves) (Hong 2012). Reducing the production of these energy sources can and will have significant impacts on many arctic states and communities. At the same time, the reduction of these resources plays a crucial role in the mitigation against climate change and the future sustainability of the planet. Not surprisingly, the responses of state and non-state actors have resulted in different narratives, what I will argue are "security" narratives, reflecting the tensions between economies, energy and the environment (climate change).

The Arctic is not a homogenous region, defined by one set of characteristics. This is the case with regards to the level of dependency upon extractive industries (particularly mining and petroleum) or upon natural resource dependent economies (Huskey et al. 2014). Although subjected to different challenges and conditions due to different political and resource management systems, a common feature for many of the arctic states is the importance of oil, gas and coal to operation of formal economies. These industries are also among those that are targeted for necessary reduction in the Paris COP 21 agreement as they contribute heavily to the carbon emissions attributed to climate change.

In this chapter I examine some of the different narratives that have framed the debates or discussions about oil and gas extraction in arctic states. The narratives articulate perceptions of values and survival that can be further analysed through three different "categories" of security: energy, economic and environmental security. After briefly explaining what we mean by these categories, I discuss the use of some of the dominant narratives to support petroleum extraction that have been used in Canada and Norway, two states heavily reliant upon oil and gas revenues, comparing these narratives to critiques about the environmental impacts of the oil and gas industry and subsequent calls for reductions in this industry.

4.2   Energy, Environment, Ethics, and Security in the Arctic

The concept of security is, at its core, about freedom from care or worry. It is about reducing or eliminating fear. The security concept further relies on the interplay between five elements: actors, values, practices, survival, and the future. The actors can range from the state to individuals, and values can range from the material (physical wellbeing) to the immaterial (identity). The values that are relevant to security are those values relevant to our (human) survival. The values that are relevant for survival are often understood within particular categories – such as values we prioritize around economy, identity, food, health, energy or the environment. Fear, and indeed fear for our future survival, has played an increasing role in climate change narratives (O'Brien et al. 2010), but can also be found in justifications for continued oil and gas extraction.

Security in the arctic context had been traditionally dominated by a state-based, militarized security perspective reflecting the tensions of the Cold War (Heininen 2004; Hong 2012). However, the recognition of the importance of the environment to the survival of states and communities came increasingly to the fore in the 1980s, and has played a pivotal role in our understanding of arctic security since the 1990s (Hoogensen Gjørv et al. 2014). The meaning of environmental security has ranged from reflecting the importance of protecting natural resources for the security of the state, to protecting the environment for human security (human wellbeing), to protecting ecosystems for their own sake (Barnett 2003; Dalby 2002, 2014). The human security literature has enhanced our understanding of environmental security by taking a "bottom-up" perspective that prioritizes individuals and communities over the state. Environmental security thus has gravitated towards a better understanding of the relationships between human communities and their natural surroundings, and the ways in which human beings are dependent upon a thriving and diverse environment (O'Brien et al. 2010). Environmental security focuses on communities identifying threats to their existence and lifestyles, and their capacities to mitigate against or adapt to these threats where need be, in concert with other actors (state, industry, NGOs) (Barnett et al. 2010; O'Brien et al. 2010; Hoogensen Gjørv et al. 2014).

Energy security reflects some of the earlier iterations of environmental security, insofar as it embodies the important role of natural resources to the security of the state. It has evolved to become a concept in its own right, engaging its own contestations and debates (Yergin 2006; Sovacool 2011). The production of, and necessity for, energy affects state security and policy, including the economic viability of the state. Energy security definitions and debates are also increasingly affected by the demands of electorates/populations, both in countries more accustomed to constant and consistent energy availability, but also in countries that have not enjoyed regular access and require more reliable access to resources in order to ensure increased development, employment, and living standards. Energy security has no one fixed definition, but as noted by Löschel and others, the "reference to physical energy availability, energy prices, and their volatility is made in most definitions" (Rabinow

1991: 1665). Cherp and Jewell (2014: 415) refer to the practice of understanding energy security through the "four As": availability, accessibility, affordability, and acceptability.

A thus far less-employed (but no less relevant) term is "economic security", which has been acknowledged in the field of security studies, though not extensively explored. The human security literature has addressed the concept to a degree (UNDP 1994). Economic security has broadly been referred to as economic wellbeing, and pertains to potential threats against finances or resources needed to function and thrive in society, for individuals and states alike (Carlarne 2009). For many northern communities, economic security emanates from opportunities arising from energy industries, binding the survival (and thus security) of the individual, community and state together through a reliance on one dominant industry. Economic security is not the same as energy security, but they are clearly linked. Energy security depends on affordable prices, but these cannot be so low so as to burden the industry and make it unprofitable (therefore making the industry possibly unreliable as it is unsustainable). At the same time, economic security relies on the potential to maintain or increase wealth and, therefore, would benefit from higher prices, but not to the degree that people and states look to other resources for their energy supplies because the price for the dominant energy resource is too expensive and thus inaccessible.

Security is about what we *say* and *do* to ensure that that the things we value most survive into the future. Indeed, what those values are informs what we believe are integral to our survival in the future (hence the classic focus on the state — we value the state on the assumption that the state is integral to our survival). Values and the future expectations that they embody are often based on a mix of knowledge (how and what we claim to "know") and ethics (what we believe is "right"). The debates regarding climate change and non-renewable energy sources are contentious and have included controversial competing claims to knowledge in attempts to reveal the "truth" behind the potentially devastating or relatively benign (depending on the knowledge claim) impact of non-renewable energy development. Ethics address our "everyday" practices, which are guided by a set of principles or values that tell us what is right or wrong in a given context, but also include a systematic philosophical inquiry regarding the foundation and the nature of those principles or values (Hutchings 2010: 6). As some ethicists would point out, we determine what is good or bad based on what we believe we know will be the consequences of a particular act in a given context (Floyd 2010, 2011).

In climate change debates, the ethical is often associated with the acknowledgement of the danger, as well as the threats associated with climate change and a willingness to do something about it (as such, acting ethically). By contrast, the unethical is associated with resistance to or the prevention of "politically challenging decisions and actions from being taken to avoid dangerous climate change." (O'Brien et al. 2010: 10). Others, however, argue for a different ethics, which claims other values like economic sustainability and access to critical energy sources that are deemed integral to community, state or human survival.

4.3 Securing Lifestyles/Threatening Lives?

Only the most obvious commodity in all this is the petroleum that literally fuels most of contemporary civilization and the car-driving inhabitants of contemporary cities. It is this industrial urban society, now spreading to all corners of the globe, that is ironically threatening everyone's human security… Because that global economy, the increasingly artificial context of our lives, is what is changing the biosphere and threatening to unleash the kind of dramatic climate disruptions that worried the Pentagon consultants in 2003; industry and consumer lifestyles are now the cause of environmental threats. (Dalby 2008: 271)

Practices in energy security, in particular the extraction of conventional (oil well) and non-conventional non-renewable oil resources (bitumen-based oil from the oil sands in Alberta, Canada), have generated considerable debate. The value we place on energy and the way in which we address the "ethics" of energy feature heavily in these debates (Kristoffersen 2014). The production and consumption of non-renewable resources, which are central to energy production in general, play a significant role in the negative effects of human-induced climate change. Climate change, in turn, can and will have increasing implications for the existence of many human societies (and ecosystems) as we currently know them, threatening those futures.

The tension between the exploitation of natural resources, particularly non-renewable resources, and the desire to continue the lifestyles that depend on this exploitation has led to a provocative discursive trend. The assumptions behind exploitation/extraction have been often linked with doing "wrong" or doing "right" with regards to the environment, the future, and our future expectations. This trend involves constructing a framework of ethics around energy issues, often as an ethical response to climate change critiques.

Making climate change a "security" issue has been difficult (though not for lack of effort), because it has been difficult to argue that climate change poses an immediate, existential threat, particularly against the state (Mayer 2012). The concept of security has been often framed in temporal terms, such that threats need to be immediate and existential (Buzan et al. 1998). The threats of climate change have not fit well into that framework. Climate change is more commonly recognized as a security issue, but the time dimension is fluid rather than immediate. We do not know how far in the future we will feel the most serious effects, but we (in this case, the international community backed by scientists) are conducting research that gives us a good idea. Climate change science is heavily knowledge dependent but additionally relies on claims about our priorities (values). The claims carry a moral as well as a scientific weight, such that we have a moral obligation to work to reduce climate change inducing emissions, for our own wellbeing and human security, as well as for our children and their children (Klein 2014). Energy producers as well as political actors (local and national regulatory agencies and governments) are subjected to social and political pressures to satisfy both scientific as well as ethical considerations. This pressure, in turn, appears to have encouraged the emergence of the "ethical oil" position that, though first popularized in the Canadian context, is relevant in the Norwegian context as well.

4.4   "Ethical Oil," Ethics of Oil: From Canada to Norway

> But the efforts of anti-oil sands lobby have consequences: every attack on Alberta's indus-
> try further risks people's jobs, their retirement portfolios, and their peace of mind. And,
> more to the point, every barrel of oil not produced in the oil sands means one more, less
> ethical barrel produced in some OPEC dictatorship (Levant 2010: 163–164)

The Alberta oil sands (also known as the "tar sands") illustrate well the competing claims for environmental, energy and economic security. They are grounded in
concurrent competing claims about knowledge, which in turn reflect a set of priorities and values that different actors hold and link with a vision of the future. The
purpose of this section of the chapter is not to determine whose claim is "right", but
rather to identify how these claims work in relation to each other and become part
of a debate about security.

The tensions between ethics, values and science were well illustrated in a debate
in Canada in 2010 regarding the oil sands.[1] The book *Ethical Oil: The Case for
Canada's Oil Sands* by Ezra Levant (2010) had, for a period of time, inspired the
use of the term "ethical oil" as a potential way of combating the critique waged
against the oil sands. His work also inspired a website: www.ethicaloil.org. In *Tar
Sands: Dirty Oil and the Future of a Continent* (2010) Andrew Nikiforuk argues
against the oil sands, invoking research on climate change as support for an anti-oil-
sands perspective. These two books do not encapsulate the entire debate, but they do
illustrate the direction the debate has taken.

Nikiforuk, a free-lance journalist located in Calgary, Alberta, presented arguments rooted in various studies that demonstrated the dangers of developing the oil
sands project. His core message on climate change pertained to the link between
increasing oil and gas production, and the impact this production has on the climate,
starting with the claim that despite the $6 million CAD the Canadian government
has invested in climate change programs, they had yet at that time to meet any targets for reducing greenhouse gases. More importantly, the author gathered evidence
demonstrating the effects that climate change has had on the Canadian Subarctic
and, in particular, the Athabasca region where the oil sands are located. Increasing
temperatures of 1 °C have already started a trend towards drying up reclamation
sites, and it is predicted that another increase by 1 °C will exacerbate this trend by
reducing rainfall and causing increased rates of evaporation. Salt and acid levels in
water sources are also expected to increase. Citing a 2006 report by Canada's
Environment Commissioner Johanne Gélinas, Nikiforuk (2010) claimed that
Canada has experienced a 51 % increase in greenhouse gases over a 15 year period,
to which the oil sands had made a significant contribution.

Nikiforuk further argued that the extraction of unconventional oil leads to double
or triple the carbon emissions of conventional sources, as unconventional sources

---

[1] The labeling of this non-conventional energy source sets up the ethical battleground: proponents
refer to the source as the oil sands whereas challengers refer to it as the tar sands. The latter has a
negative connotation. For consistency I have referred to them as "oil sands" but not to reflect of my
view on their ethical role in energy development.

like oil sands require more fossil fuels in the extraction process and cause more emissions during the later processing phases. Energy expert Alex Farrell at University of California, Berkeley notes: "When we face tradeoffs between economics, security and environment, the environment often ends up getting the short end of the stick" (quoted in Nikiforuk 2010: 129). The ethical position reflects a valuing of the environment over the convenience, economic benefit, abundance and the perceived reliability of the energy source. Nikiforuk's claims are encapsulated in an environmental security narrative, whereby he claims that the existence of human communities as well as ecosystems is threatened.

Levant challenged perceptions of the "ethical", by presenting arguments that, in his view, debunked the core of environmental concerns against unconventional oil extraction. He argued that charges of unethical environmental practices waged against corporations were and are misplaced, especially when weighed against other core values. While the author acknowledged problems around water usage, forest depletion, and increased emissions of CO2, he argued that there are other issues worth considering, such as the extent to which corporate actions support gender equality, promote peace, and fight terrorism. Levant's criticism was rooted, in part, in the subjective ways that moral choices are being made. For him, morals reflect biases and "political fashions" (Levant 2010: 54). He challenged the sources from which anti-oil sands ethical knowledge and standards were drawn, noting that even "emails from individuals – no matter their expertise, bias or vendetta" could be used to measure corporate ethical practices, particularly those of oil and gas industries.

Levant endeavoured to expose cooperation between different actors, such as financial institutions, indigenous communities, and environmental groups, that he claimed work together to generate a climate-change narrative of fear and ethics. He said that these actors attempted to draw investors towards "ethical investments" which he asserted are suspect money-making arrangements taking advantage of misguided environmental values rather than being based on sound science and truth: "If you're a fair trade coffee-drinking, Prius-driving, Green Party-voting recycler who dabbles in vegetarianism, you've found your fund" (Levant 2010: 86). Levant further criticized the financial institutions that promote ethically "green" investment funds by showing the links that some of the same organizations have to the oil and gas industry. He criticized the financial incentives and disincentives in the form of tax relief for meeting carbon emission targets, or fines for not doing so.

Levant's arguments reflected a security perspective rooted in the values of market economics, and further uses human rights ethics to back up his position (Levant 2010). His approach was to reveal inconsistences in arguments and practices based on environmental security and then argue how economically important industries additionally stand on ethically firm ground (ie: not a part of conflict zones, or human rights abuse-based countries). He criticized environmentally-focused think tanks such as the Canadian Pembina Institute as being anti-economic and anti-business, as well as overtly political, implying that his own position was apolitical and based on perceptions of a "value-free" economics. He further claimed that attempts to meet the demands of such politicized organizations would "hobble" the Alberta economy, risking jobs, peace of mind, and retirement portfolios. In other words, climate change measures that would curtail the production of fossil fuels (and thereby carbon emissions) would threaten the economic security and wellbeing of Albertans.

The term "ethical oil" itself has appeared to have fallen out of use for the most part – indeed, some have even argued that the term has been less than helpful, minimizing real environmental concerns while simplifying important and complex arguments about the benefits of continued oil and gas extraction in terms of economics, technology, and human rights (Findlay 2012). However, the logic embedded within the term are still very present, and not just in Canada. In many respects the logics of oil and gas extraction discourses in Norway have long reflected similar arguments as the "ethical oil" position, though the language and tactics have been more nuanced compared to the rather polarized way in which the debate in Canada was framed.

Following COP 21 the Norwegian government was challenged by a number of environmental groups to increase measures that would reduce Norway's reliance on fossil fuel industries and ensure a reduction of CO^2 emissions (NRK 2015). The response by the government was a clear signal that production would not be reduced, using arguments not unfamiliar to those used in "ethical oil" narrative: Norway needs to continue to produce oil and gas because, otherwise, oil and gas will be extracted with much higher emissions from other (read: ethically problematic) parts of the world. Furthermore, Norwegian oil and gas is a preferable choice to the "klimaversting" (climate offender), coal. In other words, Norwegian oil and gas production, it is argued, provides an environmentally positive solution to climate change challenges, and indeed will contribute to meeting the COP 21 goals of reduced overall emissions (NRK 2015).

The language of "drilling for the environment" which the above reflects is not new (Kristoffersen 2014). Kristoffersen has examined the narratives and practices of the Norwegian government on oil and gas extraction since the mid-2000s. The Norwegian government has openly fronted a national self-image as an environmental champion, claiming to be a frontrunner in carbon taxes, trade quotas, and cooperation with industry to meet environmental challenges (Regjeringen Stotenberg II 2008). At the same time the Norwegian government, with little difference in approach between labour and conservative governments, has actively pursued continued petroleum extraction, in close cooperation with industry (Kristoffersen and Young 2010). As Kristoffersen and Young note, resistance from civil society environmental agencies was muted, as many environmental were absorbed into the state apparatus throughout the 1990s. By coopting the environmentalist message, the Norwegian state blurred the lines between independent environmental advocacy and state-based economic and energy interests. This process nevertheless allowed the state to employ the ethics of environmentalism to support its policies.

Norway's contributions to the reduction of CO^2 emissions is currently less focused upon alternatives to oil and gas, and more focused upon using science to find technological solutions to making oil and gas extraction "cleaner." At best, Norway follows a hybrid solution where there are no predicted timelines for reducing or eliminating dependence on hydrocarbons while exploring possible renewable options. The reliance on Norwegian technology means not only cleaner extraction of petroleum resources in the north, but also an ethical claim for a better type of oil (cleaner) than that produced in the global South, where such technology is not available or used (Kristoffersen and Young 2010). Norway's contribution to reduced carbon emissions is thus seen in light of overall global responsibility, rather than focusing specifically on Norway's responsibility. This means that even though emissions in Norway appear to have risen in 2015, the Norwegian approach is to look at its emissions in relation to global efforts and, therefore, claim an environmentally-responsible image due to its relative contribution to global emission reductions (Mathismoen and Færaas 2016).

Improvements in technology also work in tandem with climate change and melting ice in the Arctic, whereby access to resources theoretically increase. Norway and other arctic states can continue to extract oil and gas resources from their arctic regions when technology allows (though for the time being it is not profitable to do so because of technological costs as well as the low price of oil). Indeed, the melting ice represents opportunity and access to resources rather than a threat to human societies, and is still a discourse that is on the table (Mathismoen and Færaas 2016; Støre 2016). At some point the cost-benefit analysis could turn again in favour of a more rigorous pursuit of resources in the high Arctic. The oil and gas industry is still a major economic contributor to the Norwegian economy, and although Prime Minister Erna Solberg claimed that Norway is pursuing building alternative sectors for employment and supporting the economy, there is no indication that this means a concurrent downsizing of petroleum extraction (NRK 2015). There is no apparent desire to predict when oil and gas extraction will come to an end either (Mathismoen and Færaas 2016). As is the case with Canada, Norwegian oil and gas will be here for the foreseeable future, according to their national leaderships. In early 2016 Norwegian government had released 56 new extraction permits in the Barents Sea, Norwegian Sea, and the North Sea with the expectation that the petroleum industry will continue exploration and extraction and provide employment in the years to come (Storholm 2016). In Norway, economic and energy security based on oil and gas consumption still play a primary role in the country's vision for survival in the future. Environmental security is not neglected, but is addressed through secondary measures (rather than directly by curtailing emission-generating activities) by making extractives "cleaner" or more ethically palatable.

4.5 Conclusion

At their core, the debates about non-renewable resources focus upon differing values and perceptions about the future. Economic security is crucial to the survival of states and their communities: this is true for both market and formal economies as well as informal and traditional economies. It is not surprising that states like Canada and Norway are engaging in debates, simplistic or more nuanced, that reflect tensions between visions of ensuring the survival of current expectations and lifestyles dependent on economic and energy development, and the survival of the environment as we know and depend on it. These are two different visions of what "we" (as people part of our local, national or global communities) believe should survive. Norway and Canada may continue to operate as if hoping that these often conflicting visions of the future can survive together, moving towards hybrid solutions that slow down the rejection of fossil-fuel energy resources. Framing the debate in terms of security includes a power struggle between different actors that brings a voice and action to differing values within a given context.

Energy, environmental and economic security can have complementary and competing priorities, depending on the values embodied within these concepts. Energy and economic security have had a close relationship, particularly for export countries where their reliance on oil and gas incomes sustains a particular politics and way of life. In this case there is little room for the prioritization of environmental security, particularly when framed around climate change. As noted by Øverland, "global climate policy has largely been a failure, and energy consumption as well as greenhouse gas emissions have continued to soar" (Øverland 2015: 1). The demands for a strong and stable economy dependent upon the sale and/or consumption of non-renewable energy resources have thus far outweighed the fears surrounding the consequences of climate change on human wellbeing. Alternatively, environmental security that prioritizes a more rigorous climate policy and thus prioritizes limited temperature increases will demand enormous changes in the current energy sector. This could have destabilizing effects on the energy sector in at least the short term while renewable energy sources have time to dominate energy supply and demand (Rapp 2015; Øverland 2015).

Do we want to maintain lifestyles that rely heavily on non-renewable energy but which can and will lead to significant impacts upon the environment, not least a changing climate? This implies a shift towards adaptation, whereby our security depends on our ability to adapt and adjust as the changing climate affects our lives, with the hope that our lifestyles are still relatively "maintained". Or are we more concerned about ensuring that we live in an environment which is as stable and predictable as possible? What are we willing to do to ensure that our relationship to the environment upon which we depend will have positive effects for the future? In this case, policy and practice would embrace mitigation efforts in an attempt to lessen or reduce potential changes, if not reverse them. It can be said, therefore, that these continuing debates revolve around contesting and competing understandings of *security* based on ethics, values and knowledge.

References

Barnett, J. (2003). Security and climate change. *Global Environmental Change, 13*, 7–17.
Barnett, J., Matthew, R., & O'Brien, K. (2010). Global environmental change and human security. An introduction. In R. Matthew, J. Barnett, B. McDonald, & K. O'Brien (Eds.), *Global environmental change and human security* (pp. 3–32). Cambridge, MA: MIT Press.

BBC News. (2015). COP21: Paris climate deal is 'best chance to save planet'. *BBC Science and Environment*. 13.01.16.

Buzan, B., Wæver, O., & de Wilde, J. (1998). *Security: A new framework for analysis*. London: Lynne Rienner.

Carlarne, C. (2009). Risky business: The ups and downs of mixing economics, security, and climate change. *Melbourne Journal of International Law, 10*(2), 439–469.

CBC News. (2015). 'Historic' Paris climate deal adopted: Agreement criticized for imposing no sanctions on countries that fail to reduce emissions. *CBC News World* 13.01.16.

Cherp, A., & Jewell, J. (2014). The concept of energy security: Beyond the four As. *Energy Policy, 75*, 415–421.

Cook, J., Nuccitelli, D., Green, S. A., Richardson, M., Winkler, B., Painting, R., Way, R., Jacobs, P., & Skuce, A. (2013). Quantifying the consensus on anthropogenic global warming in the scientific literature. *Environmental Research Letters, 8*(2), 1–7.

Dalby, S. (2002). *Environmental security*. Minneapolis: University of Minnesota.

Dalby, S. (2008). Environmental change. In P. D. Williams (Ed.), *Security studies: An introduction* (pp. 260–273). London/New York: Routledge.

Dalby, S. (2014). Rethinking geopolitics: Climate security in the anthropocene. *Global Policy, 5*(1), 1–9.

Findlay, M. H. (2012). Please stop calling it "ethical oil". GlobePolitics (24 April 2012) http://www.theglobeandmail.com/news/politics/please-stop-calling-it-ethical-oil/article4101409/

Floyd, R. (2010). *Security and the environment: Securitization theory and US environmental security policy*. Cambridge: Cambridge University Press.

Floyd, R. (2011). Can securitization theory be used in normative analysis? Towards a just securitization theory. *Security Dialogue, 42*(4–5), 427–439.

Heininen, L. (2004). Circumpolar international relations and geopolitics. In N. Einarsson, J. N. Larsen, A. Nilsson, & O. R. Young (Eds.), *Arctic human development report*. Akureyri: Stefansson Arctic Institute.

Hong, N. (2012). The energy factor in the Arctic dispute: A pathway to conflict or cooperation? *Journal of World Energy Law and Business, 5*(1), 13–26.

Hoogensen Gjørv, G., Bazely, D. R., Goloviznina, M., & Tanentzap, A. J. (Eds.). (2014). *Environmental and human security in the Arctic*. London: Routledge.

Huskey, L., Mäenpää, I., & Pelyasov, A. (2014). Economic systems. In J. N. Larsen & G. Fondahl (Eds.), *Arctic human development report: Regional processes and global linkages* (pp. 151–183). Copenhagen: Norden.

Hutchings, K. (2010). *Global ethics: An introduction*. Cambridge: Polity Press.

Klein, N. (2014). *This changes everything: Capitalism vs. climate*. New York: Simon & Schuster.

Kristoffersen, B. (2014). *Drilling oil into arctic minds? State security, industry consensus and local contestation*. PhD dissertation, UiT The Arctic University of Norway.

Kristoffersen, B., & Young, S. (2010). Geographies of security and statehood in Norway's 'Battle of the North'. *Geoforum, 41*(2010), 577–584.

Levant, E. (2010). *Ethical oil: The case for Canada's oil sands*. Toronto: McClelland & Stewart.

Mathismoen, O., & Færaas, A. (2016). Klimaministeren kan ikke love at det blir klimakutt i Norge de neste to årene (Climate minister cannot promise that there will be an climate-related cuts in Norway in the next two years). *Nyheter Innenriks: Politikk* (January 17, 2016) http://www.aftenposten.no/nyheter/iriks/politikk/Klimaministeren-kan-ikke-love-at-det-blir-klimakutt-i-Norge-de-neste-to-arene-8316053.html

Mayer, M. (2012). Chaotic climate change and security. *International Political Sociology, 6*, 165–185.

Nikiforuk, A. (2010). *Tar sands. Dirty oil and the future of the continent*. Vancouver: Greystone Books.

NRK. (2015). *Møter krav om norsk oljestopp med kald skulder* (Responds to demand for Norwegian oil-stop with a cold shoulder). 13 December 2015. Retrieved January 13, 2016, from nrk.no/nordland.

O'Brien, K., St. Clair, A. L., & Kristoffersen, B. (Eds.). (2010). *Climate change, ethics and human security*. Cambridge: Cambridge University Press.

Oreskes, N. (2004). Beyond the ivory tower: The scientific consensus on climate change. *Science, 306*(5702), 1686.

Øverland, I. (2015). *Future petroleum geopolitics: Consequences of climate policy and unconventional oil and gas. Handbook of clean energy systems*. J. Yan. Online, Wiley.

Rabinow, P. (Ed.). (1991). *The Foucault reader: An introduction to Foucault's thought*. London: Penguin.

Rapp, O. M. (2015). En grønn verden er mulig – men prisen er enorm (A green world is possible – but the price is enormous). *Aftenposten*, 11 August 2015. http://www.aftenposten.no/fakta/innsikt/En-gronn-verden-er-mulig--men-prisen-er-enorm-8062129.html

Regjeringen Stotenberg II. (2008). Norsk miljøpolitikk i tråd med OECDs anbefalinger (Norwegian environmental policy in line with OECD's recommendations). Regjeringen.no.

Sovacool, B. K. (2011). Introduction. In B. K. Sovacool (Ed.), *The Routledge handbook of energy security* (pp. 1–42). Abingdon: Routledge.

Støre, J. G. (2016). Is this a new era for the Arctic? Global Agenda 5 January 2016. http://www.weforum.org/agenda/2016/01/is-this-a-new-era-for-the-arctic/

Storholm, L. (2016). 56 nye utvinningstillatelser på norsk sokkel (56 new extraction permits on the Norwegian continental shelf). *High North News* 19 January 2016. http://www.highnorthnews.com/56-nye-utvinningstillatelser-pa-norsk-sokkel/

UNDP. (1994). *Human development report 1994: New dimensions of human security*. New York: United Nations Development Programme.

Yergin, D. (2006). Ensuring energy security. *Foreign Affairs, 85*(2), 69–82.

Chapter 5
Sustainable Security in the Arctic and Military Cooperation

Michał Łuszczuk

Abstract The comprehensive transformation occurring in the Arctic has strengthened and broadened concerns regarding the state of regional security and its impact on the international system. While arctic security in the post-Cold War period has been perceived and examined according to both traditional (hard security) and critical perspectives, this chapter explores a different approach, namely the concept of sustainable security. Such an approach complies with the multidimensional, multilevel and multilateral specificity of the arctic security environment, which is heavily influenced by both the geopolitical and climatic features of the region. The concept is then reconsidered as a potential analytical tool for examining today's arctic security, and the crucial role of cooperation between militaries in pursuing this sustainable security in the Arctic is discussed.

Keywords Arctic security • Sustainable security • Military cooperation

5.1 Introduction

When we try to consider all the manifold developments taking place nowadays in the circumpolar North, we can be overwhelmed by their aggregation, intensity and rapidity. Consequently, these attributes sometimes overshadow one further fundamental feature of the transformation going on in the Arctic, namely the interconnectedness of many processes (Dalby 2009). This aspect is very crucial when we try to understand and advance both sustainabilities and security in the region.

This chapter, building on the concept of sustainable security to analyze the composition and dynamics of arctic security, draws special attention to the assessment of military cooperation among the arctic states and its meaning for peaceful, safe, and prosperous development in the region, as outlined in the Arctic Council's

M. Łuszczuk (✉)
Maria Curie-Skłodowska University, Lublin, Poland
e-mail: mluszczuk@gmail.com

© Springer International Publishing Switzerland 2017
G. Fondahl, G.N. Wilson (eds.), *Northern Sustainabilities: Understanding and Addressing Change in the Circumpolar World*, Springer Polar Sciences,
DOI 10.1007/978-3-319-46150-2_5

'Vision for the Arctic' statement (Arctic Council 2013b). It is argued that the armed forces of the arctic states, beyond their responsibilities for dealing with all 'hard security' challenges, are also best equipped with both the monitoring instruments and the capabilities to operate in such a vast and harsh region.

The chapter begins with a brief discussion of the transnational nature of the security risks in the Arctic and the question of the essence of the (re)militarisation taking place in the region. The concept of sustainable security is then presented and reconsidered as a potential analytical tool for examining today's arctic security environment. Finally, after a short presentation of the scope and examples of military cooperation in the Arctic (which had been developing quite well until 2014, when tensions in relations between the West and Russia beyond the Arctic increased), the conclusion speculates on the role of cooperation between militaries in pursuing sustainable security in the Arctic.

5.2   Complexities of Arctic Security

Most of the environmental and social processes transforming the arctic region at the beginning of the twenty-first century are truly transnational, which means that their origins and range respect neither national boundaries nor traditional ideas of territorial sovereignty (ACIA 2004; AHDR II 2014; Stępień et al. 2014). Accordingly, even if not all of them can be considered as so-called 'transnational threats',[1] in general they pose a challenge to the traditional and even not-traditional (critical) understandings of state security and its military dimension – and ultimately, to international relations in the region (Conley et al. 2012; Huebert et al. 2012; Kraska 2011; Zellen 2013).

To some degree, this situation is not unique to the Arctic. Since at least the end of the Cold War, numerous transnational developments, particularly new types of uncertainties and risks encountered in different parts of the world, have led to a process of redefining the nature of (in)security in international relations (Baylis 2008; Hoogensen 2012). As some researchers argue (Munck 2009), the redefinition of security encompasses both its 'widening' (which could be understood primarily as the securitization of non-military threats) and its 'deepening' (which means discussing new levels and objects of security). According to Dupont (2001:2), "stretching the boundaries of conventional thinking about security" has mainly been induced by a new class of threats that include "complex, interconnected and multidimensional,

---

[1] There are five broad categories of transnational threats that may challenge human security, national governance and interests, and international stability. These include: transnational crime, transnational terrorism, international migration flows, disease and international pandemics, and global environmental degradation and climate change (Smith 2000). Transnational threats can also be described as events or phenomena whose scope goes beyond borders, and whose dynamics are significantly (but not necessarily exclusively) driven by non-state actors, activities (e.g., global economic activates or trends), or forces (e.g., environmental pollution, climate change (Cockayne and Mikulaschek 2008).

non-military transnational issues" that moved "from the periphery to the centre of the security concerns of both states and peoples."

The evolution of the understanding of security in the arctic region follows this global pattern to some extent. However, there are many peculiarities related to both the climatic and geopolitical characteristics of the region (Berkman and Vylegzhanin 2013a; Tamnes and Offerdal 2014). These aspects are closely interconnected with the heterogeneity, vulnerability, and resilience of the social-ecological subsystems of the Arctic and the role of the region as a whole in the Earth's climate system. They also distinguish the Arctic and make the region's security truly comprehensive, indivisible, and ever-evolving (Chater and Greaves 2014; Exner-Pirot 2013). Of course, this kind of environment is extremely difficult and demanding for the arctic states that are trying to preserve their national interests, develop the Arctic's potential to their greatest benefit, and secure themselves against a wide range of threats, including transnational ones. In fact, addressing these challenges and providing adequate capabilities for effective action still remains the main political and economic challenge in probably all arctic states. The arctic states are at the forefront among all stakeholders in the Arctic, not only because of their territorial presence in the region and their undisputed rights to the economic resources or geopolitical potential of the region, but primarily due to their responsibilities and needs in managing and/or mitigating the consequences of the multidimensional transnational transformation facing the Arctic (Bailes and Heininen 2012).

The arctic states achieved collective agreement and coordinated action once before, at the end of the 1980s, when a surge of initiatives for international northern circumpolar cooperation in chosen fields was launched, following the 1987 Murmansk Speech by former Soviet President Mikhail Gorbachev (Åtland 2007; Keskitalo 2007). Scientific research and environmental protection in particular benefited from this cooperation. The new post-cold war thaw quickly expanded in the form of the Arctic Environmental Protection Strategy (AEPS), signed by the eight arctic states in 1991, and then transformed into a set of new institutionalized intergovernmental forums such as the Barents Euro-Arctic Region (BEAR) in 1993, the Arctic Council (AC) in 1996, and the Arctic Military Environmental Cooperation (AMEC), also in 1996 (Heininen 2013). It is important to note that the factor connecting all these initiatives was environmental protection (or rather, environmental degradation), which became a part of the international agenda of northern cooperation (Heininen 2013). This agenda provides an excellent example of the new security discourses and concepts – primarily critical ones such as environmental and human security (Shepherd 2013) – also being applied with regard to the arctic region (Daveluy et al. 2011; Heininen 2013; Hoogensen et al. 2009).

The Arctic experienced many constructive effects of the demilitarization processes, global governance, and institutionalization of arctic nations, all of which replaced the clashes of the Cold War era (Bergman Rosamond 2011). However, it should also be kept in mind that "the Arctic story is one of marginality, centrality, securitisation and desecuritisation, militarisation and demilitarisation" all taking place simultaneously (Palosaari and Möller 2004:255). Palosaari and Möller (2004) argue that militarization is still a key concept in the Arctic, even though the majority of existing security challenges and threats to the region are of a non-military character. Despite the widespread institutionalization of the Arctic and the effective soft-law regime supporting international collaboration, the region has not entirely escaped the geopolitical tensions emerging from the competitive use and control of navigation routes, natural resources, and sovereignty claims, just to indicate the most obvious contentious issues (Berkman and Vylegzhanin 2013b). In this context, how can growing activities in the military domain be understood?

5.3 Arctic (Re)Militarisation – What Is It For?

The coastal states of the Arctic Ocean have followed one model in their arctic strategies, admitting that their main focus is safeguarding sovereignty over their arctic territories and securing a fair share in the exploitation of the Ocean's resources, while the non-coastal arctic states seem to prefer mainly the development of international cooperation (Haftendorn 2011). They all have also committed themselves to using hydrocarbons responsibly in order to avoid destroying the highly fragile arctic environment, and to ensure the well-being of indigenous people in the Arctic; "in so doing, they try to blend military preparedness with enhanced cooperation" (Haftendorn 2011:339).

A review of current and projected military forces in the arctic region indicates that the process of modernization and the creation of new capacities to address new challenges is associated with the environmental, economic, and political changes anticipated in the region, rather than constituting a response to major threat perceptions (Wezeman 2012). Conventional military forces specifically adjusted to the harsh arctic environment were projected to remain rather local in range, especially given the size of the arctic region, and would remain in some cases very much below Cold War levels (Wezeman 2012).

It bears pointing out that following the reduction of political and military tensions in the 1990s, some military capabilities are being restored or redeployed in the Arctic. In many instances, these capabilities are defensive in nature and linked to intensified activities surrounding either the extraction of raw materials or new 'soft' security issues. Soft capabilities help to address climate change, cybercrime, search and rescue, disaster response, and humanitarian assistance. As has already been highlighted, due to the extreme weather conditions, it is primarily the military or coast guard assets (plus a limited number of the icebreakers capable of operating in the Arctic Ocean) that tend to be able to safely operate under arctic conditions. It is even argued that "in order to avoid the possibility of a costly and dangerous shift of military resources to the north, it would be sensible for constabulary forces to assume some of the burden for security provision to enable greater economic activity" (Le Mière and Mazo 2013:101). Taking into account that two agreements reached in 2011 and 2013 under the auspices of the Arctic Council have focused on issues of maritime safety and security (Search and Rescue – SAR), it is not surprising that they have also encouraged greater constabulary engagement among the arctic states.

The Agreement on Cooperation on Aeronautical and Maritime Search and Rescue in the Arctic (Arctic Council 2011a) was approved on the 12 May 2011, during the Seventh Ministerial Meeting of the Arctic Council organized in Nuuk in Greenland (Takei 2013). While this Agreement was designed to become "a measure to strengthen the Arctic Council's ability to face the changing environmental and political conditions in the region" (Kao et al. 2012:832), it also provided grounds for a closer cooperation among SAR entities – both military and paramilitary – in terms of having the practical ability to perform such rescue tasks (Exner-Pirot 2012; Łuszczuk 2014; Wood-Donnelly 2013). Simultaneously, it can be described as a typical search and rescue agreement that focuses more on coordination and cooperation rather than on upgrading and/or providing new capacity (Rottem 2014). In addition to presenting and explaining the areas of responsibility of the individual states in the region, the Agreement also "calls for carrying out joint SAR exercises and training, and facilitating expeditious cooperative responses to search and rescue situations" (VanderZwaag 2014:324). Following the SAR Agreement, several activities were already undertaken, including the first Arctic Council Search and Rescue Table Top Exercise (Whitehorse, Yukon, on 4–6 Ocotber 2011) (Arctic Council 2011b), as well the live exercise "SAREX" near Greenland's coasts in 2012 and in 2013 (Kudsk 2012). Although, the usefulness of the Agreement in strengthening national SAR capabilities still remains in question, as the Agreement openly informs that its implementation 'shall be subject to availability of relevant resources" (VanderZwaag 2014), it addresses the military or paramilitary character of developments specific to this region.The second binding agreement negotiated under the Arctic Council's auspices is the Agreement on Cooperation on Marine Oil Pollution Preparedness and Response in the Arctic. It is also crucial for the development of regional cooperation in the domain of safety (Arctic Council 2013a). It was developed by a task force, established in Nuuk in 2011and co-chaired by Mr. Karsten Klepsvik (Norway), Ambassador Anton Vasiliev (Russian Federation), and Ambassador David Balton (United States) – over the course of a series of meetings held between October 2011 and October 2012. The arctic states endorsed this Agreement in Kiruna, Sweden on15 May 2013, at the Eighth Ministerial Meeting of the Arctic Council (Trigatti et al. 2014).

This agreement obligates each party to keep up a national system for responding quickly and effectively to oil pollution incidents and to establish a minimum level of pre-positioned oil spill combating equipment, a program for training, and other tools and resources, in cooperation with the oil and shipping industries (Nordtveit 2015). The agreement refers to any oil spill incident, regardless of the cause of the incident. The states are also obliged to monitor the areas under their jurisdiction, and as far as feasible, the adjacent international areas. According to the provisions, each party can request assistance from the other states when an oil spill occurs. The agreement contains also a provision for reimbursement of costs and the establishment of operational guidelines for cooperation (Nordtveit 2015).

This Agreement has also been perceived as an important step forward in arctic state cooperation in preparing for an increase in oil and gas and shipping activities that are expected to occur in the coming years (Johnstone 2015; Trigatti et al. 2014).

However, it is critisised "for making implementation subject to the capabilities of the parties and availability of relevant resources" (VanderZwaag 2014:325).

Even if these two agreements (may) play an important role, and the Arctic is a region with exceptional potential to foster collaborative relations between states, it is true that this region still lacks any form of comprehensive security architecture, which may result from an overly fragmented understanding of arctic security (Le Mière and Mazo 2013). What is more, while the concept of sustainable development in the region has enjoyed widespread acceptance for more than two decades, this concept is not linked to security issues on any official level. Could this be changed and on what basis? In addressing these questions it might be worth considering the concept of sustainable security.

5.4 Sustainable Security

The concept of sustainable security was originally conceived as a response to the international position of the United States and American foreign policy in the wake of the presidency of George W. Bush (Smith 2008). It was spearheaded by critics of the old-fashioned, but erstwhile dominant, yet short-sighted understanding of national security and threats against American security: "Sustainable security, in short, can shape our continued ability to simultaneously prevent or defend against real-time threats to America, reduce the sweeping human insecurity around the world, and manage long term threats to our collective, global security" (Smith 2008:2). In its original incarnation, the concept of sustainable security combined three approaches: (1) national security, or the safety of the United States; (2) human security – well-being and safety of people; and (3) collective security – the shared interests of the entire world (Smith 2008).

The concept was soon developed and expanded towards a more universal application (Voigt 2009). It gained stronger and more nuanced foundations, highlighting today's urgent need for a more comprehensive and effective understanding of the interconnectedness of transnational risks, which can only be effectively tackled simultaneously. Although it would seem easier to deal with these challenges separately, in reality it is not possible because none of these issues can be solved in isolation from the others; addressing them requires a truly holistic approach (Voigt 2009).

According to this concept, a contemporary understanding and implementation of security demands more than the ability and capabilities to protect and defend the nation state. It should be stressed that ensuring security in today's world requires a particular focus on sustainable conditions for security, which means that sustainable development is a security imperative (Voigt 2009).

The relationship between sustainable development and security is reciprocal and mutually reinforcing – neither can be severed from the other. While sustainable development is probable only in a politically secure environment, stable and durable peace depends on sustainable conditions for the creation of long-term sustainable livelihoods; the provision of basic services, the protection of life-sustaining ecosystems services, and the sound management of natural resources (Voigt 2009).

Unfortunately, the interaction of climate change with ongoing environmental degradation has the potential to exacerbate a wide range of conflicts in different dimensions, scales, and levels. Consequently, efforts to build a basis for long-term peace and human security are undermined. Linking sustainable development and state security challenges the traditional separation of security and development in many ways, but if successful, can be extremely constructive. Making this connection could be valuable in helping to outline more adequate policies and laws aimed at creating secure and effective conditions. As Voigt (2009:181) points out: "Nature, society and security are interdependent. Policies and measures based on this understanding are likely to be more effective and secure than those that dis-embed people from nature." Achieving sustainable security is both a means to an end and an end in itself. National security aims to ensure the ability of states to protect their citizens from external aggression. Environmental security provides favorable conditions for all livelihoods. Human security focuses on the management of threats and challenges that affect people everywhere – inside, outside, and across state borders. Sustainable security combines all of these aims into one model (Smith 2008). Although at present the concept is still rather ambiguous, it offers an additional, suitable approach to analyzing of the security in the Arctic.

5.5 Sustainable Security in the Arctic Based on Military Cooperation

The notion of sustainable security as an analytical framework, or even as a conceptual foundation for the emerging arctic security system, is an interesting proposal, especially in comparison to more established approaches or concepts that have been at the forefront of narrative analysis in recent years. It is all the more relevant given that there is rather a common agreement that the changing security environment in this region indeed covers many different aspects: national security; economic developments; investments in human capital; and ecological sustainability and social welfare (Huebert et al. 2012; Kraska and Baker 2014). Nevertheless, there are many uncertainties about certain practical aspects of incorporating this concept into the framework of current activities of the arctic states, whether they are considered individually or in the form of regional cooperation, such as that facilitated by the Arctic Council. One of these uncertainties is the issue of military cooperation in the Arctic, which seems to be extremely important in large part due to the particular engagement of the respective armed forces of each country in this harsh region (Hilde 2013; Le Mière and Mazo 2013).

Of course, the Arctic was used as a strategic space or a theatre for military operations for decades during the Cold War (Griffiths 1992). It is imaginable that it could be used in this way once again, especially if the recently observed Russian military resurgence in the Arctic becomes a part of Moscow's broader repertoire of actions aimed at restoring Russia's position in the world (Olesen 2014). This scenario might not become reality as long as a cooperative dynamic continues (Käpylä and Mikkola 2013). Indeed, previous military cooperation developed in several configurations between the 1990s and 2014. Examples include the Norwegian-Russian 'POMOR' exercises in the Barents region; cooperation between Norway and NATO combined with partners from the Partnership for Peace program, in the form of the 'Cold Response' exercises; the annual Operation Nanook in the Canadian North, with the participation of Denmark and the United States; and the Canadian and American Tri-Command Framework for Arctic Cooperation. These exercises provide good examples of the potential for joint military activities that can lead to the 'defense diplomacy' (Exner-Pirot 2012) or even more institutionalised forms of cooperation (Heininen 2013; Schaller 2014).

Additionally, two new forms of circumpolar collaboration between militaries in the North have emerged recently, both of which deserve attention. Their purpose is to discuss broader perspectives rather than a singular focus on the military aspects of arctic security. In June, 2011, the American European Command and the Norwegian Defense Forces co-organized the first meeting of the Arctic Security Forces Roundtable (ASFR) in Oslo as part of their bid to establish a forum for cooperation on climate change issues. The ASFR participants are Canada, Denmark, Finland, France, Germany, Greenland, Iceland, Netherlands, Norway, Russia, Sweden, United Kingdom, and the United States. The ASFR is an annual general officer level event expected to provide a collaborative opportunity for security forces around the Arctic to discuss safety and security issues and possibilities for expanding co-operation in the region.

With the theme 'Military Adaption to Climate Change', the roundtable focused on promoting collaboration to address challenges related to managing security forces in the Arctic, with an emphasis on support for environmental protection, infrastructure, joint exercises and training, and maritime domain awareness. During the meeting, one of the key discussion points was the establishment of better coordination of current standard operating procedures across all arctic security stakeholders (Schissler 2012). The ASFR took place again in 2012 in Bodø, Norway, and in 2013 in Naantali, Finland (Schissler 2012). More than 50 representatives from 11 countries (not including Russia) participated in the latest meeting, which was held in Sortland, Norway in 2014. During this event participants discussed an array of issues including the impact of climate change, trade and economic issues, search and rescue challenges, disaster relief operations, and other security related matters. While the most recent ASFR event, hosted by Iceland in May 2015, gathered representatives from 11 states, it is worth noting that another meeting, planned for mid-2014, where the military leaders of AC member states would have discussed how to cooperate when working on peaceful tasks like SAR, had to be postponed due to political tensions (Foughty 2014). It would have been the third in a series designed primarily to reaffirm agreements made in the Arctic Council on S&R and cooperative oil-spill response (Bailes 2015).

In contrast to the open format of the ASFR, a second initiative – launched by Canada in 2012 and called the Northern Chiefs of Defense (CHOD) meetings – has been structured as an unofficial assembly of the Chiefs of Defense and senior military officials from the eight states that are members of the Arctic Council. However, it has no direct link to the Council, which, it should be remembered, has no mandate to discuss 'hard security' issues. The goal of the Northern CHOD meetings is to offer the attendees an informal opportunity to conduct direct multi- and bilateral discussions focused on issues of interest to the North, with a view to strengthening the common understanding of the practical challenges associated with operating in the region. The meetings also aim to develop a fuller appreciation of the unique challenges each of the countries face, particularly when supporting civilian authorities in such areas as emergency response. The Northern CHOD met in 2012 in Goose Bay, Canada, and in 2013 in Ilulissat, Greenland; the 2014 meeting was postponed due to tensions and strained relations with Russia. While the meetings discussed above were rather informal and retained a working character, their importance stems from the provision of opportunities to establish new, direct contacts, exchange information and plans and, last but not least, offer a highly specialized forum to develop the joint capabilities of states, both in accordance with their political preferences and their real needs.

5.6 Concluding Remarks

The development of regional cooperation around transnational risks in the arctic region will depend on many factors. At the end of the day, however, the way in which arctic and non-arctic stakeholders will react in the face of old and new challenges is the shared responsibility of all interested international actors. This responsibility will prove to be effective if it is supplemented by mutual trust developed on the basis of the implementation of existing agreements and good cooperation in all forums. Furthermore, a collective, or at least shared, understanding of regional security among the arctic and non-arctic states is more than welcome. If sustainable security is to become a helpful approach in this regard, it would be useful to develop it in relation to the already existing forms of military and paramilitary cooperation in the Arctic. Even if temporary difficulties or crises may arise, in longer term these forms of collaboration may yet prove their mettle.

Acknowledgement This chapter was drafted as part of a research effort supported by Poland's National Centre for Science's post-doctoral fellowship under the grant DEC-2011/04/S/HS5/00172.

References

ACIA. (2004). *Arctic climate impact assessment*. Cambridge: Cambridge University Press, from http://www.amap.no/documents/download/1057

AHDR-II (2014). *Arctic human development report: Regional processes and global linkages* (TemaNord No. 2014:567). Copenhagen: Nordic Council of Ministers.

Arctic Council. (2011a). Agreement on cooperation on aeronautical and maritime search and rescue in the Arctic. *Arctic council*. Retrieved from http://www.arctic-council.org/index.php/en/document-archive/category/20-main-documents-from-nuuk?download=73:arctic-search-and-rescue-agreement-english

Arctic Council. (2011b). First Arctic Council SAR exercise in Whitehorse, Canada. Retrieved from http://www.arctic-council.org/index.php/en/environment-and-people/oceans/search-and-rescue/209-sar-exercise-whitehorse

Arctic Council. (2013a). Agreement on cooperation on marine oil pollution, preparedness and response in the Arctic. Retrieved from http://www.arctic-council.org/index.php/en/document-archive/category/425-main-documents-from-kiruna-ministerial-meeting?download=1767:agreement-on-cooperation-on-marine-oil-pollution-preparedness-and-respons-in-the-arctic

Arctic Council. (2013b). Vision for the Arctic, from Arctic Council Secretariat. Retrieved from http://www.arctic-council.org/index.php/en/document-archive/category/425-main-documents-from-kiruna-ministerial-meeting?download=1749:kiruna-vision-for-the-arctic

Åtland, K. (2007). *The European Arctic after the Cold War: How can we analyze it in terms of security?* Kjeller: Norwegian Defence Research Establishment.

Bailes, A. J. K. (2015). A new arctic chill? Reactions in the North to new tensions with Russia, from Scottish Global Forum. Retrieved from http://www.scottishglobalforum.net/alyson-bailes-arctic-chill.html#_ftn3

Bailes, A. J. K., & Heininen, L. (2012). *Strategy papers on the Arctic or High North: A comparative study and analysis*. Reykjavik: University of Iceland.

Baylis, J. (2008). The concept of security in international relations. In H. G. Brauch, Ú. O. Spring, C. Mesjasz, J. Grin, P. Dunay, N. C. Behera, et al. (Eds.), *Hexagon series on human and environmental security and peace. Globalization and environmental challenges* (pp. 495–502). Berlin/Heidelberg: Springer.

Bergman Rosamond, A. (2011). *Perspectives on security in the Arctic area* (DIIS Report: 2011:09). Copenhagen: DIIS.

Berkman, P. A., & Vylegzhanin, A. N. (2013a). *Environmental security in the Arctic Ocean. NATO science for peace and security* (Series C, Environmental security). Dordrecht: Springer.

Berkman, P. A., & Vylegzhanin, A. N. (Eds.). (2013b). *NATO science for peace and security series C: Environmental security. Environmental security in the Arctic Ocean*. Dordrecht: Springer.

Chater, A., & Greaves, W. (2014). Arctic. In J. Sperling (Ed.), *Handbook of governance and security* (pp. 123–147). Cheltenham: Edward Elgar Publishing.

Cockayne, J., & Mikulaschek, C. (2008). Transnational security challenges and the United Nations: Overcoming sovereign walls and institutional silos, from International Peace Academy. Retrieved from http://www.ipinst.org/media/pdf/publications/westpoint.pdf

Conley, H. A., Toland, T., & Kraut, J. (2012). *A new security architecture for the Arctic: An American perspective : A report of the CSIS Europe program*. Washington, DC: Center for Strategic and International Studies.

Dalby, S. (2009). *Security and environmental change*. Cambridge/Malden: Polity.

Daveluy, M., Lévesque, F., & Ferguson, J. (2011). *Humanizing security in the Arctic*. Occasional publications series: no. 68. Edmonton: CCI Press.

Dupont, A. (2001). *East Asia imperilled: Transnational challenges to security. Cambridge Asia-Pacific studies*. Cambridge: Cambridge University Press.

Exner-Pirot, H. (2012). Defence diplomacy in the Arctic: The search and rescue agreement as a confidence builder. *Canadian Foreign Policy Journal, 18*(2), 195–207.

Exner-Pirot, H. (2013). What is the Arctic a case of? The Arctic as a regional environmental security complex and the implications for policy. *The Polar Journal, 3*(1), 120–135.

Foughty, P. (2014). US, Norway co-host 4th annual Arctic security forces roundtable, from U.S. European Command: http://www.eucom.mil/media-library/article/26802/us-norway-co-host-4th-annual-arctic-security-forces-roundtable

Griffiths, F. (1992). *Arctic alternatives: Civility or militarism in the circumpolar North.* Canadian papers in peace studies: 1992 no. 3. Toronto/Downsview: Science for Peace; S. Stevens.

Haftendorn, H. (2011). NATO and the Arctic: Is the Atlantic alliance a cold war relic in a peaceful region now faced with non-military challenges? *European Security, 20*(3), 337–361.

Heininen, L. (2013). "Politicization" of the environment: Environmental politics and security in the circumpolar north. In B. S. Zellen (Ed.), *Northern lights series: no. 15. The fast-changing Arctic. Rethinking arctic security for a warmer world* (pp. 35–55). Calgary: University of Calgary Press.

Hilde, P. S. (2013). The "new" Arctic – The military dimension. *Journal of Military and Strategic Studies, 15*(2), 130–153.

Hoogensen, G. (2012). Security by any other name: Negative security, positive security, and a multi-actor security approach. *Review of International Studies, 38*(04), 835–859.

Hoogensen, G., Bazely, D., Christensen, J., Tanentzap, A., & Bojko, E. (2009). Human security in the Arctic – Yes, it is relevant! *Journal of Human Security, 5*(2), 1–10.

Huebert, R., Exner-Pirot, H., Lajeunesse, A., & Gulledge, J. (2012). *Climate change & international security: The Arctic as a bellwether.* Arlington: The Center for Climate and Energy Solutions.

Johnstone, R. L. (2015). *Offshore oil and gas development in the arctic under international law: Risk and responsibility* (Queen Mary studies in international law, Vol. 14). Dordrecht: Kluwer Academic Publishers.

Kao, S.-M., Pearre, N. S., & Firestone, J. (2012). Adoption of the arctic search and rescue agreement: A shift of the arctic regime toward a hard law basis? *Marine Policy, 36*(3), 832–838.

Käpylä, J. & Mikkola, M. (2013). Arctic conflict potential: Towards an extra-Arctic perspective, from The Finnish Institute of International Affairs: http://www.fiia.fi/assets/publications/bp138.pdf

Keskitalo, C. (2007). International region-building: Development of the arctic as an international region. *Cooperation and Conflict, 42*(2), 187–205.

Kraska, J. (Ed.). (2011). *Arctic security in an age of climate change.* Cambridge: Cambridge University Press.

Kraska, J., & Baker, B. (2014). Emerging Arctic security challenges. Retrieved October 14, 2014, from Center for a New American Security: http://www.cnas.org/sites/default/files/publications-pdf/CNAS_EmergingArcticSecurityChallenges_policybrief.pdf

Kudsk, H. (2012). First live Arctic search and rescue exercise – SAREX 2012, from Arctic council. Retrieved from http://www.arctic-council.org/index.php/en/environment-and-people/oceans/search-and-rescue/620-first-arctic-search-and-rescue-exercise-sarex-2012

Le Mière, C., & Mazo, J. (2013). *Arctic opening: Insecurity and opportunity* (Adelphi series, Vol. 440). Abingdon: Routledge.

Łuszczuk, M. (2014). Regional significance of the Arctic Search and Rescue Agreement. *Rocznik Bezpieczeństwa Międzynarodowego, 8*(1), 38–50.

Munck, R. (2009). Globalization and the limits of current security paradigms. In D. Grenfell & P. James (Eds.), *Rethinking globalizations: Vol. 15. Rethinking insecurity, war and violence. Beyond savage globalization?* (pp. 33–43). London/New York: Routledge.

Nordtveit, E. (2015). Arctic council update. In M. H. Nordquist (Ed.), *Center for oceans law and policy* (Freedom of navigation and globalization, Vol. 18, pp. 139–149). Leiden: Brill Nijhoff.

Olesen, M. R. (2014). After Ukraine: Keeping the Arctic stable. Retrieved from http://www.diis.dk/files/media/publications/publikationer_2014/diis-policybrief-keeping-the-arctic-stable-web.pdf

Palosaari, T., & Möller, F. (2004). Security and marginality: Arctic Europe after the double enlargement. *Cooperation and Conflict, 39*(3), 255–281.

Rottem, S. V. (2014). The Arctic Council and the search and rescue agreement: The case of Norway. *Polar Record, 50*(03), 284–292.

Schaller, B. (2014). Confidence- & security-building measures in the Arctic: The organization for security & co-operation in Europe as a role model for the area? *Arctic Yearbook, 3*, 434–452.

Schissler, M. O. (2012). Arctic nations meet to discuss communication, maritime domain awareness strategy, from Headquarters USEUCOM. Retrieved from http://www.eucom.mil/media-library/blog%20post/24109/arctic-nations-meet-to-discuss-communication-maritime-domain-awareness-strategy

Shepherd, L. J. (2013). *Critical approaches to security: An introduction to theories and methods.* London/New York: Routledge.

Smith, P. J. (2000). Transnational security threats and state survival: A role for the military? *Parameters, 30*(3), 77–91.

Smith, G. (2008). In search of sustainable security: Linking national security, human security, and collective security to protect America and our world. Center for American Progress. Retrieved from http://cdn.americanprogress.org/wp-content/uploads/issues/2008/06/pdf/sustainable_security1.pdf

Stępień, A., Koivurova, T., & Kankaanpää, P. (Eds.). (2014). *Strategic assessment of development of the Arctic.* Rovaniemi: Arctic Centre of the University of Lapland for the European Commission.

Takei, Y. (2013). Agreement on cooperation on aeronautical and maritime search and rescue in the Arctic: An assessment. *Aegean Review of the Law of the Sea and Maritime Law, 2*(1–2), 81–109.

Tamnes, R., & Offerdal, K. (Eds.). (2014). *Routledge global security studies. Geopolitics and security in the Arctic: Regional dynamics in a global world.* London/New York: Routledge.

Trigatti, L., Bjerkemo, O.-K., & Everett, M. (2014). Agreement on Cooperation on Marine Oil Pollution Preparedness and Response in the Arctic. *International Oil Spill Conference Proceedings, 2014*(1), 1485–1496.

VanderZwaag, D. L. (2014). The Arctic Council and the future of Arctic Ocean governance: Edging forward in a sea of governance challenges. In T. Stephens & D. VanderZwaag (Eds.), *Polar oceans governance in an era of environmental change* (pp. 308–338). Cheltenham: Edward Elgar Publishing.

Voigt, C. (2009). Sustainable security. *Yearbook of International Environmental Law, 19*(1), 163–196.

Wezeman, S. T. (2012). *Military capabilities in the Arctic.* Solna: Stockholm International Peace Research Institute from SIPRI.

Wood-Donnelly, C. (2013). The Arctic Search and Rescue Agreement: Text, framing and logics. *The Yearbook of Polar Law Online, 5*(1), 299–318.

Zellen, B. S. (Ed.). (2013). *Northern lights series: no. 15. The fast-changing Arctic: Rethinking arctic security for a warmer world.* Calgary: University of Calgary Press.

Chapter 6
Measuring Community Adaptive and Transformative Capacity in the Arctic Context

Matthew Berman, Gary Kofinas, and Shauna BurnSilver

Abstract Adaptive capacity (AC) plays a prominent role in reducing community vulnerability, an essential goal for achieving sustainability. The related concept, transformative capacity (TC), describes a set of tools from the resilience paradigm for making more fundamental system changes. While the literature appears to agree generally on the meaning of AC and TC, operational definitions vary widely in empirical applications. We address measurement of AC and TC in empirical studies of community vulnerability and resilience, with special attention to the problems of arctic communities. We discuss how some challenges follow from ambiguities in the broader vulnerability model within which AC is embedded. Other issues are more technical, such as a confounding of stocks (capacity) with flows (time-specific inputs or outcomes). We view AC and TC as forms of capital, as distinct from flows (i.e., ecosystem services, well-being), and propose a set of sequential steps for measuring the contribution of AC and TC assets to reducing vulnerability. We demonstrate the conceptual application in a comparative analysis of AC in two arctic Alaskan communities responding to an increase in the price of fuel. The comparative case study illustrates some key empirical challenges in measuring AC for small arctic communities.

Keywords Vulnerability • Resilience • Fuel price • Alaska

M. Berman (✉)
Institute of Social and Economic Research, University of Alaska Anchorage,
Anchorage, AK, USA
e-mail: matthew.berman@uaa.alaska.edu

G. Kofinas
Institute of Arctic Biology and School of Natural Resources and Extension, University of
Alaska Fairbanks, Fairbanks, AK, USA
e-mail: gary.kofinas@alaska.edu

S. BurnSilver
School of Human Evolution and Social Change, Arizona State University, Phoenix, AZ, USA
e-mail: shauna.burnsilver@asu.edu

© Springer International Publishing Switzerland 2017
G. Fondahl, G.N. Wilson (eds.), *Northern Sustainabilities: Understanding and Addressing Change in the Circumpolar World*, Springer Polar Sciences,
DOI 10.1007/978-3-319-46150-2_6

6.1   Introduction

Arctic communities are coping with multiple forces for rapid change, including climate warming and associated environmental effects, land-use change, fluctuating oil prices, and potential deep cuts in public spending. Because residents of arctic communities can and do move to improve well-being (Howe et al. 2013), the ability to adapt to change is a necessary condition for sustainability (Turner et al. 2003). What resources do communities have available to them to help meet the challenges of adaptation and sustainability? How can one determine which of these resources makes a difference in outcomes? Studies of community adaptation to environmental change generally group community assets for responding into one of two types. *Adaptive capacity* (AC) refers to assets that help communities cope with change, and plays a prominent role in the vulnerability literature (Turner et al. 2003; Adger 2006; Hovelsrud and Smit 2010). *Transformative capacity* (TC) – a related concept in the literature on resilience of social-ecological systems (SES) – helps communities change to adapt (Kofinas et al. 2013; Wilson et al. 2013; Pike et al. 2010; Folke et al. 2009; Gallopin 2006).

Different authors in the large and growing literature on vulnerability and SES resilience generally use similar sets of words when they define AC or TC. The apparent consensus dissolves, however, when they apply these terms to analyze case studies of vulnerability and resilience in communities and regions. Scholars often tailor the empirical definition to their individual cases, with their unique configurations of environmental risks, historical conditions, and forces for change. Although this customization enables a rich description of a case, it greatly limits opportunities to compare cases – either over time in the same place or across different places or systems – to address analytical objectives. Comparative empirical research is essential for evaluating strategies to increase AC and/or TC to reduce community vulnerability or increase resilience.

This chapter addresses the measurement of adaptive and transformative capacity in empirical studies of community vulnerability and resilience, with special attention to the problems facing arctic communities. We discuss how some challenges to comparing measures of community adaptive capacity follow from logical ambiguities in the broader vulnerability model within which AC is embedded. Other issues are more technical, such as a confounding of stocks (capacity) with flows (time-specific inputs or outcomes). We view AC and TC as capital stocks, distinct from flows (i.e., ecosystem services, well-being) (Kofinas et al. 2013). We propose a set of sequential steps for measuring the contribution of AC and TC to reducing vulnerability, and demonstrate their conceptual application in a comparative analysis of AC in two arctic Alaska communities responding to an increase in the price of fuel.

6.2 Adaptive Capacity and Vulnerability

The concept of adaptive capacity is deeply embedded in the vulnerability paradigm from which it emerged. Understanding AC therefore requires a clear comprehension of the vulnerability paradigm. According to a leading proponent, vulnerability describes "states of susceptibility to harm, powerlessness, and marginality of both physical and social systems … guiding normative analysis of actions to enhance well-being through reduction of risk" (Adger 2006, p. 268). The definition proposed by the Intergovernmental Panel on Climate Change (IPCC) highlights the normative nature of vulnerability: "Vulnerability: the propensity or predisposition to be adversely affected" (Field et al. 2012).

The vulnerability of a social-ecological system (SES) is widely portrayed as emerging from the intersection of three essential components: exposure, sensitivity, and adaptive capacity: *exposure* refers to the extent, duration, and intensity of external forces of change affecting the system; *sensitivity* denotes the degree to which a given exposure affects or modifies the system; *adaptive capacity* (AC) is the ability of a system to accommodate change and variability without causing harm (Adger 2006; Gallopin 2006). A SES that is highly vulnerable to climate change, for example, would be potentially sensitive to substantial harmful effects from relatively modest changes in climate, with little ability to adapt (McCarthy et al. 2001).

A large number of theoretical and empirical studies use the term 'vulnerability' to frame problems of environmental risk in rural communities. Although the authors of these studies generally associate vulnerability with exposure, sensitivity, and adaptive capacity, and use the same language to define these terms, their views of the processes that generate vulnerability often diverge (Callo-Concha and Ewert 2014; Pike et al. 2010; Gallopin 2006; Smit and Wandel 2006).

Callo-Concha and Ewert (2014) discuss a model of vulnerability in which sensitivity is the outcome from the interaction of exposure (an external process) with adaptive capacity (an internal process). Their AC included feedback through a manager's perception of risk (Fig. 6.1). On the other hand, Smit and Wandel (2006) represent exposure, sensitivity, and adaptive capacity as a nested hierarchy of over-

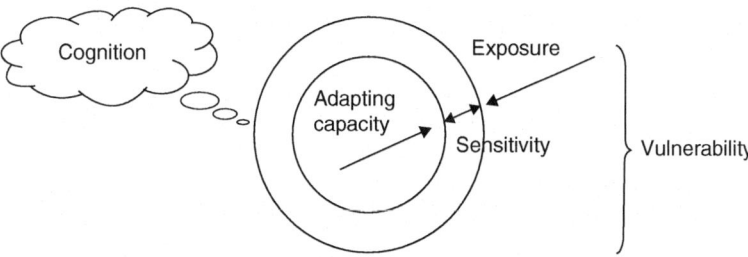

Fig. 6.1 Ontology of vulnerability: interaction of exposure, adaptive capacity, and sensitivity (From Callo-Concha and Ewert 2014)

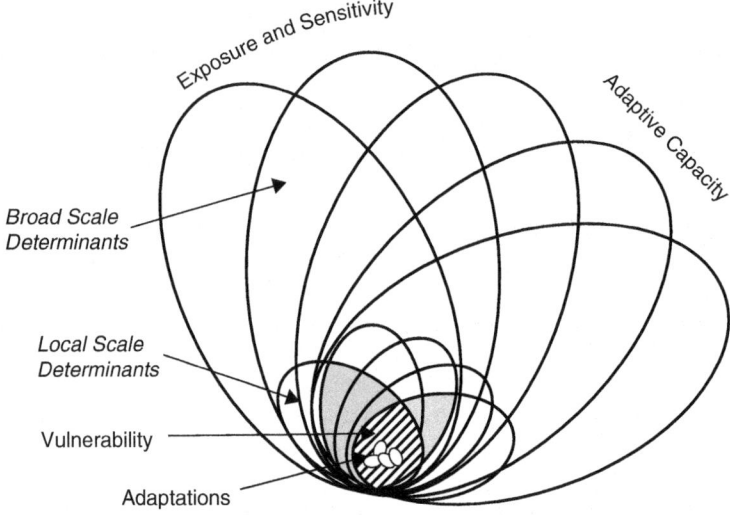

Fig. 6.2 Nested hierarchy model of vulnerability (From Smit and Wandel 2006: 286)

lapping forces driven by potentially interdependent processes (Fig. 6.2). They consistently mention "exposure and sensitivity" together as a single phrase, as if the two terms were a coupled construct.

Turner et al. (2003) discuss the role of vulnerability analysis in sustainability science, distinguishing adaptive capacity from the *capacity to cope* or respond to environmental hazards – the concept that others generally call adaptive capacity. Turner et al.'s (2003) conceptualization of adaptive capacity enables the SES to change and potentially restructure after a disturbance. Other authors, generally those approaching SES adaptation from the resilience paradigm, refer to *transformative capacity* (TC) as the capacity to reorganize in more fundamental ways, such as by changing key institutional arrangements (Kofinas et al. 2013; Wilson et al. 2013; Folke et al. 2009). To reduce confusion, we adopt Wilson et al.'s terminology, using AC to refer to the capacity to reduce the likelihood of harm generally, and TC to refer to the capacity for the SES to restructure more fundamentally.

Vulnerability, as it has developed in the environmental risk literature, has proved a useful concept to describe different aspects of susceptibility to harm. However, the widely varying interpretation of the three components – exposure, sensitivity and adaptive capacity – and how they interact to generate that susceptibility arise from the fundamental limitations of the vulnerability paradigm as a model for grounding empirical work. Models strategically simplify reality to help illuminate how things work, demonstrate cause and effect, and predict the future: for example, if A and B occur then C is a likely outcome. A good model should generate testable hypotheses; one should be able to evaluate the validity of a model by testing these hypotheses with historical data. The vulnerability model as developed in the literature falls short in three fundamental ways.

First, the model often produces no measurable outcomes (Nelson et al. 2010a). The description of vulnerability typically refers to an inferred likelihood of an imagined adverse future state, rather than to a current or historical state. No specific outcomes are defined as adverse. As Smit and Wandel (2006:286) put it:

> This conceptualization broadly indicates the ways in which vulnerabilities of communities are shaped. It does not necessarily imply that the elements of exposure, sensitivity and adaptive capacity can or should be measured in order to numerically compare the relative vulnerability of communities, regions or countries.

Second, the three-component vulnerability model contains contradictory logic. Specifically, it logically confounds sensitivity with adaptive capacity. A SES possessing greater AC could reduce its sensitivity to specific environmental hazards through greater preparedness and diversification. A related circularity of reasoning arises when researchers attempt to measure AC using income, harvests, or some other current SES outcome. While widespread (Nelson et al. 2010b; Adger et al. 2004; Yohe and Tol 2002; Kliskey et al. 2008; Sietchping 2007), such a practice is logically inconsistent. Since vulnerability means that adverse outcomes are likely (Field et al. 2012), using any of these outcomes to measure AC means only that bad outcomes facilitate bad outcomes.

More generally, the vulnerability model as developed in the literature confounds drivers, processes, and outcomes. For example, while exposure is typically envisioned as a purely external force for change, a number of scholars have described internally generated exposures, such as demographic or economic instability, or civil conflict (Young 2009; Turner et al. 2003; Denevan 1983). The vague empirical character of vulnerability, combined with the ambiguous role of adaptive capacity, prevents the model from generating testable hypotheses. Without empirically testable hypotheses, the concepts of vulnerability and adaptive capacity remain vague generalities that are difficult to observe or refute. Measurements of AC in empirical applications emerge as idiosyncratic and *ad hoc*: useful for framing individual grounded case studies of communities and regions, but offering little potential for comparative research or generalization.

6.3   Measuring the Role of AC and TC in Vulnerability and Resilience

Adger (2006) emphasizes the importance of consistent frameworks for measuring vulnerability that enable socially relevant quantitative and qualitative insights. Gallopin (2006: 302) suggests that the first step towards consistency was for researchers to agree on the definitions of "harm" and "transformation." Keeping the empirical objectives at the forefront, we begin by defining vulnerability as the probability that a specific change in an external driver (exposure) causes one or more measurable adverse outcomes to occur. The adverse outcomes would often be defined best as threshold levels for one or more measures of well-being, such as

Vulnerability

Resilience	High	Low
High	Poverty trap	Stable, high-performing
Low	Refugee risk	Opportunistic

Fig. 6.3 Characterization of social-ecological systems with respect to vulnerability and resilience

employment, income, resource harvests or nutrition. Vulnerability outcomes represent a balance between the disruptive effects of the exposure and the mitigating effects of coping activities aided by adaptive or transformative capacity. In this framework, sensitivity and vulnerability are closely related. Sensitivity refers to an incremental change in an outcome per unit change in the external driver, taking into account the effects of AC and TC. Vulnerability is the cumulative effect on the likelihood of harm.

Despite the frequent practice in the vulnerability literature of conflating resilience with negative vulnerability (Cutter et al. 2008), we follow the SES resilience literature (Kofinas et al. 2013; Walker et al. 2004) and define resilience as the likelihood that the local SES maintains its key elements of structure and function after an exposure. This definition of resilience makes it a positive rather than normative concept. However, the literature on transformative capacity is normative, often placing TC as the subset of AC that promotes SES reorganization to avoid adverse outcomes (Wilson et al. 2013; Folke et al. 2009). In our view, vulnerable communities can be embedded in resilient or non-resilient SESs (Fig. 6.3). AC acts to mitigate the effects of exposures on activities contributing to well-being, reducing vulnerability generally (Fig. 6.4). TC focuses on enhancing the potential for innovative activities or new livelihoods.

The processes by which AC and TC work to reduce vulnerability can be multiscalar and complex. Households may adjust livelihood activities to maintain flows of ecosystem services, or adjust the disposition of harvests through sharing relationships or organized food storage to avoid low consumption outcomes. They may also invest in infrastructure to manage environmental change, or move to a safer place (Berman 2013; Agrawal 2010). Communities may undertake collective actions such as formal redistribution, taxation and public borrowing, community moves, or public investments to harden infrastructure.

We propose, therefore, a procedure in five sequential steps, as summarized in Table 6.1, for the empirical analysis of the roles of AC and TC in vulnerability. The first step is to define the specific locally relevant exposure or exposures. Next, the specific indicators of vulnerability should be defined as "vulnerability of *yyy* to *xxx*," where *yyy* represents the probability that a particular SES outcome; *yyy* crosses a threshold defined as adverse, after exposure to *xxx*. Resilience, if included, would likewise be defined as "resilience of the SES to *xxx*." The third step would be to

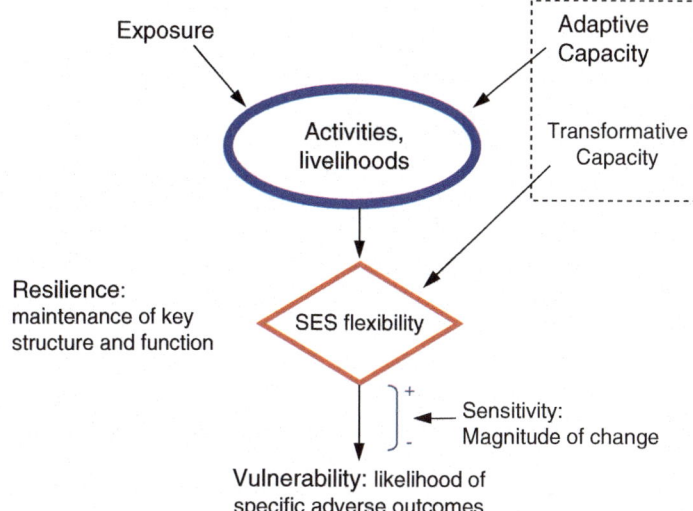

Fig. 6.4 Empirically tractable model of vulnerability and resilience

Table 6.1 Stepwise procedure for determining role of adaptive capacity (AC) in mitigating vulnerability

Step		Notes	Examples
1.	Define specific exposure or exposures to analyze	External drivers of change	Sea ice retreat, permafrost melt
2.	Define locally relevant adverse outcomes that could occur from the exposure(s)	Typically an indicator exceeding or falling short of a defined threshold	Food insecurity, infrastructure damage
3.	Determine locally relevant actions to avoid adverse outcomes	Actions households are taking or could take	Shift harvest times or target species, harvest sharing, harden infrastructure
4.	Define AC as assets that increase effectiveness of actions to avoid bad outcomes	Stocks, not flows; i.e., ecosystem health rather than ecosystem services	Knowledge systems, healthy ecosystems, effective institutions
5.	Assess contribution of AC to reducing sensitivity (*change* in outcome related to well-being) and vulnerability (likelihood of bad outcome)	Given the exposure pattern, is higher measured relative AC associated with responses that provide higher well-being?	Compare different places or same place over time, using historical data on exposure, AC, and well-being

consider relevant actions that households and communities could take to try to avoid adverse outcomes, including both preventive measures and coping responses. These activities might include new activities that result from SES reorganization as well as activities that are part of the existing SES structure and function.

Defining these activities facilitates the critical fourth step in the procedure, which is to define AC as the assets that help households undertake actions to avoid adverse *yyy* outcomes after exposure to *xxx* and increase their effectiveness. Different authors use various names to classify the types of AC assets. Regardless of the categories used, components of AC, including its subset TC, represent drivers of vulnerability. That is, AC components are assets (stocks) that generate the services that directly affect vulnerability outcomes, not the services (flows) themselves. The social construction of vulnerability implies that the delineation of AC should take into account the distribution of rights (entitlements) that enable people to access these resources; unequal access to AC assets leads to disparate vulnerability outcomes (Adger et al. 2003, following Sen 1981).

One may use the procedure outlined in Table 6.1 to generate hypotheses about the effects of specific components of AC and TC on community vulnerability to various drivers of change. The hypotheses may be tested with historical data on adaptations to past exposures, or used to generate predictions of future outcomes that can be tested as the future exposures occur. To illustrate the application of the model, we consider the vulnerability of two small, predominantly Alaska Native communities in northern Alaska to an increase in fuel costs. We focus on adaptive capacity within the current SES, leaving consideration of TC to future research.

6.4 Comparing AC in Two Arctic Alaska Communities

6.4.1 Setting

Kaktovik is a predominately Iñupiat community with about 240 residents, located on the coast of the Beaufort Sea on Alaska's North Slope. Venetie is a Gwich'in Athabascan community of about 170 people, located in northern Interior Alaska. Kaktovik and Venetie, while ethnically different, share a number of commonalities with respect to livelihoods. Households in both communities engage in a mixed economy combining subsistence resource harvests and cash incomes in household production, with social relationships facilitating flows of food and other resources among households. Cash incomes pay for motorized equipment, fuel to access resources (boats, snowmachines, all-terrain vehicles), and store-bought food. Both communities strongly articulate traditional values of sharing and cooperation (Caulfield 1983; Chance 1990; Kofinas et al. 2010; Kofinas et al. 2016) and are eligible for limited federal funding to operate tribal governments, and for Alaska Permanent Fund Dividends.

Kaktovik and Venetie, however, differ with respect to a number of key elements. Table 6.2 summarizes some potential indicators of AC for the two communities. As a coastal community, Kaktovik has access to the marine resources of the Beaufort Sea, as well as terrestrial resources, offering it a significant benefit of ecological diversity and opportunities to harvest large marine mammals such as bowhead and beluga whales. Venetie, unlike Kaktovik, has access to wood for heating fuel.

Table 6.2 Domains and indicators of adaptive and transformative capacity: summary comparison of two arctic communities

Capacity domain	Category	Indicator	Kaktovik	Venetie
Ecosystems	Ecological diversity	Main harvested species	Marine mammals, caribou, Dall sheep	Salmon, moose, caribou
	Ecosystem health	Porcupine caribou herd population[a]	197,000 (record high)	197,000 (record high)
Geography	Climate	Heating degree days[b]	19,763	17,280
	Remoteness	Round-trip airfare to Fairbanks	$1332 (April 2014)	$320 (April 2014)
Human capital	Formal education	Pct. of population 25+ with a high school degree[a]	74 %	47 %
	Traditional knowledge	Number of skilled hunters by age	High, all ages	High, all ages
Physical infrastructure	Housing quality	Median house value[c]	$110,400	$50,000
	Water-sewer system	Percentage of homes with plumbing facilities	90 % homes piped water/ sewer[d]	7 % homes flush/ haul, 93 % none[e]
Social Capital	Social Ties	Household degree (No. of ties/HH)[f]	44.6	29.2
Cultural capital	Language retention	Pct. of population 5+ speaking Native language[a]	41 %	68 %
Institutions	Local government	Main local authority	North Slope Borough	Tribal government
Financial Capital	Local revenue base	Per-capita taxable property value[g]	$2.4 million	–

[a] Medred (2014)
[b] National Weather Service climate data
[c] American Community Survey, 2009–2013 average
[d] Shepro et al. (2003)
[e] Alaska Dept. of Community and Regional Affairs, Rural Utility Business Advisor Program, Quarterly Report: 2015, October – December (Q2), Venetie: http://commerce.alaska.gov/cra/DCRAExternal/RUBA
[f] Kofinas et al. (2016). Household degree in Kaktovik includes whale feasts
[g] Office of the State Assessor (2015), 2014 appraisal data

As a North Slope community, Kaktovik benefits from its inclusion in the North Slope Borough (NSB). The NSB's ability to tax oil infrastructure has enabled it to improve local infrastructure and public utilities in Kaktovik relative to Venetie. The median home value in Kaktovik is more than double that in Venetie, and most homes have piped water and sewer facilities. In Venetie only 7 % of homes have indoor plumbing.

Most households in Kaktovik also include shareholders of the relatively wealthy Arctic Slope Regional Corporation (ASRC) and the Kaktovik Iñupiat Corporation (KIC). Venetie, on the other hand, has a history of strong tribal governance that has given the community somewhat more local autonomy. A key geographic difference is one of relative remoteness. Kaktovik is one of the most remote communities in Alaska: the lowest round trip cost of air travel to the nearest urban center (Fairbanks) in April 2014 was $1332 per person. Venetie is much less remote, with a round-trip airfare to Fairbanks costing $320.

6.4.2 Conceptual Application of Five Steps to Assess Vulnerability to Fuel Price Increase

Clearly, many differences in AC attributes relevant to vulnerability outcomes exist for the two communities, making a true hypothesis test infeasible. Consequently, we aim to achieve a lesser goal: generate a set of testable hypotheses for vulnerability to a single exposure – an increase in fuel price – and describe how they could be tested in a future analysis.

6.4.2.1 Step 1: Define Specific Exposure

Between the summer of 2007 and the summer 2008, world oil prices doubled, reaching a peak of over $140 per barrel in July 2008, before declining sharply as the global recession set in. The effect on fuel prices in Alaska was large and immediate. The average gasoline price in Alaska rose from about $4.55 in November 2007 to $5.97 in October 2008. The price increase for heating oil was even greater (DCRA 2008). Table 6.3 shows the effects in the two communities (DCRA 2007, 2008). Price data from the survey are not available for Venetie, so we use available information for Arctic Village, a neighboring community that shares a tribal government with Venetie and also receives all its oil and gasoline via air at comparable costs.

The prices for all fuel types in Arctic Village were much higher than in Kaktovik before the fuel increase. The North Slope Borough heavily subsidizes gasoline and provides heating oil for residential use at only the cost of delivery. Between June 2007 and October 2008, fuel prices rose substantially in both locations, although the

Table 6.3 Fuel cost changes 2007–2008

Community	Fuel product	June 2007 price[a]	October 2008 price[b]	Absolute change	Percentage change
Kaktovik	Heating oil, residential	$1.50	$2.00	$0.50	33 %
	Heating oil, commercial	$4.95	$6.95	$2.00	40 %
	Gasoline, retail	$3.45	$4.90	$1.45	42 %
Arctic Village	Heating oil, residential	$6.35	$9.00	$2.65	42 %
	Heating oil, commercial	$6.35	$9.00	$2.65	42 %
	Gasoline, retail	$7.00	$9.00	$2.00	29 %

[a]DCRA (2007)
[b]DCRA (2008)

Table 6.4 Hypothesized measurable outcomes from increase in fuel costs

Variable	Hypothesized outcome	Measurable outcome – Compare proportional change in:
Harvest	Decline in harvest at community scale	Total lbs harvested by community
Inter-household sharing of harvest	Decline in aggregate community sharing	Total lbs of flows between households, density of ties, number of households receiving
Food security	Decline in food security	Increase in the number of households reporting food insecurity using USDA measures
Migration	More out-migration. Decrease in base of highly productive adult community residents	Total population and demographics; sufficient population of youth to maintain school
Increase in need for public assistance	Increase in applications for heat subsidies	Increase no. of applications for heating program.

NSB fuel subsidies substantially reduced the impact on the residential cost of heating oil in Kaktovik. However, because the gasoline price in Kaktovik started at a lower base, the relative change in prices was similar in both communities.

6.4.2.2   Step 2: Define Locally Relevant Adverse Outcomes

We posit a number of hypotheses for measurable outcomes.

H1: The increase in fuel costs will be associated with a larger decline in community harvest per capita (lbs), a decline in food security (scale 1–10), and an increase in out-migration of households in Venetie relative to Kaktovik (Table 6.4).

6.4.2.3   Step 3: Determine Locally Relevant Actions to Avoid Adverse Outcomes

Households in both communities, but especially in Venetie, will be observed doing the following:

- Switching equipment for harvesting to save fuel, such as using more-efficient propeller driven boats instead of jet boats.
- Combining hunting trips with other households (e.g., leverage equipment and gas)
- Hunting closer to the community.
- Developing systems of communication between harvest parties that increase the efficiency of hunting effort.
- Altering sharing patterns: some households may share less of a household harvest with others (e.g., store more harvest for individual use), while others may share more (e.g., share widely and decrease impact overall).
- Distributing food more widely through communal feasts.
- Households with cash from employment subsidizing those without cash for fuel.
- Using or cultivating social networks to access additional resources.
- Developing wood heat technologies (long term – Venetie only)

Village councils in both communities, but especially in Venetie, may make appeals to regulatory (e.g., Alaska Board of Game) and governing bodies (Alaska State Legislature) to liberalize harvesting regulations for more hunting season flexibility or for supplemental funding to meet resultant shortfalls.

6.4.2.4   Step 4: Define AC Assets That Increase the Effectiveness of Actions (see Step 3) to Avoid Bad Outcomes

We discuss AC relevant to the specific exposure of increased fuel costs in the categories outlined in Table 6.2.

Ecosystems   Although significantly affected by climate change, regional ecosystems remain in a healthy condition, providing important services to both communities.

Geography   Although both communities experience severe cold temperatures, Venetie's location in the boreal forest offers wood as an alternative heating source. The use of wood, however, requires equipment, some fuel, and labor.

Infrastructure (and Equipment)   Both communities have storage capacities for fuel and air access for delivery of fuels. Persistent higher incomes in Kaktovik derived from employment opportunities offered by the NSB may result in wider distribution of harvesting equipment among hunting and fishing households compared to Venetie.

Human Capital   Kaktovik's higher levels of formal education suggest that it may have more capacity to respond to an increase in fuel cost, for example, by having trained staff personnel who can help households apply for fuel cost subsidies. Both communities have skilled hunters of all ages, although rigorous measures of knowledge assets are difficult to ascertain.

Social Capital This asset includes linkages both internal (bonding networks) and external to the community (bridging networks), enhancing the ability of individuals to act collectively. Both communities show evidence of strong and active traditions of food sharing and cooperation in harvesting (Kofinas et al. 2016), which may be critical for avoiding harvest shortfalls and distributing harvests to limit food insecurity. External bonds could translate into political power through governance institutions, as discussed below. Kaktovik households on average have more food and non-food sharing ties to other households than Venetie households (Table 6.2), which in part reflects the additional cooperative effort involved in organizing labor and distributing meat and *muktuk* from bowhead and beluga whaling.

Cultural Capital One measure of cultural capital is the use and retention of Native languages (ASI 2010). Recent census data show that language retention is higher in Venetie than in Kaktovik (Table 6.2). However, it is not clear that the retention of language is relevant for responding to a fuel cost increase. More important is the cultural traditions of sharing and cooperative harvesting, included in the section on social capital. Both communities have broad participation in food harvesting, with similar ratios of the number of harvesters to the number of pounds harvested (Kofinas et al. 2016).

Institutional and Financial Capital Venetie's tribal government owns 1.8 million acres of land outright. However, the region lacks marketable resources and the tribe has no rights to tax activities on its land even if substantial economic activity did exist there. The borough status of the North Slope gives it the ability to generate extraordinary wealth from the taxation of oil and gas infrastructure. The spending of tax revenues locally provides Kaktovik and other North Slope communities with a robust cash economy that, if anything, is strengthened by the rise in oil prices. Kaktovik's association with the North Slope Borough's wealth also gives it access to subsidized gasoline and home heating fuel.

On the other hand, Venetie has pursued other opportunities to reduce fuel costs. The tribal government is actively exploring the feasibility of using wood chips to heat public buildings. For a number of years, Citgo, the U.S. subsidiary of the Venezuelan national oil company, has operated a social program that offers households a voucher to purchase 100 gal of home heating oil annually. During the 2007–2008 fuel cost spike, Citgo contributed $5.3 million to native non-profit organizations to purchase fuel for more than 12,000 households (USA Today 2006). Although many villages in Alaska declined to participate, Venetie was among the communities that accepted the offer.

6.4.2.5 ˙ Step 5: Assess Contribution of AC to Reducing Vulnerability

AC in many categories is lower in Venetie. Kaktovik's access to NSB fuel subsidies reduced the absolute change in household fuel costs there, although the percentage increase in prices was similar in both places. Although some data do exist for many of these aspects of AC, evidence supporting the quantification of the effects of fuel cost increases on harvests, food security, and community population is limited. Brinkman et al. (2014) reported that increases in fuel costs over 10 years reduced the number of hunting trips taken and length of trips in Venetie and seven neighboring villages. However, the authors also noted that households were switching equipment types to save fuel.

The population declined by 18 % between the 2000 and 2010 censuses in both communities. However, a number of factors could be causing this trend, and it is not necessarily linked to increases in fuel prices. The federal/state heating fuel assistance program is available only for low-income residents. With a poverty rate of 14 %, Kaktovik had far fewer households that qualified than Venetie, which had a poverty rate of 40 % (American Community Survey 2009–2013 average), making it difficult to compare changes over time in the two communities.

6.5 Discussion

The difficulty in documenting harm, given this example of a salient and documented exposure, suggests the empirical challenges to assessing contributions to vulnerability retrospectively, let alone prospectively. Community level data are rarely available at the time scale (annual) required to test the hypotheses summarized in Table 6.4 regarding total resource harvests, flows among households, and food insecurity. The Arctic Social Indicators (ASI) project (ASI 2014, 2010) attempted to define and measure indicators of well-being in six domains that could be used to compare communities around the Arctic. However, data for many ASI indicators – local renewable resource harvests, for example – are not available at the community scale on a regular basis, if at all.

The ASI case highlights the tension between generic measures of AC that may be widely available for comparative purposes versus locally relevant measures specific to the exposure and vulnerability of interest. A number of composite measures of vulnerability and/or adaptive capacity have been developed to compare a set of places with respect to common exposures (Brose 2015; Cutter et al. 2008). However, such a prefabricated approach to assessing AC blends together indicators of qualitatively different assets, which typically vary in their importance locally and address different adaptation challenges. The weights that construct the aggregate index are inherently arbitrary and problematic for the varied cultural heritages and political economies of arctic communities.

The unique situation of the contribution of the North Slope Borough to AC illustrates the limitation of quantitative measures of AC for the Kaktovik-Venetie comparison. It would be difficult to find a comparable combination of institutional and financial capital available to small communities anywhere in the Arctic or even around the world. Understanding the local context of exposure and vulnerability is critical to a meaningful assessment. The best practice will always remain one that assesses adaptive capacity, qualitatively as well as quantitatively, and describes how each component and subcomponent contributes to mitigating the specific vulnerabilities analyzed.

6.6   Conclusion

While great progress has been made in the study of vulnerability, several problems remain. These include a lack of a common conceptual approach, including serious conceptual ambiguity in use of the terms exposure, sensitivity, and adaptive capacity. This conceptual ambiguity makes it difficult to compare cases, and generate and test hypotheses about adaptive capacity. The temporal dimension of vulnerability analysis (anticipated versus measured vulnerability) has often been overlooked.

In this paper we discussed those ambiguities and provided a framework to support a more systematic and reproducible empirical assessment of vulnerability, including the ability to compare cases. As an application of the framework, we worked through the case of exposure to increased fuel prices for two northern Alaska villages. We provide a brief narrative about the increased fuel costs in 2007–2008 to contextualize the case. Our example here is limited to describing in some detail the first four of the five steps, identifying measurable (bad) outcomes, and proposing testable hypotheses. Our data and analysis are clearly incomplete. However, we generate a series of outstanding questions about this approach, which motivates further discussion.

Acknowledgement   The research for this chapter received support from the Alaska EPSCoR program, funded by National Science Foundation award # OIA 1208927 and the State of Alaska, as well as from National Science Foundation award # ICER 1342979. Support for background research was provided by The Sharing Project of the Northwest Cooperative Ecosystems Studies Unit through US Department of the Interior Bureau of Ocean Energy Management Cooperative Agreement M07AC12496.

References

Adger, W. N. (2006). Vulnerability. *Global Environmental Change, 16*, 268–281.
Adger, W. N., Huq, S., Brown, K., Conway, D., & Hulme, M. (2003). Adaptation to climate change in the developing world. *Progress in Development Studies, 3*(3), 179–195.
Adger, W. N., Brooks, N., Bentham, G., Agnew, M., & Eriksen, S. (2004). New indicators of vulnerability and adaptive capacity. Tyndall Centre for Climate Change Research Technical Report 7, January.

Agrawal, A. (2010). Local institutions and adaptation to climate change. In Mearns, R., Norton, A. (Eds.), *Social dimensions of climate change: Equity and vulnerability in a warming world* (pp. 173–197). Washington, DC: International Bank for Reconstruction and Development (The World Bank).

ASI. (2010). *Arctic social indicators: A follow-up to the arctic human development report* (J. N. Larsen, P. Schweitzer, & G. Fondahl, Eds.). Copenhagen: Nordic Council of Ministers.

ASI. (2014). *Arctic social indicators ASI-II implementation* (J. N. Larsen, P. Schweitzer, & A. Petrov, Eds.). Stockholm: Nordic Council of Ministers. http://www.svs.is/en/home/10-all-languages-content/22-the-arctic-social-indicators-project-asi-5

Berman, M. (2013). Modeling regional dynamics of human-rangifer systems: A framework for comparative analysis. *Ecology and Society, 18*(4), 43. http://dx.doi.org/10.5751/ES-05535-180443.

Brinkman, T., Maracle, K. B., Kelly, J., Vandyke, M., Firmin, A., & Springsteen, A. (2014). Impact of fuel costs on high-latitude subsistence activities. *Ecology and Society, 19*(4), 18. doi:10.5751/ES-06861-190418.

Brose, D. A. (2015). *Developing a framework for measuring community resilience.* Washington, DC: National Academies Press.

Callo-Concha, D., & Ewert, F. (2014). Using the concepts of resilience, vulnerability and adaptability for the assessment and analysis of agricultural systems. *Change and Adaptation in Socioecological Systems, 1*, 1–11.

Caulfield, R. A. (1983). *Subsistence land use in upper Yukon-Porcupine communities, Alaska.* Juneau: Alaska Department of Fish and Game, Division of Subsistence.

Chance, N. A. (1990). *The Inupiat and arctic Alaska: An ethnography of development.* Fort Worth: Harcourt Brace.

Cutter, S. L., Barnes, L., Berry, M., Burton, C., Evans, E., Tate, E., & Webb, J. (2008). A place-based model for understanding community resilience to natural disasters. *Global Environmental Change, 18*, 598–606.

Denevan, W. M. (1983). Adaptation, variation and cultural geography. *Professional Geographer, 35*(4), 399–406.

Division of Community and Regional Affairs (DCRA). (2007). Current community conditions: Fuel Prices Across Alaska, June 2007 Update. Alaska Department of Community and Economic Development. August.

Division of Community and Regional Affairs (DCRA). (2008). Current community conditions: Fuel prices across Alaska, November 2008 Update. Alaska Department of Community and Economic Development. December.

Field, C. B., Barros, V., Stocker, T. F., Qin, D., Dokken, D. J., Ebi, K. L., Mastrandrea, M. D., Mach, K. J., Plattner, G.-K., Allen, S. K., Tignor, M., & Midgley, P. M. (Eds.). (2012). *Managing the risks of extreme events and disasters to advance climate change adaptation* (Special report of working groups I and II of the Intergovernmental Panel on Climate Change). Cambridge/New York: Cambridge University Press.

Folke, C., Chapin, F. S., & Olsson, P. (2009). Transformations in ecosystem stewardship. In F. S. Chapin, G. Kofinas, & C. Folke (Eds.), *Principles of ecosystem stewardship: Resilience-based natural resource management in a changing world* (pp. 103–128). New York: Springer.

Gallopín, G. C. (2006). Linkages between vulnerability, resilience, and adaptive capacity. *Global Environmental Change, 16*, 293–303.

Hovelsrud G.K., & Smit B., (Eds.), (2010). *Community adaptation and vulnerability in arctic regions.* Dordrecht: Springer.

Howe, E. L., Huskey, L., & Berman, M. (2013). Migration in arctic Alaska: Empirical evidence of the stepping stones hypothesis. *Migration Studies, 2*(1), 97–123. doi:10.1093/migration/mnt017.

Kliskey, A., Alessa, L., Lammers, R., Arp, C., White, D., Hinzman, L., & Busey, R. (2008). The Arctic Water Resource Vulnerability Index: an integrated assessment tool for community resilience and vulnerability with respect to freshwater. *Environmental Management, 42*(3), 523–541.

Kofinas, G., Chapin, F. S., BurnSilver, S., Schmidt, J., Fresco, N., Kielland, K., Martin, S., Springsteen, A., & Rupp, T. S. (2010). Resilience of Athabascan subsistence systems to interior Alaska's changing climate. *Canadian Journal of Forest Research, 40*, 1347–1359.

Kofinas, G., Clark, D., Hovelsrud, G. K., Alessa, L., Amundsen, H., Berman, M., Berkes, F., Chapin, F. S., Forbes, B., Ford, J., Gerlach, C., & Olsen, J. (2013). Adaptive and transformative capacity. In *Arctic Council, Arctic resilience interim report 2013* (pp. 73–93). Stockholm: Stockholm Environment Institute and Stockholm Resilience Centre.

Kofinas, G., BurnSilver. S., Magdanz, J., Stotts, R., & Okada, M. (2016). *Subsistence sharing networks and cooperation: Kaktovik, Wainwright, and Venetie Alaska.* Fairbanks: University of Alaska Fairbanks.

McCarthy, J. J., Canziani, O. F., Leary, N. A., Dokken, D. J., & White, K. S. (Eds.). (2001). *Climate change 2001: Impacts, adaptation, and vulnerability* (IPCC third assessment report, working group II. Impacts, adaptation and vulnerability.). Cambridge: Cambridge University Press.

Medred, C. (2014). AK beat: Porcupine Caribou Herd hits historic highs. *Alaska Dispatch,* March 17.

Nelson, R., Kokic, P., Crimp, S., Meinke, H., & Howden, S. M. (2010a). The vulnerability of Australian rural communities to climate variability and change: Part I-conceptualising and measuring vulnerability. *Environmental Science & Policy, 13*(1), 8–17. http://dx.doi.org/10.1016/j.envsci.2009.09.006.

Nelson, R., Kokic, P., Crimp, S., Martin, P., Meinke, H., Howden, S. M., de Voil, P., & Nidumolu, U. (2010b). The vulnerability of Australian rural communities to climate variability and change: Part II-integrating impacts with adaptive capacity. *Environmental Science & Policy, 13*(1), 18–27. http://dx.doi.org/10.1016/j.envsci.2009.09.007.

Office of the State Assessor. (2015). *Alaska taxable 2014.* Anchorage: Alaska Department of Commerce, Community and Economic Development, January.

Pike, A., Dawley, S., & Tomaney, J. (2010). Resilience, adaptation and adaptability. *Cambridge Journal of Regions, Economy and Society, 3*, 59–70. doi:10.1093/cjres/rsq001.

Sen, A. K. (1981). *Poverty and famines: An essay on entitlement and deprivation.* Oxford: Clarendon.

Shepro, C. E., Maas, D. C., et al. (2003). *North Slope Borough 2003 economic profile and census report.* Barrow: North Slope Borough, Department of Planning and Community Services.

Sietchiping, R. (2007). Applying an index of adaptive capacity to climate change in northwestern Victoria, Australia. *Applied GIS, 2*(3), 16.1–16.28. doi:10.2104/ag060016.

Smit, B., & Wandel, J. (2006). Adaptation, adaptive capacity and vulnerability. *Global Environmental Change, 16*, 282–292.

Turner, B. L., II, Kasperson, R. E., Matson, P. A., McCarthy, J. J., Corell, R. W., Christensen, L., Eckley, N., Kasperson, J. X., Luers, A., Martello, M. L., Polsky, C., Pulsipher, A., & Schiller, A. (2003). A framework for vulnerability analysis in sustainability science. *Proceedings of the National Academy of Sciences of the United States of America, 100*(14), 8074–8079.

USA Today. (2006). *Alaska villages reject Venezuela oil.* http://usatoday30.usatoday.com/news/nation/2006-10-09-alaska-venezuela_x.htm

Walker, B., Holling, C. S., Carpenter, S. R., & Kinzig, A. (2004). Resilience, adaptability and transformability in social-ecological systems. *Ecology and Society, 9*(2), 5. http://www.ecologyandsociety.org/vol9/iss2/art5

Wilson, S., Pearson, L. J., Kashima, Y., Lusher, D., & Pearson, C. (2013). Separating adaptive maintenance (resilience) and transformative capacity of social-ecological systems. *Ecology and Society, 18*(1), 22.

Yohe, G., & Tol, R. (2002). Indicators for social and economic coping capacity: Moving toward a working definition of adaptive capacity. *Global Environmental Change, 12*, 25–40.

Young, O. R. (2009). Institutional dynamics: Resilience, vulnerability and adaptation in environmental and resource regimes. *Global Environmental Change, 20*, 238–385. doi:10.1016/j.gloenvcha.2009.10.001.

Chapter 7
Political and Fiscal Limitations of Inuit Self-Determination in the Canadian Arctic

Umut Riza Ozkan and Stephan Schott

Abstract Our chapter examines the political and fiscal factors of collective capabilities for self-determination in three Inuit-dominated areas in Canada with active self-government regimes: the relatively new territory of Nunavut, the region of Nunavik (in Québec) and the region of Nunatsiavut (in Newfoundland and Labrador). We derive and measure important political and fiscal indicators identified by the literature (institutional independence, representation, local capacity and fiscal ability) that are based on the most recent information from the three regions. Our analysis indicates that most Inuit governments are still quite financially dependent and constrained. In terms of institutional independence, and representation, Nunavik lags behind the other regions due to a fragmented governance system with three separate regional public administration bodies that receive funding through provincial parent departments. In terms of local capacity, as measured by the proportion of Aboriginal people with university degree, the three regions lie well below Canadian and provincial averages. Nunatsiavut has significantly higher education levels than other Inuit regions; however, the percentage of Aboriginal people with university degrees is also higher in Newfoundland and Labrador than in Quebec and in Canada.

Keywords Fiscal independence • Self-governance • Sustainable development • Multi-level governance • Fiscal federalism

U.R. Ozkan
University of Montreal, Montreal, QC, Canada

S. Schott (✉)
Carleton University, Ottawa, ON, Canada
e-mail: stephan.schott@carleton.ca

© Springer International Publishing Switzerland 2017
G. Fondahl, G.N. Wilson (eds.), *Northern Sustainabilities: Understanding and Addressing Change in the Circumpolar World*, Springer Polar Sciences,
DOI 10.1007/978-3-319-46150-2_7

7.1 Introduction

A crucial component of sustainable livelihoods in the North is the ability, capacity, and capability to determine one's own destiny and to independently make decisions that will affect the life of future generations in the North. As arctic communities transition to mixed economies that consist of a wage-based sector and a subsistence sector (Usher et al. 2002), they face tremendous challenges in balancing individual decision-making capabilities at the household level without sacrificing collective capabilities (Hund 2004; Ford et al. 2006, 2008). Sufficient institutional independence and capacity give indigenous communities the power to preserve their traditional sector, which is essential for the survival of their crucial collective capabilities.

At the international level two important studies – the Study of Arctic Living Conditions (SLiCA) and the Arctic Social Indicators (ASI) Report (Larsen et al. 2010) – recognize the vital role of political/fiscal autonomy in achieving sustainable development in the polar region. SLiCA's survey questions on political resources focused on Aboriginal people's influence on wage employment, formal education, hunting and fishing, traditional education, as well as "aspirations for influence" and "expectations for influence" (Andersen and Poppel 2002: 314). Nevertheless, these SLiCA survey questions on political resources do not fully pay attention to indigenous people's political and fiscal abilities, and local capacity to influence decisions in the identified areas.

To address these issues, the Arctic Social Indicators Report introduced "fate control" as one of the six domains and identified quantifiable indicators to assess social well-being in the Arctic. Fate control refers to both the individual and collective capabilities of indigenous people's control over their own destiny. The fate control indicators that were suggested by the ASI do not, however, cover the actual influence of indigenous representatives or authorities in the determination of (renewable and nonrenewable) natural resource management rules, the limitations imposed by central governments on local or regional policy making, and the capacity of local or regional governing bodies to prepare and implement policies and regulations as important factors (Ozkan and Schott 2013). In this paper, we connect this specific literature on self-government in the polar region (Rodon 2014; Abele and Prince 2010; Alcantara and Nelles 2013; White 2002; Wilson 2008; Henderson 2004; Natcher et al. 2012) with the broader comparative decentralization and regional authority literature (Treisman 2002; Hooghe et al. 2008; Keman 2000; Sorens 2011). The latter literature typically focuses on the relative power of central and regional governments, but does not focus on the self-determination path through comprehensive land claims agreements and incremental paths to self-governance and devolution that exist in many arctic regions, especially in Canada.

We will focus on the political and fiscal factors of collective capabilities for self-determination in three Inuit-dominated areas in Canada with active self-government regimes: the territory of Nunavut, the region of Nunavik (in Québec), and the region of Nunatsiavut (in Newfoundland and Labrador). We derive and measure some political and fiscal indicators based on the most recent information from the three regions, and discuss the relevance of the results for self-determination, social well-being and sustainable development in these areas.

7.2   Self-Government Indicators

7.2.1   Literature Review

A number of studies focus on measuring political and fiscal decentralization and developing related indices. Hooghe and Marks (2001) proposed a regional governance index based on indicators in four areas: 'constitutional federalism', 'special territorial autonomy', 'role of regions in central government', and 'regional elections'. More recently, Hooghe et al. (2008) introduced a regional authority index by using the following indicators:

- *Institutional depth* (whether the regional authority is a deconcentrated and a general-purpose administration)
- *Policy scope* (which refers to the regional government's residual and regional policy powers)
- *Representation* (whether the region has an assembly with law-making power, whether the regional legislative assembly is directly/indirectly elected and whether the regional executive is selected by the assembly or directly elected)
- *Law making* authority (whether the regional government is represented in the national assembly and has extensive legislative authority in the national assembly)
- *Executive power sharing* (whether central and regional governments have routine meetings to negotiate policy and these meetings lead to legally binding decisions)
- *Fiscal control* (whether the regional governments have any power in influencing decisions related to the distribution of tax revenues)
- *Constitutional reform* (whether regional governments can influence the constitutional reform process).

In addition to these indicators, it is also important to focus on fiscal decentralization since "the pattern of revenue and expenditure decentralization" does not always overlap with the political division of power between the central and regional governments (Treisman 2006:291). Some studies focused on subnational control over expenditure and revenue mobilization by using indicators such as 'subnational share of total government spending', subnational expenditure-GDP ratio, and the share of intergovernmental transfers in sub-national revenues (Oates 1985; Davoodi and Zou 1998; De Mello 2000; Treisman 2006). To measure fiscal decentralization, Ebel and

Yilmaz (2002) used OECD data on fiscal design, which disaggregated sub-national revenues into tax revenues (own tax revenue and tax sharing[1]), non-tax revenues, and intergovernmental grants (general-purpose grants and specific grants).[2]

7.2.2 Geographical and Political Focus

In the Canadian Arctic four comprehensive land claims involving Inuit-dominated regions have been settled over the last four decades that led to the establishment of three regional governments (see Table 7.1): the Territorial Government of Nunavut, the Kativik Regional Government, and the Nunatsiavut Government.[3] The last is the latest regional government in Canada's Eastern Subarctic and is located in the province of Newfoundland and Labrador. The Labrador Inuit Land Claims Agreement (LILCA) of 2005 set a precedent by including self-government provisions within the land claim. Nunatsiavut is comprised of five communities (Nain, Postville, Rigolet, Makkovik and Hopedale), and is the first of the Inuit regions in Canada outside Nunavut to have achieved self-government. Although it remains part of Newfoundland and Labrador, the government has authority over many central governance areas including health, education, culture and language, justice, and community matters. While recognizing that this authority may not be used in practice in certain policy areas, this paper does not aim to measure 'policy discretion' exercised by the Nunatsiavut government or the other two Inuit regions' governance bodies.

Nunavut is Canada's youngest territory and occupies large portions of Canada's North. On 25 May 1993 the Nunavut Land Claims Agreement (NCLA) was signed by the Government of Canada, the Government of the Northwest Territories, and the Tungavik Federation of Nunavut (now Nunavut Tunngavik Inc.). In addition to the creation of management and advisory groups, and various financial considerations, the NCLA gave the Inuit of Nunavut title to approximately 350,000 sq. km of land, 35,257 sq. km of which included mineral rights (ITK (2015)). Nunavut consists of three regions (Kitikmeot, Kivalliq and Qikiqtalluk (formerly Baffin region)) with a total of 25 communities.

---

[1] 'Own tax revenue' refers to sub-national governments' control "over tax rate or a revenue tax base and rate", whereas 'tax sharing revenue' is defined by subnational governments' "limited or no control over the rate and base of a tax and the central government decision on how to split revenues" (Ebel and Yilmaz 2002:8).

[2] General-purpose grants are the grants that "may be allocated based on either objective criteria or central government's discretion", while "specific grants are earmarked for certain purposes, and the allocation may or may not be conditional across subnational governments" (Ebel and Yilmaz 2002:9).

[3] The establishment of the Beaufort/Delta Government in the Inuvialuit Settlement Region in the Northwest Territories, which is still in a negotiation process, is not included in our assessment of governance indicators.

Table 7.1 Inuit political organizations and their comprehensive land claims

	Inuvialuit settlement region	Nunavut	Nunavik	Nunatsiavut
Political Organization	Committee for Original Peoples' Entitlement (1970–1984)	Tunngavik Federation of Nunavut (1982–1993)	Northern Quebec Inuit Association (1971–1978)	Labrador Inuit Association (since 1973)
Land Claim Agreement and Year Claim was Submitted	Inuvialuit Final Agreement (1974)[a]	Nunavut Land Claims Agreement (1977)	James Bay and Northern Quebec Agreement (1973)	Labrador Inuit Land Claims Agreement (1977)
Agreement-in-Principle Settled	October 1978	April 1990	November 1974	June 2001
Beneficiary Corporation	Inuvialuit Regional Corporation (IRC) (1984)	Nunavut Tunngavik Incorporated (NTI) (1993)	Makivik Corporation (1978)	Labrador Inuit Development Corporation (LIDC) (1982)
Final Agreement Settled	June 1984	April 1993	November 1975	January 2005
Regional Government	Beaufort/Delta Government (TBA)[b]	Government of Nunavut (1999)	Kativik Regional Government (1976); Nunavik Government (TBA)	Nunatsiavut Government (Fall 2006)

Source: Bonesteel (2006)

[a]This agreement is also called the COPE agreement

[b]Provisions for an Inuit regional government, the Western Arctic Regional Municipality (WARM), were initially part of the land claim but were not included in the final settlement. Gwich'in and Inuvialuit are currently negotiating to establish some form of aboriginal self-government in the western Arctic

Although the region of Nunavik does not yet have an official legislative assembly, it does have three separate governing bodies: the Kativik Regional Government, the Kativik School Board, and the Nunavik Regional Board of Health and Social Services. On 5 December 2007, the province of Québec, the federal government and the Makivik Corporation on behalf of the Inuit of Nunavik signed the agreement-in-principle to formally create a new regional government in Nunavik. In a regional referendum on 27 April 2011, voters in Nunavik clearly rejected an amalgamation of all three separate bodies that would have put them under the authority of a new elected body called the Nunavik Assembly. Some of the stated reasons for the rejection was disappointment that the final agreement mostly amalgamated existing institutions without allowing Nunavimmiut (Inuit of Nunavik) more control over regional decision-making (Nunatsiaq Online 2011). Many residents hoped that Nunavik would become a more autonomous region, with the ability to make its own laws and to better protect the Inuit language and culture.

7.2.3 Analytical Framework

We concentrate our analysis on the degree of self-governing capabilities of these three regions, and how they vary in terms of relevant political and fiscal indicators of self-determination. Building upon this literature and our previous study, this chapter focuses on four areas of regional autonomy – institutional independence, representation, local capacity, and fiscal ability/dependence – to analyze individual and collective capabilities of indigenous control over their own destiny. Our previous work involved 'policy discretion', another area of regional autonomy that measures the variety of policies for which a regional government has responsibility and chooses to execute its authority (Ozkan and Schott 2013). Given the limited space, it would not be possible to conduct an in-depth examination of each policy area and determine whether the regional governments use their policy authority in the areas where they have authoritative competence.

The 'institutional independence' indicator measures the relative degree of power that the central government possesses to change or veto the policies (including related ordinances, by-laws, and/or legislation) of regional or local governing bodies at will (ranging from 0–3) (Sorens 2011:208). The institutional independence variable is rated '0' if there is no functioning general-purpose administration at the regional level; '1' if there is a deconcentrated,[4] general purpose[5] administration; '2' if there is "a non-deconcentrated, general-purpose, administration subject to central (or provincial) government veto"; and '3' if there is "a non-deconcentrated, general-purpose, administration not subject to central (or provincial) government veto" (Hooghe et al. 2008:124). We adjust this variable for the purpose of our study by specifying the policy areas (economic policy, cultural and language policy, education policy, health policy, social policy, environment policy) where three regional administrations may be subject to either federal government veto (as in the case of the Territory of Nunavut) or provincial government veto (by either the Province of Newfoundland and Labrador in the case of Nunatsiavut or the province of Québec in the case of Nunavik). The main motivation behind this adjustment is that all three selected Inuit regions (Nunavut, Nunatsiavut and Nunavik) have non-deconcentrated regional administrations; therefore, measurement of the provincial and/or federal government's control or veto over these regional administrations in specific policy areas becomes more important in terms of understanding the variation among the regional bodies (See Table 7.2).

---

[4] De-concentration refers to the central government's delegation of some of its policy responsibilities to its lower-level units (but not to separate public legal personalities such as elected public authorities like municipalities and regional administrations) (Hooghe et al. 2008).

[5] General-purpose administration refers to regional authorities responsible for more than one policy area (Hooghe et al. 2008).

Table 7.2 Self-determination indicators for the three Inuit Arctic regions

Indicators	Measurement
Institutional independence	'0' if there is no functioning general-purpose administration at the regional level
	'1' if there is a deconcentrated, general purpose administration
	'2' if there is 'a non-deconcentrated, general-purpose, administration subject to federal (or provincial) government veto in all of the six policy areas [economic policy, cultural and language policy, education policy, health policy, social policy (including income support and training policy, housing and family policy, and youth and child services), and environment policy]
	'3' if there is 'a non-deconcentrated, general-purpose, administration subject to federal (or provincial) government veto in five policy areas
	'4' if there is 'a non-deconcentrated, general-purpose, administration subject to federal (or provincial) government veto in four policy areas
	'5' if there is 'a non-deconcentrated, general-purpose, administration subject to federal (or provincial) government veto in three policy areas
	'6' if there is 'a non-deconcentrated, general-purpose, administration not subject to federal (or provincial) government veto' in two areas.
	'7' if there is 'a non-deconcentrated, general-purpose, administration not subject to federal (or provincial) government veto' in at most one area.
Representation	0: no regional legislative assembly, the regional executive is appointed by the provincial/federal government
	1: an indirectly elected regional legislative assembly with a regional executive appointed by provincial/federal government
	2: a directly elected legislative assembly with a regional executive appointed by provincial/federal government, or an indirectly elected assembly with dual executives appointed by regional assembly and provincial/federal government
	3: indirectly elected legislative assembly with a directly elected or assembly-appointed executive, or a directly elected legislative assembly with dual executives appointed by assembly and provincial/federal government
	4: directly elected regional legislative assembly with directly elected or assembly-appointed executive (Sorens 2011).
Local capacity	*Percentage of Aboriginal people with master's degrees (or equivalent)
	*Percentage of Aboriginal people with university and master's degree (or equivalent)
	*Occupational Capacity per 1000 Aboriginal People
Fiscal ability/ dependence	*Own source revenues/expenditures
	*Own source revenues-debt ratio
	*Debt limit/capita
	*Federal/provincial transfer/capita

The 'representation' measure assesses "the capacity of regional actors to select regional office holders: in the case of legislators, by direct election in the region, or failing that, indirect election by subnational office holders; in the case of an executive, by the regional assembly, or failing that, a mixed system of a regional/central dual executive" (Hooghe et al. 2008:129). The area of 'local capacity' reflects whether local authorities have the professional staff to formulate or implement public policies.[6] This indicator was selected because even if a region has authority in certain policy areas and its own fiscal resources, it would not be able to exercise this authority in the absence of adequate human capital. It can be measured by first disaggregating the public service occupations into categories such as policy analysts, teachers, social workers, health professionals, and scientists, and then calculating the ratio of these professionals per capita.

Certain international standards for 'local capacity' in different occupations exist. For instance, the World Health Organization (WHO) has set a minimum threshold of 23 health professionals (doctors, nurses, and midwives) per 10,000 people in order to fulfill essential maternal and health services (WHO 2012). In the absence of international standards, the local capacity indicator can still provide insights for comparing regions in terms of their relative local capacity in different public service occupations. For example, the number of public servants with Masters or equivalent degrees in each policy sector could be a useful indicator to explain the erosion or strengthening of local capacity in specific policy areas. To our knowledge, there is not a comparable data set for the Canadian Arctic that disaggregates public sector occupations or public servants' educational attainment. Given the lack of comparable data for the indicators identified above, we will instead rely on post-secondary education and occupation by sector in proportion to the Aboriginal population in each region.

For fiscal indicators, we measure the ratio of own source revenues to expenditures of the regional governments. We also examine the ratio of own source revenues to accumulated debt in order to examine the ability of the regional government to service debt out of local revenues and to identify potential dependence on other levels of government for debt guarantees. We then derive the accumulated or potential debt (in case of a debt limit) per capita as an indicator of the ability to accumulate debt. Finally we compute transfer payments from other governments per capita to assess the financial support from other sources.

---

[6] At first glance, this indicator seems highly correlated with education indicators of a given region; yet these professionals may not necessarily be the people who were born and earned their education in this region. Some argue that it is not sustainable to bring people from the South to fill the public service positions in the Arctic; yet it can also buy time for indigenous communities in the North to establish their human capital for the public sector, while benefiting from the expertise of professionals in policy-making and implementation processes.

Table 7.3 Institutional independence, policy discretion and representation		Institutional independence	Representation
	Nunavut	7	4
	Nunavik	3	0
	Nunatsiavut	5.66	3.6

7.3   Political and Fiscal Indicators for the Canadian Arctic

7.3.1   Institutional Independence

As stated above, all Inuit regions have non-deconcentrated regional administrations. In Nunavik's case, the KSB and the NRBHSS are specific-purpose regional bodies responsible for education and health care, respectively. The KRG is a general-purpose and decentralized body with supra-municipal powers like many of the regional county municipalities (*municipalité régionale de comté*) in the rest of Québec. Its council is composed of elected municipal representatives who are appointed by Inuit communities and the Naskapi First Nation of Kawawachikamach. The KRG is in charge of a number of areas, including regional and local economic development, housing, policing and civil security, Inuit hunting and fishing and wildlife conservation, childcare services, municipal infrastructure development, and employment training and income support. Each regional body "is responsible… to their parent provincial department" (Rodon and Grey 2009:9). To determine the institutional independence of Nunavik, we calculated the total score of the existing three Inuit regional bodies (Table 7.3). This is because all these three Inuit bodies in Nunavik were established by the James Bay and Northern Quebec Agreement and, therefore, it would be more appropriate to assess these three bodies together in this study, in order to understand the self-determination power of this region.

The regional bodies of Nunavik are subject to a veto by the provincial/federal government in five policy areas, with the exception of cultural and language policy. The Québec Ministry of Education cannot veto ordinances by the KSB pertaining to Inuit culture and Inuktitut (KSB 2015). Consequently, the Nunavik regional bodies are scored as a non-deconcentrated administration subject to provincial/federal government's control or veto in five out of six identified policy areas (at a score of 3).

The Nunavut territorial government has a general-purpose administration: it has responsibilities in many policy areas (housing policy, education policy, health policy, economic policy, social policy, and language and cultural policy, etc.). It is, however, subject to the federal government's control or veto in the area of environmental policy[7] (e.g. the protection of endangered species).[8] Therefore, the institutional independence score of the Nunavut government is 7.

---

[7] The Parliament of Canada can repeal any decision of the Government of Nunavut with regards to protection of endangered species within 45 days.

[8] For instance, in the past the Department of Fisheries and Oceans vetoed a decision of the Nunavut Wildlife Management Board regarding the turbot quota, in order to protect the species (http://www.cba.org/nunavut/pdf/NU_interimreport.pdf).

The Nunatsiavut regional government is a non-deconcentrated general-purpose administration with law-making power; however, the Inuit legislation made by the Nunatsiavut government can be overridden by federal or provincial legislation or the Law of General Application in certain policy areas such as environmental protection policy, health policy,[9] and income support policy (LILCA 2005). Since the Nunatsiavut Government is subject to a veto in only one of the three areas of social policy (income support and training policy, housing and family policy, and youth and child services), the veto power of federal/provincial government was considered as 'partial' in social policy.[10] Therefore, the institutional dependence of Nunatsiavut is scored between 5 and 6. More precisely, Nunatsiavut receives a score of 5.66 (the three social policy areas have been weighed equally; that is, 0.33 points has been assigned for each of these areas) with regards to institutional independence.

7.3.2 Representation in the Legislative Assembly

In terms of 'representation', Nunavik scores 0 since the Nunavik regional administration does not have legislative power but instead it only exercises 'administrative autonomy'. The KRG is one of 16 equivalent territories (*territoire équivalent* (TE)) in Québec with no Legislative Assembly or Council that has special legislative power or authority. In contrast, Nunavut has a directly elected legislature and assembly-appointed executive. According to the Nunavut Act (1993), each member of the Legislative Assembly is 'elected to represent an electoral district in Nunavut'. The Cabinet (or executive council) is directly selected by the members of the Assembly. Nunavut, therefore, receives a maximum score of 4.

The Nunatsiavut Legislative Assembly is composed of the ordinary members of the Assembly, the AngajukKak[11] of each of the five Inuit Community Governments, the Chair of each of the two Inuit Community Corporations (NunaKatiget ICC and Sivunivut ICC), and the President of Nunatsiavut. Currently, there are 18 members

---

[9] In the area of health policy, the Law of General Application prevails over Inuit Law in matters related to programs for (a) health promotion, injury prevention, disease prevention and control, environmental health prevention and control and environmental health; (b) public and community health care programs and services; addictions and substance abuse programs; (d) promotion of mental health wellness, prevention of mental health problems and the provision of mental health support services; and (e) premises, centres, facilities and buildings.

[10] There may be a mismatch between legal authority the Inuit region possesses and whether this authority is used in practice. For instance, Wilson et al. (2016) argue that the Nunatsiavut Government is not involved in areas like housing policy. Instead, the provincial government as well as regional quasi-governmental organizations like the Torngat Regional Housing Authority plays a key role. As stated above, it is beyond the scope of this study to ascertain the extent the regional bodies use their policy authority in the areas where they have authoritative competence ('policy discretion'). We try to capture, to some extent, this dimension of autonomy through measuring the local capacity in these regions.

[11] The equivalent of a Mayor.

in total (seven indirectly elected members and eleven directly elected members) in the Inuit Legislative Assembly. The ordinary members of the Assembly and the President of Nunatsiavut are directly elected to the Nunatsiavut Assembly, whereas the AngajukKaks and the Chairs of the Inuit Corporations are indirectly elected to the Nunatsiavut Legislative Assembly. AngajukKaks are elected to their positions in each of their communities. The three communities outside the land claim area – Happy Valley-Goose Bay, Mud Lake, and North West River – elect chairpersons for the Nunakatiget Inuit Community Corporation (Happy Valley-Goose Bay/Mud Lake) and the Sivunivut Inuit Community Corporation (North West River) (Labour Inuit Constitution 2005: 38–42). The Executive Council is composed of the First Minister (appointed by the President upon nomination by the Assembly) and other members (selected by the President upon the advice of the First Minister) (Labour Inuit Constitution 2005: 65–69).

Since the majority (approximately 63 %) of the Assembly's members are directly elected and the executive is appointed by the directly elected President, Nunatsiavut's representation score falls between 3 and 4 (=3.6).

7.3.3   Local Capacity[12]

When we examine the proportion of Aboriginal people with university degrees and with Master's degrees (or equivalent), we notice that it lies well below Canadian and provincial averages for Aboriginal people. Nunatsiavut has significantly higher education levels than other Inuit regions. The percentage of Aboriginal people with university degrees or above is also higher in Newfoundland and Labrador compared to Québec and in Canada (Statistics Canada 2013) (see Table 7.4).

Table 7.4 Education levels compared to federal/provincial averages

Region/Territory	Percentage of Aboriginal people with master's degrees (or equivalent)	Percentage Aboriginal people with university certificate, diploma or degree at bachelor level or above
Nunavut	0.34 % [(60/17395)*100]	1.43 % [(250/17395)*100]
Nunavik	0.14 % [(10/6850)*100]	1.02 % [(70/6850)*100]
Nunatsiavut	0.56 % [(10/1780)*100]	2.5 % [(45/1780)*100]
Canada	5.31 % [(53575/1008585)*100]	7.44 % [(75070/100885)*100]
Québec	5.26 % [(5700/108350)	7.66 % [(8310/108350)*100]
Newfoundland & Labrador	7.08 % [(1960/27645)*100]	9.36 % [(2590/27645)*100]

Source: Statistics Canada 2013

---

[12] Here, we created our own measures by adapting from World Health Organization's capacity indicators for health professions (WHO 2012).

In order to assess local capacity for self-governance and administration, we really need to have information about the number of Aboriginal and non-Aboriginal individuals in professional occupations. We could only identify occupation by sector of Aboriginal people in proportion to the Aboriginal population for all three Inuit self-government regions. Nunatsiavut clearly stands out again in comparison with Nunavut and Nunavik in terms of educational training for all occupations except for personal, protective and transportation services. Nunatsiavut also has significantly more occupational capacity in natural and applied sciences, natural resources, agriculture and related production occupations, and manufacturing and utilities (see Table 7.5). Nunavut stands out for its low proportion of indigenous employees in health, education, law and social, community and government services. Nunavik has a higher capacity than the other two regions in health occupations and sales and service occupations. It also has a significantly higher proportion of indigenous employees in education, law and social, community and government services compared to Nunavut. A more detailed analysis of non-Aboriginal professional capacity needs to be conducted to properly assess local capacity and determine if it meets acceptable standards and thresholds.

7.3.4 Fiscal Ability

Fiscal ability is the ability to have authority over budget decisions and to control crucial policy decisions and essential services and programmes. All of the arctic self-governments in this study depend on provincial and or federal transfer payments. The more independent regional governments are in raising own source revenues in proportion to their expenditures, the more authority they have in terms of independently controlling their budgets. Furthermore, fiscal independence creates certainty and a sustainable path to sovereignty. We therefore develop four relevant indicators to examine fiscal ability and dependence. Table 7.6 shows that there is quite a difference between the regions. Nunatsiavut has by far the greatest fiscal independence, partially due to mining agreements and sharing of royalties with the Newfoundland and Labrador government (Nunatsiavut Assembly 2013).

Nunatsiavut receives more transfer payments than Nunavik on a per capita basis (not on aggregate). The figure for Nunavik, however, is only based on the financial statements from the Kativik Regional Government (KRG), one of the three public administration bodies in Nunavik (KRG 2013). We could not find information about federal transfers or own source revenues and expenditures for the Kativik School Board or the Nunavik Regional Board of Health and Social Services. Based on KRG financial statements for 2013 (the most recent available year), Nunavik receives only one quarter of the transfer payments per capita compared to Nunavut, and has the lowest fiscal independence, since only about 8 % of expenditures are financed from local sources. Nunavik also stands out with respect to the proportion of own source revenues to total debt. It has significant debt that could not easily be paid back without substantial transfer payments, which makes the region quite

Table 7.5 Post-secondary education and occupation by sector in proportion to aboriginal population

	Capacity per 1000 Aboriginal people			Capacity per 1000 people
Post-secondary education degree by fields of study	Nunavik	Nunavut	Nunatsiavut	Canada
Education	12.858	10.095	17.241	32.30
Visual and performing arts, and communications technologies	2.074	2.839	3.831	13.29
Humanities	3.318	3.155	5.747	23.92
Social and behavioural sciences and law	13.273	9.621	17.241	43.61
Business, management and public administration	0.581	0.221	2.682	92.82
Physical and life sciences and technologies	0.83	1.42	0	16.15
Mathematics, computer and information sciences	0.83	2.524	3.831	17.52
Architecture, engineering, and related technologies	21.568	29.022	51.724	93
Agriculture, natural resources and conservation	1.244	2.208	7.663	9.5
Health and related fields	7.881	7.098	9.579	60
Personal, protective and transportation services	21.568	20.662	26.82	26.2
Occupation				
All occupations	347.988	291.009	381.226	
Management occupations	26.96	15.931	26.82	59.87
Business, finance and administration occupations	34.84	46.53	40.23	91.54
Natural and applied sciences and related occupations	6.221	6.151	13.41	38.05
Health occupations	11.613	4.259	9.579	34.24
Occupations in education, law and social, community and government services	79.22	48.896	82.375	65.50
Occupations in art, culture, recreation and sport	17.42	12.934	15.326	17.22
Sales and service occupations	99.129	79.18	84.291	132.81
Trades, transport and equipment operators and related occupations	63.044	65.3	70.881	79.13
Natural resources, agriculture and related production occupations	4.977	6.782	21.073	13.42
Occupation manufacturing and utilities	4.977	5.205	22.989	25.35

Source: Statistics Canada 2013

Table 7.6 Fiscal ability and fiscal dependence

Region/territory	Own source revenues/ expenditures	Own source revenues-debt ratio	Debt limit/capita	Federal/provincial transfers/capita
Nunavut	9.5 %[a]	71.3 %[a]	$11,107/18,050[e]	$41,127[b]
Nunavik	8.1%[c]	5.5 %	$17,689	$10,849[d]
Nunatsiavut	32.8 %	n/a	n/a	$15,767

[a] Based on 2012–2013 budget
[b] Based on forecasts for 2014–2015
[c] Nunavik receives about $12 million of revenues from local sources (KRG 2013)
[d] Nunavik receives around 20 % of transfer from the federal government and 80 % of transfer payments from the Province of Québec
[e] As of April 2015. Figures for Nunavut are based on Government of Nunavut (Department of Finance 2014), for Nunavik (KRG 2013) and for Nunatsiavut (Nunatsiavut Assembly 2013)

fiscally dependent. An agreement signed by the Government of Québec and the KRG in March 2004 to simplify the terms and conditions for transfers from various Québec government departments grants the KRG greater autonomy in allocating funds based on regional priorities. The minimum annual payments provided for in the coming years amount to $55 million of block funding (approximately 50 % of total provincial transfers) and are subject to indexation until 2028 (Gouvernement du Québec 2014).

Nunavut enjoys by far the largest per capita transfers from other levels of government, but its debt limit per capita was much more limited compared to the other two regions. In April 2015 the federal government raised the debt limit for Nunavut from $400 million to $650 million, which now brings it in line with Nunavik. Own source revenues/expenditures are still relatively low in both Nunavik and Nunavut, which is partially due to limited taxation and resource rent extraction power.

7.4 Conclusion

Our analysis demonstrates some of the political and fiscal limitations in the drive for Inuit self-determination in Canada's arctic regions. Overcoming these limitations will be a crucial step towards more autonomy, sovereignty and sustainable development. Our self-determination indicators demonstrate that there are meaningful differences between regions. Collective capabilities are essential for self-determination and self-preservation, and therefore, sustainable living in the Arctic. Our analysis indicates that most Inuit regional governments are still quite financially dependent and constrained. Nunatsiavut has made a step in the right direction, but it relies on royalties from mining (mostly from the Voisey's Bay mine) with a limited time line and rather unstable and declining commodity prices. More diversification is needed in terms of other minerals (besides nickel and copper) and in the development of other economic sectors. Nunatsiavut also does not receive the transfers per capita that Nunavut obtains. In terms of local capacity Nunatsiavut seems to have a more promising future, but all three regions lag behind national and provincial standards in almost all professions, and especially in the health field. In terms of institutional independence and representation Nunavik lags behind the other regions due to a fragmented governance system with three separate regional public administration bodies that receive funding through provincial parent departments and through special programs (Rodon and Grey 2009). Nunavik would probably benefit from a more formal, coordinated and official self-government arrangement. The region has an agreement-in-principle and is currently in negotiations to determine the details of its governance regime.

Our discussion of political and fiscal indicators in the three Inuit-dominated regions is far from conclusive. We hope it will encourage more in-depth discussion about different factors that determine self-determination capabilities in each region. The indicators we examined are important in providing information about building sustainable and independent self-governing regions. We should try to measure these indicators more accurately and consistently over time and by region, and not judge progress just by the readily available indicators such as the number of jobs created or the average income generated in a given region. Also, future research can focus on other components of the regional governance structure, especially Aboriginal corporations. It would be interesting, for example, to see the impact of these corporations on economic and social development in the Arctic.

References

Abele, F., & Prince, M. J. (2010). A little imagination required: How Canada funds territorial and northern Aboriginal governments. In A. M. Maslove (Ed.), *How Ottawa spends 2008–2009: A more orderly federalism?* (pp. 82–109). Montreal: McGill-Queen's University Press.

Alcantara, C., & Nelles, J. (2013). Indigenous peoples and the state in settler societies: Toward a more robust definition of multilevel governance. *Publius: The Journal of Federalism, 44*(1), 1–22.

Andersen, T., & Poppel, B. (2002). Living conditions in the Arctic. *Social Indicators Research, 58*, 191–216.

Bonesteel, S. (2006). *Canada's relationship with Inuit: A history of policy and program development*. Ottawa: Indian and Northern Affairs Canada. Retrieved from http://www.aadnc-aandc.gc.ca/eng/1100100016900/1100100016908.

Davoodi, H., & Zou, H. F. (1998). Fiscal decentralization and economic growth: A cross-country study. *Journal of Urban Economics, 43*(2), 244–257.

De Mello, L. R. (2000). Fiscal decentralization and intergovernmental fiscal relations: A cross-country analysis. *World Development, 28*(2), 365–380.

Ebel, R.D., & Yilmaz, S. (2002). *On the measurement and impact of fiscal decentralization* (Vol. 2809). Washington DC: World Bank Publications.

Ford, J. D., Smit, B., & Wandel, J. (2006). Vulnerability to climate change in the Arctic: a case study from Arctic Bay, Canada. *Global Environmental Change, 16*(2), 145–160.

Ford, J. D., Smit, B., Wandel, J., Allurut, M., Shappa, K., Ittusarjuat, H., & Qrunnut, K. (2008). Climate change in the Arctic: Current and future vulnerability in two Inuit communities in Canada. *The Geographical Journal, 174*(1), 45–62.

Gouvernement du Québec. (2014). Consolidated financial statements of the Gouvernement du Québec. Retrieved from http://www.finances.gouv.qc.ca/documents/Comptespublics/en/CPTEN_vol1-2013-2014.pdf

Government of Nunavut (Department of Finance). (2014). Budget 2014–2015 Fiscal and Economic Indicators. Retrieved February 1, 2015, from http://www.gov.nu.ca/sites/default/files/files/Finance/Budgets/2014-15%20budgets/2014-15_Fiscal_and_Economic_Indicators_EN.pdf

Henderson, A. (2004). Northern political culture?: Political behaviour in Nunavut. *Études/Inuit/Studies, 28*(1), 133–154.

Hooghe, L., & Marks, G. (2001). Types of multi-level governance. *European Integration Online Papers (EIoP), 5*(11), 1–32.

Hooghe, L., Marks, G., & Schakel, A. H. (2008). Operationalizing regional authority: a coding scheme for 42 countries, 1950–2006. *Regional and Federal Studies, 18*(2–3), 123–142.

Hund, A. (2004). From subsistence to the cash-based economy: Alterations in the Inuit family structure, values, and expectations. In *Annual meeting of the American Sociological Association*, Hilton San Francisco and Renaissance Park Hotel, San Francisco. (Vol. 19, p. 2008).

Inuit Tapitiit Kanatami. (2015). Nunavut land claims agreement signed, Retrieved November 12, 2015, from https://www.itk.ca/historical-event/nunavut-land-claims-agreement-signed

Keman, H. (2000). Federalism and policy performance. A conceptual and empirical inquiry. In U. Wachendorfer-Schmidt (Ed.), *Federalism and Political Performance* (pp. 196–227). London: Routledge.

KRG (2013). Kativik regional government annual report 2013, Retrieved February 1, 2015, from http://issuu.com/krgcomms/docs/krg_2013_ar?e=14751414/10540525

KSB (2015). Kativik School Board. Retrieved July 17, 2015, from http://www.kativik.qc.ca/about-kativik-school-board

Larsen, J. N., Fondahl, G., & Young, O. (2010). Introduction: Human development in the Arctic and Arctic Social Indicators. In J. N. Larsen, P. Schweitzer, & G. Fondahl (Eds.), *Arctic social indicators: Follow-up to the Arctic Human Development Report* (pp. 11–28). Copenhagen: Nordic Council of Ministers.

LIC. (2005) Labrador Inuit Constitution. Retrieved from http://www.nunatsiavut.com/wp-content/uploads/2014/03/IL%202005-02%20-%20E.pdf

LILCA. (2005) Labrador Inuit land claim agreement. Retrieved from https://www.aadnc-aandc.gc.ca/DAM/DAM-INTER-HQ/STAGING/texte-text/al_ldc_ccl_fagr_labi_labi_1307037470583_eng.pdf

Natcher, D., Felt, L., & Procter, A. (Eds.). (2012). *Settlement, subsistence, and change among the Labrador Inuit: The Nunatsiavummiut experience.* Winnipeg: University of Manitoba Press.

Nunatsiaq Online. (2011). "Nunavik votes "no" in April 27 NRG referendum", Retrieved October 5,2015, from http:// www. nunatsiaqonline. ca/stories/article/287756_nunavik_says_no_in_NRG_referendum/.

Nunatsiavut Assembly. (2013). Budget Act 2013-IL 2013-02. Retrieved February 1, 2015, from http://www.nunatsiavut.com/wp-content/uploads/2014/03/IL%202013-02%20-%20E.pdf

Nunavut Act. (1993). Statues of Canada, 1993, c. 28. http://laws-lois.justice.gc.ca/eng/acts/N-28.6/

Oates, W. E. (1985). Searching for Leviathan: An empirical study. *The American Economic Review, 75*(4), 748–757.

Ozkan, U. R., & Schott, S. (2013). Sustainable development and capabilities for the polar region. *Social Indicators Research, 114*(3), 1259–1283.

Rodon, T. (2014). "Working together": The dynamics of multilevel governance in Nunavut. *Arctic Review on Law and Politics, 5*(2), 250–270.

Rodon, T., & Grey, M. (2009). The long and winding road to self-government: The Nunavik and Nunatsiavut experiences. In F. Abele, T. J. Couchene, F. L. Seidle, & F. St-Hilaire (Eds.), *The Art of the State IV: Northern Exposure; Peoples, Powers and Prospects in Canada's North* (pp. 317–343). Montreal: Institute for Research on Public Policy.

Sorens, J. (2011). The institutions of fiscal federalism. *Publius: The Journal of Federalism, 41*(2), 207–231.

Statistics Canada. (2013). Nunavik, Nunavut and Nunavik. National Household Survey (NHS) Aboriginal Population Profile. 2011 National Household Survey. Statistics Canada Catalogue no. 99-011-X2011007. Ottawa. Released November 13, 2013. Retrieved January 16, 2015, from http://www12.statcan.gc.ca/nhs-enm/2011/dp-pd/aprof/index.cfm?Lang=E

Treisman, D. (2002). *Defining and measuring decentralization: A global perspective*. Unpublished manuscript.

Treisman, D. (2006). Explaining fiscal decentralisation: geography, colonial history, economic development and political institutions. *Commonwealth & Comparative Politics, 44*(3), 289–325.

Usher, P., Duhaime, G., & Searles, E. (2002). The household economic unit in arctic Aboriginal communities, and its measurement by means of a comprehensive survey. *Social Indicators Research, 61*(2), 175–202.

White, G. (2002). Treaty federalism in northern Canada: Aboriginal-government land claims boards. *Publius: The Journal of Federalism, 32*(3), 89–114.

Wilson, G. (2008). Nested federalism in arctic Quebec: A comparative perspective. *Canadian Journal of Political Science/Revue canadienne de science politique, 41*(1), 71–92.

Wilson, G. N., Alcantara, C., & Rodon, T. (2016). Multilevel governance in the Inuit regions of the territorial and provincial north. In M. Papillon & A. Juneau (Eds.), *Canada: The state of the federation, 2013: Aboriginal multilevel governance* (pp. 43–64). Kingston: Institute of Intergovernmental Relations.

World Health Organization. (2012). Achieving the health-related MDGs. It takes a workforce! Retrieved March 15, 2012, from http://www.who.int/hrh/workforce_mdgs/en/index.html

Chapter 8
The Social Life of Political Institutions Among the Nunavik Inuit (Arctic Québec, Canada)

Caroline Hervé

Abstract This chapter proposes to analyze the nature of the relationship between Inuit of Nunavik (arctic Québec, Canada) and the political organizations that rule their collective lives. It intends to examine what role and place people assign to these organizations in the context of self-governance in Nunavik. Using a relational approach, in order to take into account the singular ontology reflected in Inuit discourses and practices, this chapter shows that Nunavik Inuit do not see themselves separated from political institutions, but rather in a continuous relationship with them. Their attempt to achieve political autonomy in the region is not designed to attain complete separation with the Canadian federal government and with the Québec provincial government, but to establish a new relationship with them, where they would be recognized as equal partners. This chapter is based on long-term fieldwork in Nunavik communities and on research in governmental archives, which aimed to analyze power relationships among Inuit in Nunavik. It reveals that the relationships between people and the political entities are structured in the same way as their interpersonal relationships. Nunavik Inuit endowed their political institutions with agency and assigned them a specific place in their web of life.

Keywords Self-governance • Political institutions • Relational perspective • Ontology • Nunavik

8.1 Introduction

In the northern landscape of self-governance, Nunavik (arctic Québec) distinguishes itself by the difficulties it has encountered in attaining political autonomy in spite of its rich experience in regional administration. The location of the region within the province of Québec, imposing its supremacy in the beginning of the 1960s, and the

C. Hervé (✉)
Interuniversity Centre for Aboriginal Studies and Research, Québec, QC, Canada
e-mail: caroline.herve@ciera.ulaval.ca

© Springer International Publishing Switzerland 2017
G. Fondahl, G.N. Wilson (eds.), *Northern Sustainabilities: Understanding and Addressing Change in the Circumpolar World*, Springer Polar Sciences, DOI 10.1007/978-3-319-46150-2_8

discrete but still effective responsibility of the federal government have shaped a unique path to self-governance. The vigorous political dissensions between Nunavimmiut at the end of the 1970s, pacified today but still sensitive, and the effort to overcome them in a democratic way created a complex field of interrelated causes for Nunavik's complicated political history.[1] This chapter continues this reflection, proposing to leave aside institutional logics and rely instead on ontological rationalities.

To that purpose, the chapter analyzes how Nunavimmiut[2] cohabit with the political institutions that rule their collective lives today. The objective here is to use an ontological perspective to read in between the lines of their discourses and to understand the properties and societal placement Nunavimmiut attribute to federal and provincial governmental institutions. This will provide an opportunity to better understand where Nunavimmiut situate political institutions into their web of life.

Following Blaser et al. (2010), we use a relational perspective as an interpretative grid. The ontological approach incorporates the cultural anthropological trend of combining the description of variations in cultural representations and practices with grasping the substance of multiple worlds "where basic inferences are made about the kinds of beings the world is made of and how they relate to each other" (Descola 2014:273). The objective of this perspective is not only to understand the nature of the link between human beings but also their relationship with non-human entities. Western ontology divides, classifies, and isolates entities (Latour 1991); whereas in indigenous cosmopolitics, another "mode of being" is privileged (Descola 2014). Indigenous cosmologies are built upon relationships with the word that surrounds them, visible or immaterial:

> The cosmologies, socialities and historicities of such peoples include non-human agencies (and persons), in others words, their worlds are predominated by an ethos of inclusiveness and co-existence (where positionality and negotiation prevails), as opposed to the ethos of exclusiveness characteristic of the modern constitution and ontology (where relativism and hegemony prevail). (Poirier 2008:76)

Nowadays, few social scientists ask for the recognition of nonhumans as political actors or as an issue in politics (Blaser et al. 2010; De la Cadena 2010; Stengers 2010).

Concerning Inuit, it is now clear that they see themselves in constant relationships with other entities such as ancestors, animals, spirits and other nonhumans (Ouellette 2000; Laugrand and Oosten 2009; Laugrand 2013; Dupré 2014; Pernet 2014). To better understand Inuit socio-political organization, therefore, it is important to take into account their relationships with the world, visible or immaterial, that surrounds them (Oosten et al. 2001). L. Koperqualuk (2011) has started to explore the importance of religion in the current political life in Nunavik, an interesting work that should be continued. The objective of this chapter is however not

---

[1] For details on Nunavik political history, see Duhaime (1992) and Rodon (2009).

[2] Inhabitants of Nunavik.

to examine the place of nonhumans in the political arena in Nunavik but to analyse the place Nunavimmiut attribute to their political institutions in their web of life.

This chapter draws on the research from my Ph.D. dissertation in anthropology which analysed the dynamics of power relationships among Nunavik Inuit. Through a reflexive fieldwork approach, this research pointed out a high prevalence of cooperative practices among Nunavik Inuit and showed how sharing is central, not only in social interactions, but also in the bonds with nonhuman entities, such as spirits or political organizations, and in the shaping of power relationships (Hervé 2013). This research is the result of more than one and a half years of fieldwork in four communities located in the three main regions of Nunavik and a few weeks in three other communities.[3] The fieldwork was supplemented by research in federal and provincial archives in order to analyze the relationship between Inuit and various governments during the last century.[4] For this chapter, two community consultations were analyzed: the Neville-Robitaille Commission (NRC 1970)[5] and the Nunavik Commission (NC 2000).[6] Social media sites, such as the Facebook page "Nunavik and the Nunavik Regional Government", were particularly rich in Nunavimmiut perspectives about the most recent political events in their region. A self-distancing reflection was useful to identify the power games and identity issues that came up in the ethnographic research, especially when analyzing power relationships (Leservoisier 2005; Hervé 2010).

8.2   Relational Power in Inuit Nunavik

Even though the Québec and Canadian governments brought their respective administrations to the region in the early 1960s and imposed new positions of power and new perspectives of leadership on the Nunavimmiut, numerous aspects of their informal social regulation are still effective today. In Nunavik, like in other parts of *Inuit Nunangat* (Inuit territory), social harmony is highly valued (Briggs 2000). Even when a person has a role in leadership, numerous social and linguistic mechanisms tend to counterbalance this asymmetrical relationship.

---

[3] The communities where I spent most time were Ivujivik, Puvirnituq, Inukjuak, and Kuujjuaq. My position in the field varied from sharing the life of Nunavik families for a few months to living as a *Qallunaaq* (a white) in the North and having my own house. I conducted 100 interviews with Inuit and non-Inuit. For further details concerning the methodology of my doctoral research, please refer to Hervé (2013).

[4] In this text, LAC will refer to Library and Archives Canada (files Department of Indian Affairs/ District of the Administrator of the Arctic). BANQ will refer to Bibliothèque et archives nationales du Québec (files Direction générale du Nouveau-Québec).

[5] The Neville-Robitaille Commission was mandated in 1970 to consult Nunavimmiut about the transfer of the region from federal to provincial responsibility.

[6] The Nunavik Commission held a public consultation in each community of Nunavik in 2000 with the mandate to collect Nunavimmiut' views on self-governance and to prepare the future negotiations toward a regional government in the region.

Outside the hierarchy imposed by modern institutions, it is clear that the social order has been regulated through relationships (Damas 1963; Rouland 1979; Rasing 1994; Hervé 2013). Nunavimmiut choose their power figures based on the amount of food, equipment, relationships, knowledge or wealth possessed by each individual in relation to the needs of the community. Power figures are not only those who possess wealth, but also those who redistribute it; their capacity to share and help others makes them respected and influential. When Nunavimmiut talk about the way in which they came to power, they always explain that they did not claim it, but that the power came to them. They emphasize the fact that power was given to them by another group, person or entity, either in a family setting, in the political field, or in religious life (Hervé 2013).

Power in Nunavik is strongly relational. Power figures can change depending on context and collective needs, unless they accumulate different kind of wealth and impose themselves as permanent leaders (Saladin d'Anglure 1967). As opposed to stratified societies, in which individuals are classified and separated into hierarchies, relationships of power create solidarities within Inuit societies; people are not separated, but instead reunited through cooperative practices (Hervé 2013). Sharing or helping is also a way of reinforcing one's belonging to a group: "You are helping me, so you are part of my *ila-*[7]", explained an elder in Inukjuak.[8] Helping or sharing is a social obligation and there is strong pressure upon people to respect this principle. Nunavimmiut explain that they cannot say "no" to their elders, which may include older siblings, parents, grand-parents, or the oldest members of their extended family, when they are requesting help.[9] Those who do not receive any help from others are considered outcasts (Hervé 2013).

There is no word in Inuktitut to say "to order" or "to obey", but in Nunavik people use the word *maliktuq* (to follow). This word is used in all the domains of the social life: following game, following a leader or the government, following the law, and following the *Ten Commandments* (Therrien 1997). It is recognized that following someone puts another in a position of power; however, Nunavimmiut emphasize the fact that each individual is, first and foremost, independent and master of their own life. Taamusi Qumaq defines the choice to follow someone as a component of the person: "Inuk: He possesses a breath, an intelligence, a soul, a practical sense and he may follow what he appreciates or he can decide not to follow" (Qumaq 1991). A young Inuk, explaining the relationship between young people and elders, emphasized this subtle distinction: "*Well they command not so much it is just that we follow them.*"[10] People also use the term *naalaktuq* (listening attentively to someone) or *angaajuq* (to say "yes"); the latter often being used in the context of a

---

[7] *Ila-* is a radical always added with a suffix, which expresses the flexible concept of "the part of a whole". *Ilagiik* can refer to blood relatives, companion in a hunting, travel or work setting (Schneider 1985:61).

[8] Workshop of the CURA « Inuit Leadership and Governance in Nunavut and Nunavik: Life Stories, Analytical Perspectives, and Training » (Inukjuak, March 2011).

[9] Interview with Saviarjuk Usuarjuk, Ivujivik, 12/08/2009.

[10] Interview with Lucassie, Ivujivik, 04/05/2010.

cooperative relationship. All these terms, used to designate the action of obeying a request or an order, convey the importance of the free will. Those who follow have decided to do so; those who listen can decide on their future actions; those who say "yes" could have said "no". People used to say that this ability to choose is what distinguishes humans from dogs, who must follow their master.

Analyzing this dynamic helps us to identify the ambiguous nature of the concept of power.[11] Power imposes itself on the group, but it is controlled by many social constraints. It is at the same time respected and contested. There is a singular dynamic between the individual and the group. If this dialectic tension between the individual and the group can also be found in western societies, it is not grounded the same way. As Ingold (1986:130) explains:

> The difference, I believe, between the individuality of Western man and that of the hunter-gatherer is that whereas the former is conceived as an autonomous agent prior to his entry into social relations, the autonomy of the latter has its foundation in his incorporation, through generalised relations of sharing, in an unbounded social collectivity. This commitment to the whole does not so much limit as underwrite the expression of individual autonomy.

8.3 Preserving Social Harmony with Political Organizations

When examining Nunavimmiut discourses about their political institutions, it is manifest that the importance of social harmony is not limited to interpersonal relationships but applies to political entities as well. This is obvious when Nunavimmiut talk about self-governance in their region. In several indigenous groups, the western concept of political autonomy which is understood as the separation of one group from another, is not well received or understood (Blaser et al. 2010). The same can be observed in Nunavik where the idea of having a self-government is not supported by everybody. During the different consultations held in the region in the last 40 years,[12] a few sceptics voiced their doubts concerning the need to set up self-government in their region (NC 2000, Inukjuak). One of the most important aspects emphasized by Nunavimmiut was the preservation of good relationships with and between political entities and ensuring that everybody participates in the decision-making process.

For Nunavimmiut, preserving good relationships with federal and provincial governments and political organizations is possible if these institutions consult with the communities on a regular basis to understand their needs and views on collective matters. During the different public consultations held in Nunavik, participants criticized the provincial and federal governments for having made decisions in

---

[11] This "ambiguity of power" has already been highlighted by political anthropologists concerning African chieftainships (see Balandier 1967; Muller 1980; Adler 1982).

[12] The Neville-Robitalle Commission in 1970; the Nunavik Constitutional Committee (mandated to develop a strategy for political autonomy by consulting first Nunavimmiut in 1988–1989); and the Nunavik Commission in 2000.

previous decades without consulting them. It was only in 1959 that Inuit representatives were invited, for the first time, by the federal government, to speak about their needs and concerns during a meeting of the Eskimo Affairs Committee.[13] Even when regional representation was encouraged and regional and municipal structures were developed in the 1960s, people felt alienated by the lack of communication with governmental administrations. This criticism has been strong over the past 40 years, and directed at all levels of government (Anon. 1984; see also Nunavik 2011).

The Nunavimmiut's criticisms of political organizations concern concrete aspects of democratic life. Throughout the negotiation processes for self-government in Nunavik, Nunavimmiut expressed their will to choose their representatives, to decide on the negotiation procedures, and even on the referendum rules. In 2000, when the commissioners came in Kuujjuarapik to listen to people's views on self-government, the mayor, Robbie Tookalook, complained about the fact that people did not choose the representatives. The visitors explained that they were not negotiators but only commissioners (NC 2000, Kuujjuarapik). In Ivujivik, people criticized the fact that everything was already decided (NC 2000, Ivujivik). In 2011, before the referendum on Nunavik regional government, the same complaints were expressed and people regretted the lack of communication from the negotiation team (Rogers 2011).

The lack of consultation is not only seen as a dysfunction of the political entities; it is also seen as the source of a personal suffering. The feeling of wellness is deeply linked to the importance of being connected with others. During the Nunavik Commission consultations in 2000 in Inukjuak, a woman regretted publicly having been ignored by one of the regional organizations:

> Whenever Makivik comes around, or when they have their annual meetings, I have never been the subject of them turning their heads toward me. Yet I am the subject of their meetings, because I am a member of theirs. I have reached middle adulthood in this condition (of isolation). I wonder if this condition will continue to exist while my daughter grows up. (NC 2000, Inujuak)

During the Neville-Robitaille and Nunavik Commissions, several people even expressed the feeling of being treated like dogs because they were not consulted by the Québec and federal governments: *"The two governments went ahead without consulting with the Eskimos and Indians–that I consider prejudice, making proposals without consulting. It is like we were treated like dogs,"* complained Silassie Cookie, representative from Kuujjuarapik, in 1970 (NRC 1970, Kuujjuarapik). As was mentioned previously, the dog is considered by Inuit as an animal with no independence which is always following her or his master. When they are not invited to participate in political debate, Nunavimmiut feel they are ignored, poorly treated, and perceived as animals. For Nunavimmiut, a true self-government should guarantee consultation and the full participation of people during each step of the political life of the region.

---

[13] Before they were not considered ready to discuss the future of their region (see LAC, Arctic Conference, Report of the opening session, 1957).

8.4 Political Autonomy: Renegotiating Its Place in a New Relationship

A good government, like a good leader, is defined by Nunavimmiut as the one who shares resources with others (Dorais 2001; Hervé 2013). In the context of negotiating for self-government in Nunavik, people assert that the main task of the new government would be to help them and especially the needy such as elders, by sharing money, materials, or knowledge (NC 2000, Ivujivik; NC 2000, Salluit). Some Nunavimmiut also ask governments to pay for their fuel or even their phone bills. They pressure governmental institutions to share their wealth. If some perceive this as Nunavimmiut giving over their responsibilities for managing their collective life, it can also be understood as the sign of a strong and persistent cultural pattern.

Many times during the last decade Nunavimmiut spoke about the government, either provincial or federal, as a human, when they refer to it as *ataatak* (father) (Filotas 1984). But this characterization implies a relationship of power: people have to "follow" the political institutions and to "say yes" to his requests. This is obvious within the context of political dissensions in Nunavik, especially in the conflict between the Northern Quebec Inuit Association (NQIA) and Inuit Tungavingat Nunamini (ITN).[14] When ITN members who rejected the *James Bay and Northern Quebec Agreement* (JBNQA) talk about their counterparts who signed the agreement, they explained that NQIA members did not dare to say "no" to the government:

> Maybe they thought they couldn't say "no" to the government. They thought the government is so powerful. It is because it is so powerful that we have to agree everything with what the government says. [...] All we could have to say: "yes, yes, yes" to the government. "We follow you whatever you say." [...] This agreement is going to exist all the time and the government is going to have a power, power, power. Always, always. And their power will be increasing, increasing, increasing until we finish. Sure, that is what it is. (E. Sallualuk, Interview, 8 June 2010)

The moral would be that people say "yes" to political entities as an Inuk says "yes" to their oldest sibling, their parents, or the one they decide to follow because they possess needed or coveted material or immaterial goods. As Uitangak put it in his unpublished essay on Nunavik political history: "the Inuit have always followed the law of the others' society and because the Agreement [JBNQA] is law, one has to follow, like it or not." (Uitangak 1994:24).

Rather than talking about separation from the state, Nunavimmiut view political autonomy as a process of establishing a new relationship with it. They do not want to "follow" anymore; rather they want to be considered equal partners. To achieve that new relationship, Nunavimmiut feel that they must master new knowledge and

---

[14] The NQIA was engaged in the negotiations of the *James Bay and Northern Quebec Agreement* and some of its members signed the controversial agreement in 1975. The same year, the villages of Puvirnituq, Ivujivik and half of the inhabitants of Salluit decided to create an association (ITN) to oppose the agreement. They were mainly rejecting two paragraphs that planned the extinguishment of the Nunavimmiut's territorial and aboriginal rights.

skills. When Nunavimmiut started to talk about political autonomy within the co-op movement that was developed in the region in the 1960s, it was recognized that learning and mastering relevant knowledge was the first step to being able to run a government (BANQ 1971). But, as long as the Nunavimmiut do not have financial or technical strength, they think that the federal and provincial governments must continue to assist their communities (BANQ 1971). This position, however, is not perceived as one of dependence but as a form of progress towards autonomy for Nunavimmiut. The possession of new knowledge and wealth gives them the authority to elevate themselves as an equal partner to the governments and to impose their opinion; to say "no", and eventually, to refuse to follow.

Nourished by new knowledge and a political awareness within the co-operative movement, some Inuit started to claim political autonomy and expressed it officially for the first time during the Neville-Robitaille commission in 1970 (Simard and Duhaime 1981). Afterwards, a few of them founded ITN and dared to reject the *JBNQA* (Qumaq 2010). If saying "no" to the government was still inconvenient in the 1970s and 1980s, it changed in the 2000s. During the debates surrounding the referendum of April 2011, on the Final Agreement for the creation of the Nunavik Regional Government (NRG), the "no" group found a new legitimacy. In the public debate, two groups emerged, a "yes group" and a "no group". The latter criticized the fact that the political promise contained in the agreement was not strong enough, proposing only to amalgamate the existing regional organizations and forgetting to preserve Inuit culture. The electorate's rejection was complete: 66 % of the residents of Nunavik voted "no". Before the vote, some people expressed their relief to be able to say "no": "*Alianatuuq* [pleasant], the word "NO" has never sounded so good" posted an Internet user on the Facebook group page for Nunavik and the Nunavik Regional Government (Nunavik 2011). In 2011 it seemed easier for Nunavimmiut to oppose the provincial or federal governments. They were mastering important skills and knowledge that gave them more confidence. Self-governance was therefore a project easier to imagine and to materialize for a larger majority.

8.5   Conclusion

This chapter shows the importance of considering political autonomy in Indigenous peoples from a relational perspective and continues the discussion initiated by Blaser et al. (2010) in their book on globalization and autonomy. In Nunavik, it is clear that the idea of political autonomy as a complete separation from the federal and provincial governments is not well perceived. Since the 1960s, Nunavimmiut have expressed their determination to preserve a good relationship with the Québec and federal governments, and their desire to receive more services. Thus, for Nunavimmiut, achieving self-governance means negotiating a new position in their relationship with these governments (i.e. being considered as an equal partner). This echoes what Poirier highlighted when she pinpointed the importance of "position" and "negotiation" in the relational ontologies of indigenous peoples (Poirier

2008:76). In this new relationship, Nunavimmiut still continue to receive government services, but on the basis of a complementary exchange of wealth, skills, and knowledge.

The relationships between Nunavimmiut and political institutions are structured the same way as their interpersonal relationships. Social harmony is of same importance in relationships between people and with political institutions. It is expected that a good government helps people in need in the same way that a good leader shares their wealth with others. Governments, as power figures, are respected and followed. Personal or political autonomy is achieved when mastering relevant skills or possessing wealth. In that sense, political institutions are not seen as separated from the people, but in continuity with them, and they take a particular place in the Nunaviummiut web of life, allowed by the very flexible nature of the concept of *ila-* (whole) in Inuktitut.

Should political institutions be considered as nonhumans? The response to this question would need more extensive research. What is certain is that Nunavimmiut see similarities between humans and political institutions, which evokes their animist nature (Descola 2014) and that they see themselves in a continuous relationship with them. "A government, while looking like one creature, contains a great deal of elements" said an Inukjuamiuq in 2000 (NC 2000, Inukjuak). Political entities are thus endowed with agency, with a form of animacy, a concept that Ingold defines as follows: "it is the dynamic, transformative potential of the entire field of relations within which beings of all kinds, more of less person-like or thing-like, continually and reciprocally bring one another into existence" (Ingold 2006:10). The importance of taking into account nonhumans in the analysis of indigenous politics opens in the meantime the necessity to reintroduce relational ontologies in our own perspective on and practices of politics.

To ensure a sustainable politics in Nunavik, the different levels of government should take into account the importance accorded by Nunavimmiut to the development and preservation of long-term relationships with and between political institutions. Creating a trusting relationship with the people is possible if the different actors make an effort to consult and strengthen collaboration at all levels. But meaningful consultations means more than just listening; it means incorporating Nunavimmiut ideas into policies. Self-governance in Nunavik will have to integrate new tools to ensure democratic participation if they want to fully satisfy the people. Under these conditions, Nunavimmiut will feel respected and fully recognized as political partners.

Acknowledgements I want to thank the several organizations which financially supported my doctoral research: Social Sciences and Humanities Research Council of Canada (SSRHC), Institut Polaire Français Paul-Émile Victor (France), SSRHC CURA Inuit Leadership and Governance in Nunavut and Nunavik, Consulat Général de France à Québec, the Ministère Québécois des Relations Internationales, the Faculty of Social Sciences and the Department of Anthropology of Université Laval, and the Northern Scientific Training Program (Canada). I also want to thank the National Science Foundation for its financial support for my participation in ICASS 2014 in Prince George, Canada.

References

Adler, A. (1982). *La mort est le masque du roi: La royauté sacrée des Moundang du Tchad*. Paris: Payot.

Anon. (1984). General conclusions based on information from the Task Force's consultation trip, *Taqralik*, May: 21. Retrieved from http://collections.banq.qc.ca/ark:/52327/2165381

Balandier, G. (1967). *Anthropologie politique*. Paris: Presses universitaires françaises.

BANQ [Bibliothèque et Archives nationales du Québec], Compte-rendu de la reunion des agents de la DGNQ (21–23 mars 1967), Minutes of a meeting between community council members and Gilles Massé (18 février 1971). Serie E78, S30.

Blaser, M., De Costa, R., McGregor, D., & Coleman, W. D. (2010). Reconfiguring the web of life: Indigenous peoples, relationality, and globalization. In M. Blaser et al. (Eds.), *Indigenous peoples and autonomy. Insights for a global Age* (pp. 3–26). Vancouver: UBC Press.

Briggs, J. (2000). Conflict management in a modern Inuit community. In P. P. Schweitzer et al. (Eds.), *Hunters and gatherers in the modern world: Conflict, resistance and self-determination* (pp. 110–124). New York: Berghahn Books.

Damas, D. (1963). *Igluligmiut kinship and local groupings: A structural approach*. Ottawa: Department of Northern Affairs and National Resources.

De La Cadena, M. (2010). Indigenous cosmopolitics in the Andes: Conceptual reflections beyond "politics". *Cultural Anthropology, 25*(2), 334–370.

Descola, P. (2014). Modes of being and forms of predication. *HAU: Journal of Ethnographic Theory, 4*(1), 271–280.

Dorais, L.J. (2001). Maqainniq et kiinaujaliurutiit. Un village inuit dans le Québec d'aujourd'hui, Globe : Revue internationale d'études québécoises, *4*(1), 53–70.

Duhaime, G. (1992). Le chasseur et le minotaure : itinéraire de l'autonomie politique au Nunavik. *Études/Inuit/Studies, 16*(1–2), 149–177.

Dupré, F. (2014). La fabrique des parentés. Enjeux électifs, pratiques relationnelles et productions symboliques chez les Inuit des Îles Belcher (Nunavut, Arctique canadien). Dissertation, Université Laval.

Filotas, G. (1984). Impressions from the Arbvitarneq. Manuscript, Retrieved from http://www.makivik.org/fr/public-documents/

Hervé, C. (2010). Analyse de la position sociale du chercheur : des obstacles sur le terrain à l'anthropologie reflexive. *Les Cahiers du CIÉRA, 6*, 7–26.

Hervé, C. (2013). "On ne fait que s'entraider." Dynamiques des relations de pouvoir et construction de la figure du leader chez les Inuit du Nunavik (XXe siècle-2011). PhD dissertation, Université Laval.

Ingold, T. (1986). Comment. In A. Béteille et al. Individualism and equality [and comments and replies]. *Current Anthropology, 27*(2), 129–130.

Ingold, T. (2006). Rethinking the animate, reanimating thought. *Ethnos: Journal of Anthropology, 71*(1), 9–20.

Koperqualuk, L. (2011). *Puvirnitumiut religious and political dynamics*. Master's Thesis, Université Laval.

Latour, B. (1991). *Nous n'avons jamais été modernes*. Paris: La Découverte.

Laugrand, F. (2013). Les Inuit face aux changements climatiques et environnementaux. *Communication, 31*(2). Retrieved January 23, 2015, from http://communication.revues.org/4458; doi:10.4000/communication.4458

Laugrand, F., & Oosten, J. (2009). *Inuit shamanism and Christianity in the Canadian Arctic. Transitions and transformations*. Montréal/Kingston: McGill-Queens University Press.

Leservoisier, O. (2005). *Terrains ethnographiques et hiérarchise sociales. Retour réflexif sur la situation d'enquête*. Paris: Karthala.

Muller, J. C. (1980). *Le roi bouc émissaire. Pouvoir et ritual chez les Rubuka du Nigeria central*. Québec: Serge Fleury.

NC. (2000). Hearing transcriptions of the Nunavik Commission. (A copy of the transcript can be consulted at Makivik Society, Quebec City.)

NRC. (1970). Hearing transcriptions of the Neville Robitaille Commission. (A copy of the transcript can be consulted at Makivik Society, Quebec City.)

Nunavik (2011, April 26). Nunavik and the Nunavik regional government. Facebook Group. Consulted on https://www.facebook.com/groups/189933581040333/?fref=ts

Oosten, J., Laugrand, F., & Rasing, W. (2001). *Interviewing Inuit elders. Perspectives on traditional law*. Iqaluit: Nunavut Arctic College.

Oullette, N. (2000). *Tuurngait et chamanes inuit dans le Nunavik occidental contemporain*. Master, Université Laval.

Pernet, F. (2014). *La construction de la personne au Nunavik. Ontologie, continuité culturelle, et rites de passage*. Dissertation, Université Laval.

Poirier, S. (2008). Reflections on indigenous cosmopolitics–poetics. *Anthropologica, 50*, 75–85.

Qumaq, T. (1991). *Inuit Uqausillaringit. Ulirnaisigutiit.*. Québec: Association Inuksiutiit Katimajiit and Avataq Cultural Institute.

Qumaq, T. (2010). *Je veux que les Inuit soient libres de nouveau*. Québec: Les Presses de l'Université du Québec.

Rasing, W. (1994). 'Too many people': In *Order and non-conformity in Iglulingmiut social process*. Nijmegen: Katholieke Universiteit Faculteit der Rechtsgeleerdheid.

Rodon, T., & Grey, M. (2009). The long and winding road to self-government: The Nunavik and Nunatsiavut experience. In F. Abele et al. (Ed.), *Northern exposure: Peoples, powers, and prospects for Canada's North* (pp. 1–26). The art of the state, vol. 4. Montréal: Institute for Research in Public Policy.

Rogers, S. (2011). Nunavik debates new government on Facebook. Nunatsiaq News, 8April. Retrievedfromhttp:// www. nunatsiaqonline. ca/ stories/ article/080411_nunavik_debates_ new_government_on_facebook/

Rouland, N. (1979). Les modes juridiques de solution des conflits chez les Inuit. *Études/Inuit/Studies, 3*.

Saladin d'Anglure B. (1967). L'organisation sociale traditionnelle des Esquimaux de Kangiqsujuaq (Nouveau-Québec). Dissertation, Université Laval.

Schneider, L. (1985). *Ulirnaisigutiit. An Inuktitut-English dictionnary of northern Quebec, Labrador and eastern arctic dialects*. Québec: Presses de l'Université Laval.

Simard, J.-J., & Duhaime, G. (1981). Praxis autochtone et stratégies techno-bureaucratiques. L'épisode de la consultation de l'hiver 1970 au Nouveau-Québec; ses tenants et ses aboutissants. *Recherches amérindiennes au Québec, 9*(2), 115–132.

Stengers, I. (2010). Including nonhumans in political theory: Opening the Pandora's box? In B. Braun & S. J. Whatmore (Eds.), *Political matter: Technoscience, democracy, and public life* (pp. 2–22). Minneapolis: University of Minnesota Press.

Therrien, M. (1997). Inuit concepts and notions regarding the Canadian justice system. In D. Brice-Bennett et al. (Eds.), *Legal glossary/Glossaire juridique* (pp. 250–275). Iqaluit: Nunavut Arctic College.

Uitangak, J. (1994). *Walk! Don't walk! Squaring off the land of the Inuit. An essay on the historical and political developments of Nunavik*. Manuscript.

Part II
Challenges to Sustainability

Chapter 9
Gendered Consequences of Climate Change in Rural Yakutia

Liliia Vinokurova

Abstract A survey of rural communities in North and Central Yakutia provides an opportunity to identify and analyze the social consequences of natural disasters related to climate change, with a focus on the gender dimension. It was found that in situations of natural disasters archaic patterns of gender behavior tend to appear. The social anxiety of rural residents differentiates by gender. The shrinking of "living space" is noted by representatives of both sexes: the reduction of the areas of hayfields, pastures, deterioration of the transport and communications infrastructure under the influence of climatic changes. The men, mostly middle-aged and older, are concerned about the prospect of the loss of traditional occupations. Men consider it necessary to preserve traditional occupations and are more oriented to stay in ancestral lands. The women of these age groups are more concerned about the threats to health and safety associated with the effects of climate change. Women are more willing than men to change either location or occupation or both. The gender and age gap in life strategies is a serious threat to the safety (sustainability) of the rural communities of Yakutia.

Keywords Yakutia • Social consequences of climate change • Gender manifestation • Natural disasters • Sustainability

9.1 Rural Yakutia

Yakutia, also known as the Sakha Republic, is located in northeastern Siberia and is the largest constituent unit in the Russian Federation. Established to acknowledge the Sakha (Yakut) people, it is also likely the largest ethnic province in the world. In

L. Vinokurova (✉)
Institute for Humanities Research and Indigenous Studies of the North, Siberian Branch, Russian Academy of Sciences, Yakutsk, Sakha Republic (Yakutia), Russian Federation
e-mail: lilivin@mail.ru

© Springer International Publishing Switzerland 2017
G. Fondahl, G.N. Wilson (eds.), *Northern Sustainabilities: Understanding and Addressing Change in the Circumpolar World*, Springer Polar Sciences, DOI 10.1007/978-3-319-46150-2_9

Table 9.1 Population of Yakutia: Urban and rural (according to official state censuses)

Year	1939	1959	1970	1979	1989	2002	2010
Total	413.8	487.4	666.7	851.8	1094.1	949.3	958.5
Urban	111.5	238.3	375.7	521.9	732.0	610.0	614.5
Rural	302.3	249.1	291.0	329.9	362.1	339.3	344.0

Source: Chislennost' 2004:10; Results of 2010 All-Russia population census. Retrieved from: http://www.gks.ru/free_doc/new_site/perepis2010/croc/Documents/Vol1/pub-01-04.pdf

the twenty-first century, its rural areas remain "ethnic niches" for its indigenous peoples: the Sakha, Evenki, Even, the Dolgan, Yukagir and Chukchi.

A unique cultural characteristic of rural Yakutia, compared to rural areas in other Russian regions, is a relatively high persistence of traditional indigenous economic activities that are based on the use of natural resources, especially land and water. This phenomenon presents fertile ground for both historical studies and the monitoring of the current status of these activities. Surprisingly, social processes in rural Yakutia remain relatively understudied. From a historiographical perspective, over the past 50 years, it has been rural economic activities that have received most scholarly attention. Important social aspects of rural life, such as the social well-being of the people, their daily life, perceptions of the environment and of global peace have not been the subject of scientific interest.

Rural Yakutia underwent a long series of administrative and economic reforms in the second half of twentieth century. As a result, the way of life of the rural population changed and the network and structure of settlements has been transformed. According to the 1979 Census, one-third of the population in Yakutia was concentrated in large villages, and only 726 rural settlements existed, whereas in the early 1960s there were more 4400 rural settlements.

The Soviet campaign to eliminate "unpromising villages" affected Yakutia as well as other regions. Traditional winter and summer settlements located on the lands of state farms disappeared. Villagers feared the consolidation of their settlements and population transfers to larger centers which were mandated by the Soviet authorities. For them, it meant not only economic costs and damage, but also social costs, including a break with their heritage, a "loss of roots"(Vinokurova 1993; Filippova 2007).

In the 1970s, as a result of in-migration by employees of the mining industry, rural residents became a minority in the population of Yakutia. Rural population decline was not yet due to a mass migration from rural areas or urban centers: indeed, the population of the villages was increasing due to natural "causes" (i.e. higher birthrate) as shown in Table 9.1.

According to the 2010 Census, the 586 rural villages of Yakutia had a total population of 344,000. Women represented 53 % of the overall population of the Republic, and 51 % in rural areas. The residents of rural communities live in difficult conditions, and are trying to adapt to a significant number of internal and external challenges. For example, Yakutia is rich in water resources, and most of the population resides close to the region's thousands of lakes and rivers. This reality is

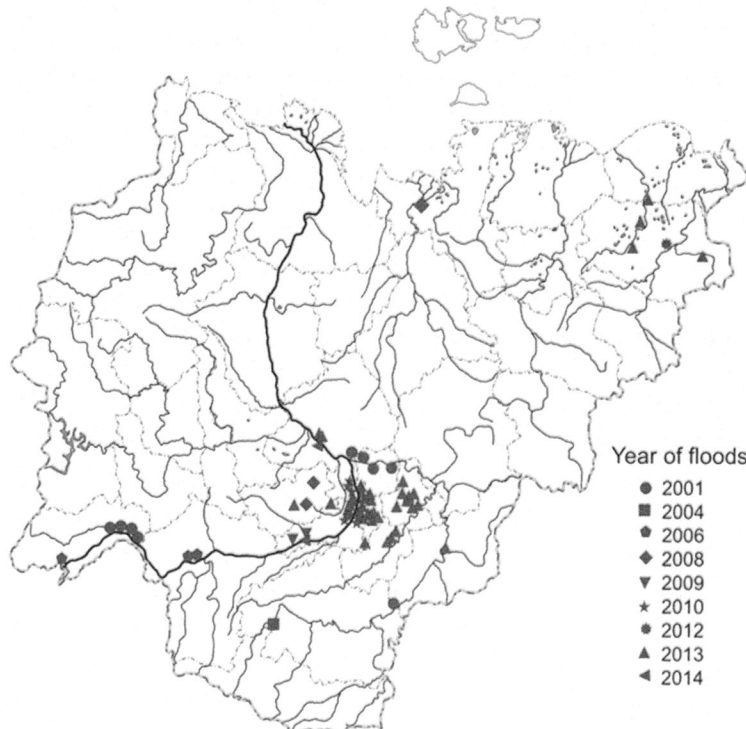

Year of floods

● 2001
■ 2004
✿ 2006
◆ 2008
▼ 2009
★ 2010
✹ 2012
▲ 2013
◄ 2014

Fig. 9.1 Data on occurrence of floods (Source: Filippova 2009:345; data on 2012, 2013, 2014. Retrieved from: http://www.mchs.gov.ru; www.ria.ru; www.newsruss.ru; www.ykt.ru; www.sakh-news.ru; www.yakutia.info.ru)

both an economic advantage for rural settlements and a serious natural hazard. Most of rural population resides within the basins of major rivers such as the Lena, Vilyui, Aldan, and Kolyma. Since most of the communities are located in forested areas along the riverbanks, these communities are potentially exposed to both floods and wildfires. We have mainly studied the impact of regular floods (Fig. 9.1) on rural communities, although wildfires are also an important feature of rural life.

It was in this historical context that the rural population experienced the collapse of the Soviet system. At the end of the Soviet period, rural life in Yakutia found itself in 'reverse development', almost as if going backwards in history. During field work in the 1990s, I observed a virtual return to the age-old ways of subsistence hunting and fishing, as well as the revival of previously abandoned fishing and agricultural areas such as small farms and old homesteads (Vinokurova et al. 2004). The traditional knowledge and adaptive skills of the ancestors demonstrated their lasting value. Traditional Sakha cattle and horse breeding, and reindeer herding in the arctic areas, along with fishing and hunting, saved the rural population from starvation. The land with its natural resources was their only refuge. Therefore, the 'health' of the land is a vital issue in today's rural communities.

The period of economic upheavals in the 1990s created many challenges for rural Yakutia. The prolonged multi-decade recession in the agricultural sector, exacerbated by the collapse of the Soviet farming system, resulted in high rural unemployment rates. Consequently, outward migration from rural to urban areas within Yakutia still remains high. Unemployment and poverty are painful realities for rural residents, especially for those who live in remote and small villages. They have no possibility to join or benefit from the rapidly changing economic situation and have been particularly affected by the breakdown of the social service system. Naturally, these processes include gendered dimensions, and affect relations between men and women. Thus, the Soviet-period gender contract, whereby patriarchal attitudes were the backbone of the family despite the broad participation of women in the state's economy and social life, was destroyed. In the post-Soviet period, new gender behaviors emerged in rural communities, with each gender assuming behavior models associated with the opposite gender.

One of the characteristics of this ongoing shift is that rural women are playing an increasingly active role in (public) economic life; for instance, by starting their own businesses. Women are assuming more prominent roles in local government and in community work, as well as in non-governmental and non-commercial organizations. By the 2000s, there were more than 300 women's community groups in rural Yakutia. Women have demonstrated the ability to meet the challenges of the free market, often more so than men. This was especially true in the 1990s (Vinokurova et al. 2004). This gender shift also manifested itself in the transformation of male social identity in Yakutia, including social behavior and life strategies (Vinokurova 2010).

9.2   Gender Aspects of the Consequences of Climate Changes

Recently, the sustainability of rural indigenous communities in Yakutia has been closely connected to climate change. Climate change entails increased seasonal fluctuations, changes in the environment and, in particular, more frequent natural disasters (e.g., floods, major forest fires). These directly impact the economy and social life of the rural population. In this article, my attention will focus on the devastating floods that have occurred in the countryside.

This chapter covers research carried out in rural regions of Yakutia from 2001 to 2014. The research team collected historical, statistical, sociological and visual data in the Verkhoaynskiy, Olekminskiy, Ust-Aldanskiy and Khangalasskiy *Ulusy* (counties; singular *ulus*) of the Republic of Sakha (Yakutia); areas that have experienced numerous floods. We have also used the results of field studies in the Tattinskiy and Megino-Khangalasskiy Ulusy for comparative analysis. We conducted special surveys and analyzed their results. Personal communications and interviews with rural residents, and their oral stories, have added new perspectives and ideas to our study, and have been of significant interest. Interviewing hunters and horse breeders has been an especially rewarding experience, as has listening to the incredible life

Table 9.2 Perception of social and natural phenomena by rural residents, in %

Most of all you and your family are concerned the following social and natural phenomena	Verkhoyansky ulus		Olekminsky ulus	
	Men	Women	Men	Women
Alcoholism, drug addiction	37.5	47.1	36.4	79.3
Health issues	18.75	23.5	18.2	10.3
Natural disasters (floods, forest fires)	43.75	70.6	54.5	31.0
Environmental conditions	18.75	11.8	27.3	20.7
Climate change	62.5	82.4	27.3	44.8

Source: Author's field work, 2001–2014

stories of local old-timers. In total, we analyzed more than 80 questionnaires and dozens of interviews and opinions from four focus groups. We realize that our results are preliminary because ours is one of the first studies of the impacts of climate change in a region of rural Russia. The social consequences of climate change in Yakutia are only starting to attract the attention of researchers (Filippova 2011; Crate 2013a; Vinokurova 2014).

Residents of rural communities acknowledge the fact of global climate change. Although the implicit and explicit hazards of natural disasters and the problems caused by climate change are not their priority, rural women, especially in the middle and old-age groups, noted climatic challenges as one of the threats to private or family safety. Women in the villages who were surveyed perceived the effects of climate change in the form of natural disasters and/or the destruction of communications as a threat to their families, especially to their children. And it is only after this, that they mention the effect of disasters on the economic problems and sociocultural environment of their settlements. Of course, they are acutely aware of the negative social and economic consequences of climate change.

Community concern with climate change is due to the various social and economic problems specific to Russia, and especially Yakutia with its severe climatic conditions. Among the acute problems in rural areas our respondents noted the following:

- Unemployment;
- The lack of social supports and guarantees for welfare;
- Threats for private safety due to criminals, alcoholism, and drug abuse; and
- Health care needs of adults and children.

Threats from nature and climate are mentioned only after other, more immediate challenges. As indicated in Table 9.2, opinions on climatic changes vary by ulusy, and likely depend on the time that the survey is conducted: we visited Verkhoyanskiy Ulus directly after a flood. Natural disasters, including huge fires during the summer of 2014, which came very close to some of the villages in which we carried out research, have long posed as a considerable threat and socio-economic stressor. Against this backdrop, the consequences of climate changes are perceived by the majority as inevitable.

The number of the floods in rural and traditional areas is significant. From the moment when a preliminary flood warning is issued, until the emergency advisory warning is lifted, and often beyond that point, rural residents face harsh and often life-threatening conditions. This is particularly the case when a rise in the water level is combined with snow storms. In the event of rapid flooding, when the warning is often late or non-existent, residents can only save themselves, not their property.

It is interesting to observe gender differences in the perception of external threats to well-being and health. Women are concerned about such risks more than men. For example, in the area affected by major floods in the Verkhoyanskiy and Olekminskiy Ulusy, the proportion of women who consider alcoholism, drug addiction, and the threat of climate change as risks is higher than that of men. The survey for this project was carried out in the village of Kyllakh, in the Olekminskiy Ulus, a village moved to a different place due to regular flooding (Filippova 2011). Most of the local women recognized alcoholism and drug abuse as the most important social threats. It is possible that their position as wives and mothers of families that suffer from these social evils manifests itself in their assessment of risk. But in the Verkhoyanskiy and Olekminskiy Ulusy the number of women who perceive climate change as causing restlessness and anxiety is higher than men.

Among rural men, we found that the latent, or non-obvious, impacts of climate change are experienced through the local environment, and are observed in greater detail by those members of the communities who spend more time outdoors. Such members, who are predominantly men, frequently raise concerns over the future of traditional ways of life. There appears to be one implication of this concern: our surveys have shown a high propensity of men who want to leave the village because of regular natural disasters; about half of the male respondents in the survey planned to leave due to frequent floods. Interviews and private conversations highlighted the following factors: (1) male young and middle-age groups expressed their emotions following the stressful experiences they underwent to save their families and property; (2) young men (18–35 years of age) often had long planned to leave and when provided with monetary compensation related to flooding they could implement their plan.

Women in the Verkhoyanskiy and Olekminskiy Ulusy expressed their willingness to adapt to the impacts of climate change, even given the increase of its negative effects. The percentage of females willing to adapt to even the most powerful effects of climate change is 4–5 % higher than the percentage of men, in both ulusy. There is a clear difference among different age groups in terms of the emotional perception of the visible damage of the surrounding landscape such as constant high water levels in the reservoirs, the formation of new thermokarst lakes due to melting permafrost, and waterlogged farmlands. It is a painful and sad situation for older members of the rural population, but for younger groups, such destruction is considered commonplace. Young men and women view these phenomena as inevitable parts of rural life. Nevertheless, they consider the current living conditions in the countryside unsatisfactory and intend to build a different life for themselves and their children.

In those regions of Yakutia that have experienced floods, we collected and generalized the gender behavior of adult men and women during disasters and found that responses to disasters such as floods and wild fires were gendered. We observed that during a flood, men are engaged in the traditional roles of rural males, evacuating women and children, rescuing and protecting property and possessions, evacuating cattle, removing pets, and monitoring the situation to ensure public safety. Women, on the other hand, take care of children and the elderly, help in the evacuation of possessions, and cook for their families, neighbors and refugees, and volunteers. In times of natural disasters in rural settlements of Yakutia, therefore, both men and women revert to their traditional roles.

With regards to emotional reactions to natural disasters, there is obviously a huge amount of stress. We note, however, that in traditional communities residents rarely display emotions. It appears that some even have difficulties expressing how they feel about the event. We have observed that sometimes women are more vocal, while men are traditionally more reserved.

In the aftermath of a flood, there is a lot of cleanup that needs to be done. Following a disaster, women and teenagers take part in making an inventory of property while men's activities are characterized by organizing meetings in order to mobilize people into action. The men also rescue livestock, repair buildings and fences, and generally attempt to maintain public safety and security. Women initiate social charity work, set up meetings with local authorities, and get the teenagers involved in the relief work where possible.

It is obvious that a disaster not only mobilizes, but also unites and consolidates, the inhabitants (both women and men) of rural communities. Healthy adult men participate in public works while women exhibit traditional behavioral patterns by taking care of children and organizing impromptu public dining rooms for volunteers and refugees. Drunkenness and crime practically disappear during this general mobilization; even the boundaries of the social hierarchy are temporarily lifted. We found that during disasters that villagers with different social status were very tolerant to each other and worked together for common safety. Such behaviour ended after the disaster was over.

Figure 9.2 demonstrates financial losses following floods. The year 2001 was an outlier because the flood in that year affected a major town, Lensk, as well as rural communities. But generally, the trend is towards the increasing economic costs of flooding.

Aside from the official data, it is important to note that there are the real human impacts that such disasters have on individuals and families in rural and remote areas. The loss of homes, livestock and possessions not only has an immediate impact on people but it also affects their long-term plans.

Gender behavior of rural residents during and after natural disasters also varies by age group and social status. Natural disasters frequently serve as a catalyst for migration and relocation of young and educated people who have a better chance of succeeding in a new environment. Notably, young rural families (in their twenties–mid thirties) who obtained some form of financial aid after the flood are most likely to leave. Young rural women also wish to relocate to a larger settlement or a big city.

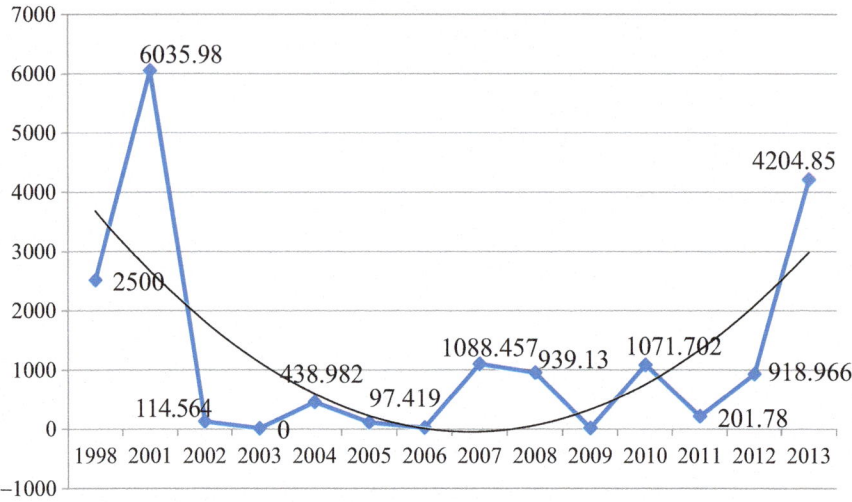

Fig. 9.2 Flood losses in Yakutia, 1998–2013 (millions of rubles) (Source: Otchety ispolnitelnykh organov gosudarstvennoi vlasti Respubliki Sakha (Yakutia) ob itogakh deyatelnosty za 2012, 2013, 2014 gody. [Reports of the executive authorities of the Republic of Sakha (Yakutia) on the results of operations for 2012, 2013, 2014]. – Retrieved from http://www.sakha.gov.ru/)

By age 25 these women typically have completed secondary school, and vocational training or even higher education. They often aspire to career or marriage choices that are typically associated with city life. For some, their motivation is ensuring better access to education for their children, and/or access to better social and health services.

The members of the 35–45 age group appear to be more established within their existing community; they are less mobile and have more commitments. They are likely to move only if offered a substantial increase in income and standards of living. People of an older age bracket find it more challenging to abandon established households and their existing social status within the community. Currently age group proportions in rural Yakutia are balanced (Fig. 9.3); therefore we suppose that migration will remain on an existing level.

Field observations reveal clear shifts in traditional gender roles. Earlier in rural society women were more conservative, more oriented to preserve old traditions of life. Currently men are more likely than women to attempt to preserve traditional ways of life, choosing to hold on to traditional occupations such as hunting and fishing. Interestingly women are more likely to explore alternative roles both in terms of employment and lifestyles. They play an increasingly active role in economic life by starting up their own micro-businesses and by adopting more prominent roles within local government.

Men are more commonly concerned about maintaining their social status and are reluctant to move, while women seem to be more open to the challenge. In this regard, older women are motivated by the desire to help their children and their

Age groups in rural Yakutia

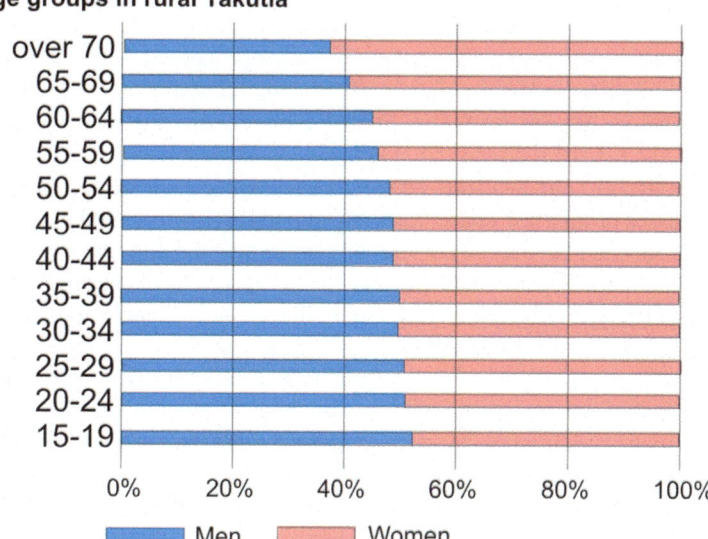

Fig. 9.3 Age groups proportion in rural Yakutia, 2010 Census (Source: Results of 2010 All-Russia population census. Website: http://www.gks.ru/free_doc/new_site/perepis2010)

Table 9.3 Prolonged climatic change impacts on traditional economy and mode of life in rural Yakutia (from the view point of residents, in %)

	Tattinsky ulus	Olekminsky ulus	Verkhoyansky ulus
"Will not impact"	Men – 5.1	Men – 0	Men – 3.1
	Women – 7.2	Women – 0	Women – 2.9
"Will possibly impact"	Men – 42.6	Men – 26.3	Men – 45.4
	Women – 45.4	Women – 17.7	Women – 47.2
"Will greatly impact"	Men – 46.1	Men – 73.7	Men – 49.2
	Women – 42.7	Women – 82.3	Women – 48.2
Difficult to answer	Men – 6.2	Men – 0	Men – 2.3
	Women – 4.7	Women – 0	Women – 1.7

Source: Author's field work, 2001–2014

families, not for individual self-realization. Another observation here is that men of the older age groups express more concerns over the health impacts of relocation, due to various relocation-related stress factors. They are afraid of being socially disconnected, isolated, and the challenges of adapting to city-life, including loneliness in a new environment. For rural seniors, who have spent most of their lives in the community, the very thought of relocation seems impossible, particularly if it is their ancestral village. They do not intend to leave their villages under any circumstances.

Generally, people in rural communities do not have access to extra sources of finance, have less access to information on existing relief programs and are not aware of their own rights, and so they do not use legal instruments to cope with the adverse impacts of floods and other natural disasters. As a rule most rely on themselves. Gender aspects of prolonged climatic change impacts have been in the focus of our interest. Table 9.3 presents rural residents' reactions to and assessment of climate change impacts on the traditional economy and the tenor of life in rural Yakutia.

The analysis of our data shows that women and men evaluate current phenomena and processes in "her/his" space. It was mentioned earlier that rural men spend more time outdoors and have more contact with wildlife and nature. However, the limits and the boundaries of the "female" and "male" spaces have changed and continue to change. Reducing the number of small villages means narrowing the traditional spaces of everyday activities for all the villagers.

The "territorial" aspect of local geography is not considered here: in other words, how the routes, distance and mobility rhythms of rural men and women have changed in the recent times. In addition to noting that that overall, local mobility of rural women – who do not migrate – has declined compared to the Soviet period, we also found that despite being more economically active, rural women are physically more isolated within their households. For example, in the Soviet period female agricultural workers were active in a large territory that included: farms; pastures; hay meadows; and winter and summer work camps. This is increasingly not the case in the post-Soviet period. This "narrowing" of living space is a challenge for the sustainability of rural communities because it means that women are less aware of the impacts of climate change.

Based on discussions with a number of rural activists, it is evident that gender bias has existed and still exists throughout Yakutia, not only in rural areas but also in urban areas. For example, in the post-Soviet period, rural women's councils (*zhensovety*) have had many discussions on rural environmental deterioration, on the social negatives associated with the narrowing of the traditional employment and on low living standards. But the male-dominated authorities mostly shrugged them off, portraying them as an expression of emotional and alarmist sentiments. In parallel with this women's activity, men who had been mid-level functionaries in the Soviet administration began to present themselves as ethnic revival activists and environmental leaders in the post-Soviet period. They introduced these problems into the public discourse, and successfully used this public resource as a start-up for their own political careers (Grigorev 2012).

Throughout the twentieth century, the bulk of women in Yakutia's villages were in low-paid categories in social production, often performing heavy physical work, in addition to house chores and raising children. For a long time women perceived this situation as the only possible life scenario. The behavior of rural women today – migration, travel, practice of obtaining various government payments and subsidies (for example, disaster relief compensation or benefits for having children) as a significant source of resources – I see as a kind of politics of escape; not only physical escape, but also mental emigration. This process is not new, but has not been researched and is rarely discussed in Yakutian society.

9.3 Conclusion

The ongoing shift in gender behavior is just one of the many complex processes occurring in the rural communities of Yakutia. The fact that it currently coincides with climate change only adds to its complexity. Coupled with these issues are inward and outward migration within the Republic, the changing ethnic structure of the population, and the third wave of industrialization and urbanization. We have only highlighted a few points about the gender shift occurring against this backdrop.

Rural males see the current and future impacts of climate change: the threat of extinction of traditional activities (hunting, fishing); and the risk that land holdings will be reduced in size, including not only personal lands, but family and tribal "ancestral lands". Older men believe that the thawing permafrost means global catastrophe and is an existential threat. Women are more concerned about the fact that the consequences of climate change affect the safety of people's lives, and the health of children and descendants. They are also concerned about the growing problem of rural unemployment, the destruction of roads and communications.

In this research area, we communicated with local authorities to share with them our thoughts and opinions. There is no global program to educate rural people about the consequences of climate change, or prevention activities related to natural disasters caused by climate change. Moreover, there is no serious attention to rural residents' own practical work in their efforts to adapt to the changing environment. In this case, the practical guide that discusses knowledge and understandings of climate change in Yakutia's Vilyui area, initiated by anthropologist Susan Crate (2013b), deserves credit.

Here we base our conclusions on the consequences of extreme situations of climate change. This work requires in-depth research concerning the gender aspects of ordinary everyday perceptions on climate change, including invisible environmental effects on villagers. Gendered experiences of such social consequences are a very attractive subject for future research

Currently there are few expectations in Russia or Yakutia regarding financial investments for any large-scale nature protection and disaster prevention programs. Therefore, women and men in rural settlements of Yakutia will be facing the social consequences of climate change for years to come. Based on years of observations, it is necessary to realize that the stability of Yakutia's rural communities is confronted with serious, multi-dimensional challenges. Not least among these are challenges in the field of gender. In the economically active age groups of rural communities there is often a discrepancy between women and men in terms of important questions of social strategy, including the assessment and planning of their own lives and the lives of their loved ones and relatives. Of course, this divergence depends on factors such as distance from major cities and to transport communications, and also the frequency and size of natural disasters.

The perceptions and adaptations of Yakutia's rural residents to climate change vary significantly. One of the important moments – the presence or absence of human will – affects the desire to challenge the most vulnerable aspects of the integrity and viability of rural communities. Gender characteristics in the evaluation of the current processes and prospects of further development of rural Yakutia significantly vary among younger and elderly rural residents alike. There is a set of challenges, among which we identify as significant threats to sustainable development, and which leads to a deep devaluation of the self-worth of traditional ways of life among young rural women and men.

Acknowledgements I would like to thank Viktoriya Filippova, Senior Researcher at the Arctic Research Section of Institute for Humanities Research and Indigenous Studies of the North, for her advice on this article and her help in producing the figures and tables. I would also like to thank my other colleagues at the Institute for their support of my long-term field work.

References

Chislennost'. (2004). Chislennost' i razmeshsenie naseleniya Respubliki Sakha (Yakutiia). Itogy Vserossyskoi perepisi naseleniya 2002 goda. Statisticheskiy sbornik. [Number and settlement of population of Republic of Sakha (Yakutia). Results of 2002 All-Russia population census. Statistical collection.] Yakutsk: Komstat RS(Ya).

Crate, S. A. (2013). Global climate change and the changing seasons: Dr. S. Crate in cooperation with Institute for Humanitarian Research and North Indigenous peoples' problems, Siberian branch of the Russian Academy of Sciences. *North – Eastern Journal of the Humanities, 2*(7), 80–90.

Crate, S. A. (with Fedorov A., Egorov P., Filippova V., Solovieva V., Ransom N., & Balls C.) (2013b). *Alamay tyyn: Byulyuu uluustarygar climat ularyytyn tuhunan uonna baar khyhalgalar* [*Precious life: About climatic changes and current challengers in Vilyiu counties*]. Yakutsk: Bichik.

Field materials of the author in rural districts of Yakutia during 2001–2014.

Filippova, V. V. (2007). *Korennye malochislennye narody Severa Yakutii v menyayushemsya prostranstve zizhnedeyatelnosty* [Indigenous peoples of Northern Yakutia in the changing space of life. The second half of 20th century]. Novosibirsk: Nauka.

Filippova, V.V. (2009) K voprosu o navodneniyakh v Yakutii [On the question of the floods in Yakutia] *Gumanitarnye nauki v Yakutii: issledovaniya molodykh uchenykh* [Humanities in Yakutia: Research of young scientists] (pp. 338–347). Yakutsk: IGIiPMNS SO RAN

Filippova, V. V. (2011). Sotsialnye vyzovy periodicheskykh navodneniy v Yakutii [Social challenges of periodic floods in Yakutia]. *Arktika i Sever, 4*, 207–212.

Grigorev, S. A. (2012) Razvitie ekologicheskogo dvizheniya Yakutii v kontse XX v. [The development of the environmental movement in Yakutia in the late XX century)] *Nauchnye problemy gumanitarnykh issledovany, 3*, 19–26.

Otchety ispolnitelnykh organov gosudarstvennoi vlasti Respubliki Sakha (Yakutia) ob itogakh deyatelnosty za 2012, 2013, 2014 gody. [Reports of the executive authorities of the Republic of Sakha (Yakutia) on the results of operations for 2012, 2013, 2014]. Retrieved from http://www.sakha.gov.ru/

Results of 2010 All-Russia population census. Retrieved from: http://www.gks.ru/free_doc/new_site/perepis2010

Vinokurova, L. (1993) *Kadry selskogo khozaystva Yakutii. 1961–1985.* [Agricultural workers of Yakutia. 1961–1985.] Yakutsk: Yakutsk Scientific Centre.

Vinokurova, L. (2010). Yakutia's men today: Widowing wives and longing for life? *Anthropology of East Europe Review, 28*(2), 131–153.

Vinokurova, L. (2014) The consequences of climate change in Yakutia: the socio-economic challenges for rural communities. In *Proceeding of 2nd International Conference "Global Warming and the Human-Nature Dimension in Siberia: Social Adaptation to the Changes of the Terrestrial Ecosystem, with an Emphasis on Water Environments" and the 7th Annual International Workshop "C/H2O/Energy balance and climate over boreal and arctic regions with special emphasis on eastern Eurasia", 8–11 October 2013* (pp. 27–29). Yakutsk: Kyoto.

Vinokurova, L. I., Popova, A. G., Boyakova, S. I., Mayrikaynova, E. T. (2004). *Zhenshina Severa: poisk novoy socialnoy identichnosty* [Northern Woman: in the search of a new social identity]. Novosibirsk: Nauka.

Chapter 10
Activating Adaptive Capacities: Fishing Communities in Northern Norway

Ingrid Bay-Larsen and Grete K. Hovelsrud

Abstract We have increasingly more knowledge about the factors and processes that determine or shape adaptive capacity, but few studies that specifically address whether such capacity in fact is activated and used. In this chapter we draw on findings from a number of related empirical studies undertaken in northern Norwegian coastal communities between 2007 and 2014 to investigate whether the narrative, 'vi står han av' –"we face whatever comes", as a subjective dimension of adaptive capacity, reflects how climate change is perceived and acted upon. We argue that the narrative reflects a discourse within which fishers perceive changing weather and the need for climate adaptation. By looking at this narrative as an analytical object we are able to identify how underlying meaning and worldview among fishers structure and constitute aspects of latent adaptive capacity.

Keywords Adaptive capacity • Perception of risk • Coastal fishers • Climate adaptation • Narratives • Northern Norway • "vi står han av"

10.1 Introduction

The heightened attention to adaptation as a necessary response to climate change and the need to develop an applicable and useful understanding of adaptation for practitioners have significantly increased research on and for adaptation. Over the past decade, research on adaptation to climate change has generated critical insights into adaptation processes, potential technological solutions, and limits and barriers to adaptation that all affect the outcome of adaptation efforts. This research has highlighted the linkages between changes in climatic, environmental and socio-economic conditions, and how communities and individuals adapt to these

I. Bay-Larsen
Nordland Research Institute, Bodø, Nordland, Norway
e-mail: iby@nforsk.no

G.K. Hovelsrud (✉)
Nord University and Nordland Research Institute, Bodø, Nordland, Norway
e-mail: grete.hovelsrud@nord.no

© Springer International Publishing Switzerland 2017
G. Fondahl, G.N. Wilson (eds.), *Northern Sustainabilities: Understanding and Addressing Change in the Circumpolar World*, Springer Polar Sciences,
DOI 10.1007/978-3-319-46150-2_10

complex, interactive changes (e.g. Leichenko and O'Brien 2008; Hovelsrud and Smit 2010). Consequently, our understanding of the factors that enable, limit or trigger adaptation responses has increased proportionally.

Interestingly, while adaptation has gained solid legitimacy and focus among researchers and policy makers across societal levels, it is documented that many northern Norwegian communities do not consider climate adaptation to be a major concern (e.g. Hovelsrud et al. 2010; Amundsen 2012). People in northern Norway are accustomed to living with greater weather variability and perceive climate change to be of a lesser concern when compared to socio-economic challenges, such as outmigration, a lack of job opportunities, education, health care and recruitment to primary industries (Hovelsrud and Smit 2010; Dannevig et al. 2013; West and Hovelsrud 2010).

Recent analyses by Hovelsrud et al. (2015) argue that fishers respond to highly variable and uncertain conditions by drawing on a "combination of individual creativity and ingenuity, heroic efforts, physical and mental toughness, and time-tested knowledge, skill and experience" (for example about when and where it is safe to fish). Both climatic and societal challenges are met with the narrative "*vi står han av*" – here unpoetically and insufficiently translated as "we face whatever comes"; a narrative with deep historical roots expressing a perception of high resilience to challenging societal and environmental conditions (cf. Nesse 2008). The phrase captures a multitude of meanings applicable both historically and currently to societal and environmental (including weather) challenges. According to Nesse (2008: 155, authors' translation)

> It provides comfort to a collective which has been tested and tried. It is not I who handles hardship, it is we – together. In addition, the phrase contains a kind of humble pride in mastering. In these four words we find the brave and courageous northern Norwegian who does not hide that they are capable of something. We also find the poetic northerner who enjoys company with others and enjoys creating linguistic euphemisms with hidden and double meanings. Last but not least, we find the northern Norwegians who live with nature – a nature both known and mastered.

In this chapter we will explore the narrative "we face whatever comes" as a proxy for the perceptions and articulations of changing weather conditions in coastal fishing communities. We suggest that the underlying values, knowledge, practices and world views reflected in this narrative have a bearing on perceptions of risks and activation of adaptive capacity. The analysis is based on interviews and conversations with fishers that live in different municipalities throughout northern Norway and that are facing similar challenges in weather and resources variability, and in socio-economic and demographic conditions. The temperature in northern Norway is projected to increase by up to 4 °C by 2100, with changes in prevailing weather, a rise in sea-level, as well as changes in mean ocean and air temperatures, and precipitation (Førland et al. 2009; Hanssen-Bauer et al. 2009). Increasing ocean temperatures are expected, and it has already been observed that this increase is changing the magnitude, composition and spatial and temporal distribution of important commercial fish stocks (e.g. Drinkwater 2005; Sundby and Nakken 2008). Fishing constitutes an important historical basis for livelihoods, culture and identity across the case sites, which is clearly reflected in local narratives (Hovelsrud et al. 2010, 2015). The fisheries resources of the region are rich and include species such as winter spawning cod (*Gadhus morhea*), Atlantic halibut (*Hippoglossus hippoglossus*), capelin (*Mallotus villosus*) and herring (*Clupea harengus*).

The empirical material for this chapter is drawn from previously published studies carried out in coastal communities in northern Norway between 2007 and 2014 (West and Hovelsrud 2010; Hovelsrud and Smit 2010; Hovelsrud et al. 2010; Dannevig and Hovelsrud 2016). These studies assessed whether northern communities are particularly vulnerable to climate change and investigate the impacts and adaptive responses to multiple stresses and changes in coupled social-ecological systems including fisheries – the main focus here.

We start by presenting the theoretical framework in section 10.2 and move on to a broader discussion of the underlying structures of the narrative in section 10.3.

10.2 Theoretical and Conceptual Framework

10.2.1 Climate Adaptation and Adaptive Capacity

The concept of adaptive capacity is rooted in both the physical and social sciences and is generally referred to as the capacity of a social-ecological system to be robust to disturbance and to adapt to both exogenous and endogenous changes, whether observed or anticipated (Armitage and Plummer 2010; see also Chap. 6, this volume). Additionally, adaptive capacity is seen as shaped by rights and access to resources, equity, available infrastructure, scientific and traditional knowledge, efficient and enabling institutions, governance systems, and the distribution of benefits and costs (Adger et al. 2009; Keskitalo et al. 2011; Kofinas et al. 2013; Hovelsrud and Smit 2010; Brown and Westaway 2011). Place-based attachment is another subjective dimension and a powerful motivator for adapting to change (Adger et al. 2009; Fresque-Baxter and Armitage 2012; Amundsen 2013). Social sciences place adaptive capacity within a suite of sources, determinants or attributes such as financial, technical, social, institutional, cultural and political resources, in relation to the social processes and structures through which they are mediated (Armitage and Plummer 2010). Whether adaptive capacity fits with norms, values, world views, knowledge and preferences is also of importance (e.g. Ostrom 1998, 2011; Adger et al. 2009). The latter have consistently been overlooked in climate change research and policy with implications for adaptive processes (Wolf et al. 2013).

In this chapter, we align ourselves with the literature that describes adaptive capacity along both objective (resources, governance, education, and income), and subjective (perceived risks and feasibility of adaptation, self-efficacy, values and meaning) dimensions (Grothmann et al. 2005; Lorenzoni et al. 2007; O'Brien and Wolf 2010; Kuruppu and Liverman 2011; Wolf et al. 2013). By connecting weather and climate change to world views, meaning, belief systems, culture and values and

Fig. 10.1 Conceptualization of how objective and subjective dimensions together form adaptive capacity. Such capacity can potentially be latent and activated in either dimension but here we suggest that it may be latent under the objective heading and activated by subjective aspects

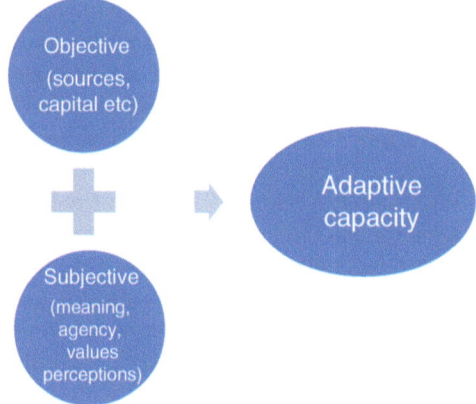

interpretations of the need to adapt, to what, by whom and in what ways (Jasanoff 2004, 2010; Hulme 2008; O'Brien and Wolf 2010), we allow for multiple understandings and narratives about adaptive needs (Amundsen 2012). Correspondingly potential endogenous limits to adaptation that may emerge from within the same goals, values, assumptions, beliefs and understandings of human society become detectable (Adger et al. 2009).

Figure 10.1 offers a schematic of the relationship between the objective and subjective dimensions of adaptive capacity. Our focus here is on one subjective dimension, exemplified by a particular narrative that has a role in whether the latent aspects of adaptive capacity are activated or not.

Our contention is that the subjective dimensions of adaptive capacity, such as values, culture, agency, cognition, perceptions, beliefs, context and place, are particularly relevant when investigating how latent capacity is activated. We therefore think of adaptive capacity as a potential attribute that must be activated to enable adaptation (Brown and Westaway 2011: 322). It is important to note that not all aspects of adaptive capacity are latent; some capacities are already in play to ensure adaptation. According to Berkes and Ross (2012: 15) the latent capacity to adapt is activated through human agency. This is also connected to psychology and cognition, such as the capacity to act independently and make free choices (Brown and Westaway 2011; Grothmann and Patt 2005). Bruner (1987: 4) recognizes that mental models are simplified representations of the world that exhibit story-like properties and that "we organize our experience and our memory of human happenings mainly in the form of narrative-stories." In the process of learning, people do not just add new information to loosely accumulated knowledge; instead they construct mental models that make sense of what they see (Kempton et al. 1995). Investigating community resilience across groups, Buikstra et al. (2010) found several such attributes of adaptive capacity, including positive outlook, core values and work ethics, community ethos, beliefs and sense of purpose. Here we consider the narrative phrase 'vi står han av' – "we face whatever comes", as rooted in context-specific experiences, practices and cognition of the actors.

Whether the latent parts of adaptive capacity are activated or not will have a bearing on how resilient or vulnerable a community is in the face of risks and perturbations. Norway as a country is described as having high adaptive capacity (or low vulnerability) to climate change and this has been found to create a level of complacency which may mask the barriers and constraints to adaptation at sectoral or local levels (O'Brien et al. 2006). In the cases presented here we explore whether the fishers' narrative of high resilience hinders long-term and needed adaptation. We surmise that these structures, internal to individual and community perceptions of climate change and adaptation policies, may obstruct the activations of latent adaptive capacity. Our exploration into the relationship between narratives and adaptive capacity will contribute to the understanding of the complex and interrelated structures that shape perceptions of risk and the need to adapt to climate change.

10.2.2 Subjective Dimensions of Adaptive Capacity: Local Perceptions as a Point of Departure

The underlying assumption of emic perspectives is that beliefs, practices and institutions can only be understood from the context within which they originate (Schmidt 2010). Focusing on narratives allows for an investigation of climate adaptation in terms of its meaning to the actors situated in a particular historical, cultural and political context. We understand the concept of narrative broadly as referring to a social process (Cortazzi 2001:384; Paschen and Ison 2014). "In narrative theory, the term 'narrative' at its most abstract is used to refer to 'structures of knowledge and storied ways of knowing" (Cortazzi 2001, quoted in Paschen and Ison 2014:1084).

Narratives are seen as stories reflecting perceptions that in turn are connected to wider discourses in society. The way people perceive and rationalize climate change and strategies for adaptation is informed, assessed and structured through broader discourses that structure and constitute their livelihood and identities. Narratives can be seen as both structures and constructs that come into existence through the communication of sentient agents. The emic position of narrative analyses takes place on the basis of individuals' or groups' shared perceptions, embedded in language and communication, and not on the basis of true, real-world-observations which are found in the literature to be a trigger of climate adaptation (Dannevig et al. 2013). Narratives and discourses are rather investigated as communication, text, frame, myths, collective memories, stories, scripts, and practice (Hajer 1995; Arts and Buizer 2009).

Attention to a duality in narratives allows for discussing and revealing the underlying structures through which changing weather and adaptation policies are being perceived among local actors. Through narrative analyses, the subjective dimensions of adaptive capacity and the meaning and saliency of changes in climate and policies can be discovered. Narrative analyses thus enable us to unpack the broader discourse through which the activation of adaptive capacity is taking place. Through the '*vi står han av*' narrative we can increase our understanding of how these subjective dimensions come into play in promoting and constraining the social actors and in activating the latent parts of adaptive capacity.

10.3 "We Face Whatever Comes" – the Narrative of Northern Coastal Communities

The narrative '*vi står han av*' "we face whatever comes" conveys "optimism and determination in the face of the unpredictable social, economic and climatic conditions that are central features of life in northern, coastal communities" (Hovelsrud et al. 2015: 207). As Hovelsrud and others (2015: 202) have noted: "fishers' narratives [in general] reflect occupational values of freedom and independence, mastery of a traditional and time-tested craft, maintenance of local traditions, knowledge and identity, providing fresh, local food, creating local employment and income, and being resilient and adaptable to unpredictable conditions." These narratives reflect perceived resilience to challenges in both environmental and socio-economic conditions, and indicate that climate adaptation is not perceived as urgent or critical for fishers, despite the fact that they are engaged in a primary industry that is exposed and sensitive to climate change impacts. The expression also refers to a high confidence in their resilience and adaptability, as well as their optimism and determination to meet future challenges. We suggest that the narrative is situated within a broader societal frame and can be structured along four dimensions.

10.3.1 Sociopolitical Context

The first dimension relates to the sociopolitical conditions within which the fishers operate as economic actors. The coastal fishers partaking in the research overwhelmingly emphasize non-climatic issues such as resource management regulations and local socio-economic challenges when describing the overall vulnerability of their livelihoods (Hovelsrud et al. 2010; West and Hovelsrud 2010; Hovelsrud et al. 2015). They express more concern about the impacts of structural fisheries reforms, which have contributed to reduced numbers in fishers and boats, and with more locally pressing issues such as outmigration and general employment, than climate change (Hovelsrud et al. 2010; West and Hovelsrud 2010). While employment in the northern Norwegian fisheries has seen a steady decline over the past several decades, the industry remains important for the local economies, for cultural identity and connection to place in the case of communities (e.g. West and Hovelsrud 2010; Amundsen 2012). Therefore, the urgency of climate change and adaptation policies is perceived as less salient for fishers when compared to the direct impacts of national fisheries management policies, fluctuations in international markets and local socio-economic conditions.

10.3.2 Coastal Culture

The second dimension pertains to the identity of fishers and the role of fishing in a wider coastal cultural context. Along with Nesse (2008), we contend that the narrative *'vi står han av'* is motivated and shaped by the collective experience of living with highly variable natural climatic variability and challenging weather conditions. The plural "we" refers to a collective experience that has been tested throughout history (Nesse 2008). Coastal fisheries are an important cornerstone of employment and income, and are strongly rooted in the place where fishers live and deliver their catch. The reduction in the number of fishers and vessels therefore, has consequences for the culture, traditions and identities of coastal communities. By maintaining a traditional livelihood, providing fresh clean food, and keeping the houses lit along the coast, the fishers see themselves as stewards of their coastal culture. In northern Norway, fishers and non-fishers alike invoke the moral justifications for fishing and the responsibilities of fishers towards the economic and social welfare of their communities (Hovelsrud et al. 2015). This echoes the poetic northerners, noted by Nesse (2008), who value their community, and who understate the daily hardships with a narrative that at once acknowledges but glosses over the challenging conditions. This confirms earlier studies showing how climate adaptation is only addressed when a risk (for example climate change) is perceived as salient and relevant by those who need to adapt and when such risks are situated within past experiences and existing cultural frames (e.g. Wlezien 2005; Dannevig and Hovelsrud 2016).

10.3.3 Fishers' Identity and Knowledge – Freedom and Independence

Independence and freedom also emerge as significant identity markers from interviews with fishers (Hovelsrud et al. 2015). Fishers generally operate under highly variable and uncertain conditions, in weather, management and markets, which require an individualistic attitude and a heavy reliance on their own knowledge and experiences. Fishing practices, traditions, operations, gear, and crafts are not derived from scientific or theoretical expertise, but rather accumulated practical experiences that have provided fishers with adaptive capacity and strategies towards challenging and variable weather throughout history.

Although climate change is projected to substantially influence livelihood activities and area planning in northern Norway (Hovelsrud et al. 2010; Øseth 2010), research indicates that local perceptions of risks associated with climate change differ from the evidence researchers have found (Hovelsrud et al. 2015). Despite the perception that anthropogenic climate change does not affect fishers' livelihood vulnerability, they are well aware that their livelihood is exposed to such changes. But, climate variability and change fall within the existing and recognized high natural variability in local weather patterns and the resource base (fish stocks). When confronted with information about the possible consequences of climate change for their livelihoods, fishers cited similar weather events that happened up to 100 years ago, and interpreted the projections as a part of the natural and seasonal variability with which they are highly familiar. The fishers framed discussions of the causes of past and current fluctuations, particularly in local fish stocks, in terms of wider marine ecosystem dynamics and interactions that include the effects of predation and the abundance of other fish and marine mammal stocks on survival and feeding conditions, the general health of the marine environment, natural fluctuations in ocean temperatures and chemistry, and human activities, including illegal fishing, the pollution of harbours, and overfishing (West and Hovelsrud 2010).

10.3.4 Fishers' Knowledge and Science

Although the fishers acknowledge global warming and also observe changing conditions that correlate with scientific observations and modelling of climate change, they do not readily view these as salient. The fishers recognize that the substantial shift in the distribution of cod and fish stock composition may affect what and where they can fish. But whereas marine scientists (e.g. Sundby and Nakken 2008) attribute the substantial shifts in cod distribution and fish stock composition to increasing temperatures, fishers argue that such changes are part of cyclical changes in fishing conditions. They do not readily accept that such changes are caused by anthropogenic climate change, because their collective memory and experience recall that such changes have happened in the past. Instead they frame the changing fisheries conditions within a wider marine ecosystem dynamic, impacted by both human activities and environmental changes (West and Hovelsrud 2010).

The fishers also contend that the science informing Norwegian marine resource management is at odds with their own experiences and knowledge. It may not be too farfetched to connect the lack of trust in marine science to that of scientific climate change discourse. This is supported by the argument that cognitive and moral sensibilities are shaped by peoples' discourses, which in turn influence how people interpret facts, evidence and the credibility of science (Hajer 1995). The confidence, or "humble pride in mastering" as Nesse (2008) notes, captured by the expression 'vi står han av' may explain how the challenges associated with changing natural systems are not perceived to be as urgent as those reflected in science and policy. However, the 'vi står han av' narrative does not necessarily signify a complete disconnect between scientific and local understanding of climate change. The fishers note that they continually make adjustments (either as individual entrepreneurs or in a professional capacity) to deal with climate/weather variability and change, and with socio-economic challenges. It nevertheless demonstrates the significance of the correspondence between different knowledge systems. Scientific knowledge about climate change and its production, framing, communication and uses in policy processes need to correspond to the wider discourses in coastal communities in order to be perceived as salient and urgent (e.g. (Jasanoff 2004; Wlezien 2005; Dannevig and Hovelsrud 2016). If the information and knowledge communicated by the bureaucracy and policy makers (in this case those involved in fisheries management or national climate policy) do not fit with how the individual or group perceive a concern or problem, it will likely not be acted upon (as is the case with the fishers in northern Norway).

10.4 Concluding Discussion

This chapter has explored how narratives such as "we face whatever comes" reflect fishers self-reliance, a history of pride in mastering a demanding environment, professional independence and freedom, as well as extensive knowledge about the marine environment derived from centuries of accumulated experience. The narrative structures we have suggested influence the perceptions of changing weather and determine whether and how adaptation is deemed relevant. As earlier studies of fishers have shown, anthropogenic climate change may not necessarily be a part of their perception of reality, and there is often a discrepancy between community perceptions and scientific findings regarding the urgency of climate change adaptation (e.g. Moser 2010). This discrepancy between scientific and fishers' understanding of climate change, in addition to other socioeconomic stressors, and a collective memory of previous successful responses to challenging weather, may impede climate adaptation. Climate change is not perceived as salient by the fishers, and they will therefore not initiate adaptive responses. The filters through which we interpret and understand realities, therefore, may create barriers for responding to climate change. For policy makers and scientists, it is imperative that the underlying and multiple realities within which people operate are addressed for climate adaptation to be considered necessary and successful (see Amundsen 2014).

In this chapter we have initiated a discussion of some of the potential reasons why latent adaptive capacity is not activated. Our analysis is driven by the idea that it is not sufficient to have a high adaptive capacity in terms of financial and social capital, infrastructure and resource access if these remain latent, and if there are aspects of society which thwart such activations. We suggest that there may be subjective dimensions and internal barriers associated with the questions of whether and how adaptive capacities are activated. These subjective dimensions are expressed through narratives that structure local perceptions of climate change and adaptation needs. The structuring dimensions of particular relevance in this chapter have been socioeconomic conditions, cultural and historic context, identity and confidence in knowledge systems. Altogether these structures constitute particular world views that, in turn, may either hinder or facilitate local adaptive capacity to reach its full potential.

We conclude that a focus on narratives provides a venue for understanding how people perceive a potential and somewhat intangible risk such as climate change, and a way to analyze multiple realities and the underpinnings for adaptive capacity. A better understanding of the apparent inertia in society to accept the need for climate adaptation, partially explained by the lack of salience of climate change science, may thus be gained from studying narratives. The lack of concern that is expressed through shared narratives strengthens and validates cultural values and world views. This, we conclude, requires the inclusion of narrative research and an understanding of differing values, priorities and experiences when addressing climate change and climate adaptation (Paschen and Ison 2014; West and Hovelsrud 2010). Conversely we conclude that in order for climate change to be a salient issue it "needs to be integrated into the everyday narratives that people tell about themselves and their world" (Lejano et al. 2013: 61). This means that there may likely be a need to expand the narrative. Fisheries management officials and politicians will likely increase their credibility and relevance by being attentive to the local narratives into which their information is incorporated and acted upon.

References

Adger, W. N., Dessai, S., & Goulden, M. (2009). Are there social limits to adaptation to climate change? *Climatic Change, 93*(3–4), 335–354.

Amundsen, H. (2012). Illusions of resilience? An analysis of community responses to change in Northern Norway. *Ecology and Society, 17*(4), 46.

Amundsen, H. (2013). Place attachment as a driver of adaptation in coastal communities in Northern Norway. *Local Environment, 20*(3), 257–276.

Amundsen, H. (2014). *Adapting to change – Community resilience in northern Norwegian municipalities.* PhD Thesis, The Faculty of Social Sciences, University of Oslo, Norway.

Armitage, D., & Plummer, R. (Eds.). (2010). *Adaptive capacity: Building environmental governance in an age of uncertainty.* Dordrecht: Springer.

Arts, B., & Buizer, M. (2009). Forests, discourses, institutions: A discursive-institutional analysis of global forest governance. *Forest Policy and Economics, 11*(5–6), 340–347.

Berkes, F., & Ross, H. (2012). Community resilience: Toward an integrated approach. *Society & Natural Resources, 26*(1), 5–20.

Brown, K., & Westaway, E. (2011). Agency, capacity, and resilience to environmental change: Lessons from human development, well-being, and disasters. *Annual Review of Environment and Resources, 36*, 321–342.

Bruner, J. (1987). Life as a narrative. *Social Research, 54*, 12–32.

Buikstra, E., Ross, H., King, C. A., Baker, P. G., Hegney, D., Mclachlan, K., & Rogers-Clark, C. (2010). The components of resilience – Perceptions of an Australian rural community. *Journal of Community Psychology, 38*(8), 975–991.

Cortazzi, M. (2001). Narrative analysis in ethnography. In P. Atkinson, A. Coffey, S. Delamont, J. Lofland, & L. Lofland (Eds.), *Handbook of ethnography* (pp. 384–393). London: Sage.

Dannevig, H., & Hovelsrud, G. K. (2016). Understanding the need for adaptation in a natural resource dependent community in Northern Norway: Issue salience, knowledge and values. *Climatic Change, 135*, 261–275.

Dannevig, H., Hovelsrud, G. K., & Husabø, I. A. (2013). Driving the agenda for climate change adaptation in Norwegian municipalities. *Environment & Planning C: Government & Policy, 31*(3), 490–505.

Drinkwater, K. F. (2005). The response of Atlantic cod (*Gadus morhua*) to future climate change. *ICES Journal of Marine Science, 62*, 1327–1337.

Førland, E. J., Benestad, R. E., Flatøy, F., Hanssen-Bauer, I., Haugen, J. E., Isaksen, K., Sorteberg, A., & Ådlandsvik, B. (2009). *Climate development in North Norway and the Svalbard region during 1900–2100* (Report No. 128). Tromsø, Norway: Norwegian Polar Institute.

Fresque-Baxter, J. A., & Armitage, D. (2012). Place identity and climate change adaptation: A synthesis and framework for understanding. *Wiley Interdisciplinary Reviews: Climate Change, 3*(3), 251–266.

Grothmann, T., & Patt, A. (2005). Adaptive capacity and human cognition: The process of individual adaptation to climate change. *Global Environmental Change, 15*(3), 199–213.

Hajer, M. A. (1995). *The politics of environmental discourse -ecological modernization and the policy process*. New York: Oxford University Press.

Hanssen-Bauer, I., Drange, H., Førland, E. J., Roald, L. A., Børsheim, K. Y., Hisdal, H., Lawrence, D., Nesje, A., Sandven, S., Sorteberg, A., Sundby, S., Vasskog, K., & Ådlandsvik, B. (2009). Klima i Norge 2100. Bakgrunnsmateriale til NOU Klimatilpasning. Background material for the Green Paper on Climate Adaptation. Norsk Klimasenter.

Hovelsrud, G. K., & Smit, B. (2010). *Community adaptation and vulnerability in the arctic regions*. Dordrecht: Springer.

Hovelsrud, G. K., Dannevig, H., West, J., & Amundsen, H. (2010). Adaptation in fisheries and municipalities: Three communities in northern Norway. In G. K. Hovelsrud & B. Smit (Eds.), *Community adaptation and vulnerability in the arctic regions* (pp. 23–63). Dordrecht: Springer.

Hovelsrud, G. K., West, J., & Dannevig, H. (2015). Exploring vulnerability and adaptation narratives among fishers, farmers and municipal planners in Northern Norway. In K. O'Brien & E. Selboe (Eds.), *The adaptive challenge of climate change* (pp. 194–212). Cambridge: Cambridge University Press.

Hulme, M. (2008). Governing and adapting to climate. A response to Ian Bailey's commentary on 'geographical work at the boundaries of climate change'. *Transactions of the Institute of British Geographers, 33*(3), 424–427.

Jasanoff, S. (2004). *States of knowledge: The co-production of science and social order*. Oxon: Routledge.

Jasanoff, S. (2010). A new climate for society. *Theory, Culture & Society, 27*(2–3), 233–253.

Kempton, W., Boster, J. S., & Hartley, J. A. (1995). *Environmental values in American culture*. Cambridge, MA: The MIT Press.

Keskitalo, E. C. H., Dannevig, H., Hovelsrud, G. K., West, J., & Swartling, Å. G. (2011). Local vulnerability and adaptive capacity in developed states: Examples from the Nordic countries and Russia. *Regional Environmental Change, 11*(3), 579–592.

Kofinas, G., Clark, D., & Hovelsrud, G. K. (2013). Adaptive and transformative capacity. In A. Council (Ed.), *Arctic resilience interim report* (pp. 73–93). Stockholm: Stockholm Environment Institute and Stockholm Resilience Centre.

Kuruppu, N., & Liverman, D. (2011). Mental preparation for climate adaptation: The role of cognition and culture in enhancing adaptive capacity of water management in Kiribati. *Global Environmental Change, 21*(2), 657–669.

Leichenko, R., & O'Brien, K. (2008). *Environmental change and globalization: Double exposures*. New York: Oxford University Press.

Lejano, R. P., Tavares-Reager, J., & Berkes, F. (2013). Climate and narrative: Environmental knowledge in everyday life. *Environmental Science & Policy, 31*, 61–70.

Lorenzoni, I., Nicholson-Cole, S., & Whitmarsh, L. (2007). Barriers perceived to engaging with climate change among the UK public and their policy implications. *Global Environmental Change, 17*(3–4), 445–459.

Moser, S. C. (2010). Communicating climate change: History, challenges, process and future directions. *Wiley Interdisciplinary Review–Climate Change, 1*(1), 31–53.

Nesse, A. (2008). *Bydialekt, riksmål og identitet: sett fra Bodø [City dialect, official language and identity]*. Oslo: Novus Publisher.

O'Brien, K. L., & Wolf, J. (2010). A values-based approach to vulnerability and adaptation to climate change. *Wiley Interdisciplinary Reviews: Climate Change, 1*(2), 232–242.

O'Brien, K., Eriksen, S., Sygna, L., et al. (2006). Questioning complacency: Climate change impacts, vulnerability, and adaptation in Norway. *AMBIO: A Journal of the Human Environment, 35*(2), 50–56.

Øseth, E. (2010). *Klimaendringer i norsk Arktis. Konsekvenser for livet i nord. Report 136 [Climate change in the Norwegian Arctic. Consequences for life in the North]*. Norsk Polarinstitutt. Tromsø.

Ostrom, E. (1998). A behavioral approach to the rational choice theory of collective action: Presidential address, American political science association, 1997. *The American Political Science Review, 92*(1), 1–22.

Ostrom, E. (2011). Background on the institutional analysis and development framework. *Policy Studies Journal, 39*(1), 7–27.

Paschen, J.-A., & Ison, R. (2014). Narrative research in climate change adaptation – Exploring a complementary paradigm for research and governance. *Research Policy, 43*, 1083–1092.

Schmidt, V. (2010). Taking ideas and discourse seriously: Explaining change through discursive institutionalims as the fourth 'new institutionalism'. *European Political Science Review, 2*(1), 1–25.

Sundby, S., & Nakken, O. (2008). Spatial shifts in spawning habitats of Arcto-Norwegian cod related to multi decadal climate oscillations and climate change. *ICES Journal of Marine Science, 65*, 953–962.

West, J., & Hovelsrud, G. K. (2010). Cross-scale adaptation challenges in the coastal fisheries: Findings from Lebesby, Northern Norway. *Arctic, 63*(3), 338–354.

Wlezien, C. (2005). On the salience of political issues: The problem with "most important problem." *Electoral Studies, 24*(4), 555–579.

Wolf, J., Allice, I., & Bell, T. (2013). Values, climate change, and implications for adaptation: Evidence from two communities in Labrador, Canada. *Global Environmental Change, 23*(2), 548–562.

Chapter 11
Signs of Non-recognition: Colonized Linguistic Landscapes and Indigenous Peoples in Chersky, Northeastern Siberia

Lena Sidorova, Jenanne Ferguson, and Laur Vallikivi

Abstract This paper analyses the presence and absence of local languages in the visual sphere of Chersky, a small settlement located in the Nizhnekolymsk *Rayon* (Lower Kolyma County) in the far northeast of the Republic of Sakha (Yakutia) in the Russian Federation. The linguistic landscape—the elements of language present in public space—can be seen as a reflection of the sustainability of a language and indeed the cultural identity of a group. An assessment of Chersky's linguistic landscape reveals that despite the region being home to Russian, Sakha, Eveny, Chukchi, and Yukaghir speakers, not all of these languages have a presence within the landscape. We analyze the ways in which the indigenous Eveny, Chukchi and Yukaghir languages are excluded from the linguistic landscape in favour of Russian, Sakha, and even English; these local languages are subsumed within a discourse that highlights the region's belonging not only to the Republic of Sakha (Yakutia), but to the Russian Federation as a whole.

Keywords Linguistic landscape • Lower Kolyma • Multilingualism • Indigenous peoples • Republic of Sakha (Yakutia)

L. Sidorova (✉)
Department of Culturology, North Eastern Federal University,
Yakutsk, Sakha Republic (Yakutia), Russian Federation
e-mail: lenasida@mail.ru

J. Ferguson
Department of Anthropology, University of Nevada-Reno, Reno, NV, USA
e-mail: jenannef@unr.edu

L. Vallikivi
Department of Ethnology, University of Tartu, Tartu, Estonia
e-mail: laur.vallikivi@ut.ee

© Springer International Publishing Switzerland 2017
G. Fondahl, G.N. Wilson (eds.), *Northern Sustainabilities: Understanding and Addressing Change in the Circumpolar World*, Springer Polar Sciences, DOI 10.1007/978-3-319-46150-2_11

11.1   Introduction

The social and cultural attitudes of a population form in various ways. One process of formation involves the visual and linguistic elements—or linguistic landscape—that are created and replicated by different agents within a particular socio-cultural field. This landscape is easily accessible for observation and analysis, but its convenience is not the only point of interest to researchers. A most important facet of the linguistic landscape is what it can say—what evidence it can provide—about deeper processes transpiring within a society that are not as visually apparent. The daily "consumption" of these elements of the landscape may be somewhat non-reflexive, but do both reflect and affect the cultural consciousness and values of the local population. The linguistic landscape also signals the sustainability of a particular language, and culture. We posit that the state of indigenous languages can be an indicator of a sustainable culture. Just as in nature, when deep tectonic processes become apparent through changes in the landscape, cultural changes may also become visible through language usage. If the languages of indigenous people in the linguistic landscape are diminished or absent, it can be indicative of the erosion or growing instability of a particular culture. An increased presence of an indigenous language in the landscape then would speak to the greater vitality of that language and culture.

This article is based on field research undertaken in April 2014 in Nizhnekolymsk *Rayon* (Lower Kolyma County) in the Republic of Sakha (Yakutia), Russian Federation, and focuses on what can be revealed by the linguistic landscape of Chersky, a small northern settlement. Our objective is to give a short overview of the linguistic situation in the town, in particular focusing on the scopes and spheres of the usage of the languages among indigenous peoples of the Lower Kolyma—Eveny, Chukchi, and Yukaghir. The extent to which a language is used in society, and how it is officially represented, are some of the criteria recommended by UNESCO for assessing the viability of a given language and its likelihood for continued usage. Studying linguistic landscapes not only provides the means to analyze the scope of language use and other semiotic systems in the public space, but it also allows us to access ideas and values of the local population, and understand how these values are transmitted through both language and image.

11.2   Linguistic Landscape

The term "linguistic landscape" was first used in an article by Canadian sociologists Rodrigue Landry and Richard Bourhis in 1997. In their definition this notion "refers to the visibility and salience of languages on public and commercial signs in a given territory or region" (1997:23). Visual linguistic elements such as "public road signs, advertising billboards, street names, place names, commercial shop signs, and public signs on government buildings combine to form the linguistic landscape of a given territory, region, or urban agglomeration" (Landry and Bourhis 1997:25). Pennycook (2009) and Shohamy (2006) believe that graffiti and other informal signs are equally important components of the linguistic landscape. We have included all the aforementioned elements in our analysis of the linguistic landscape of Chersky. As Gorter (2006) notes, the linguistic landscape should be broadly considered as all types of linguistic objects present in the public space.

Exploring linguistic landscapes should not be limited solely to the task of considering the use of natural human languages, but goes beyond the scope of linguistics into "the multiple semiotic systems which together form the meanings that we call place" (Scollon and Scollon 2003:12). According to Ben-Rafael et al. (2006:8) the presence of languages in the public space has a "socio-symbolic significance." Through the analysis of the linguistic landscape, one can see which languages in the community are the most prestigious and significant. Considering signs used in advertising, Hult (2009) and Shohamy (2006) identify popular ideas of multilingualism and how these are represented through the social and political roles of different languages. As noted by Sloboda (2009), the value of linguistic elements of the landscape is that signs are objects in which worldview and history are materialized; like monuments, they are sites of memory and history. We investigated the use of languages and images on various signs in Chersky in order to analyze how the linguistic landscape reflects the current linguistic situation of Nizhnekolymsk Rayon. In addition we also considered the presence of Soviet imagery in the official signs, to understand better how the linguistic landscape or index reflects such ideologies and thus can influence the cultural values of the local population (Sloboda 2009).

11.3   Why Chersky?

In Soviet historiography, the official history of what is now Nizhnekolymsk Rayon is presented as a story of constant positive "development" (*osvoenie*). The region has been geographically strategic for Russia since the time when Cossacks first arrived there in the 1640s to collect the fur tribute from the indigenous population. In the middle of the twentieth century, deposits of tin and gold were discovered, for which Nizhnekolymsk Rayon became an important transit hub. A seaport was built to the north of Chersky, and the local airport became a center for polar aviation, dispatching aircraft to Soviet drifting stations in the Arctic Ocean.

As the Soviet Union was particularly active in the exploration and exploitation of the Arctic, Chersky attracted civilian and military experts of different ethnic backgrounds from all parts of the country. Nizhnekolymsk Rayon was part of a vast system of Gulag camps until the 1950s, when it ceased to serve as the Ambarchik seaport and transit camp on the banks of the East Siberian Sea (see Samoylova 1993). Located three kilometres from Chersky, the seaport Zelenyi Mys also housed political prisoners from the 1930s to early 1950s. Later, in the 1960s, large amounts of goods and provisions passed through Zelenyi Mys on the way to Magadan,

Chukotka, and the Kolyma-Indigirka group of regions. In the late Soviet period (1970s), it became a thriving settlement with a relatively good infrastructure and direct flights to cities in central Russia. While a food and consumer goods deficit plagued Soviet society as a whole, Chersky (like some other arctic settlements) received special provisions. The older indigenous reindeer herders we spoke with recalled that time as the "golden age" for Nizhnekolymsk Rayon, as heavily subsidized reindeer herding enabled inhabitants to earn a decent living in the region.

11.3.1 The Indigenous Population of Nizhnekolysmk Rayon

Before the Cossacks arrived in the seventeenth century, what is now Nizhnekolymsk Rayon had been long inhabited by the ancestors of the Yukaghir and Chukchi (Shishlo 2000). By the early period of Russian colonization, cultural and linguistic diversity had further developed due to the growing number of Tungus (Evenki, Eveny), Sakha (also known as Yakut), and Russians (Old Settlers, or *starozhili*). By the early twentieth century, cultural contact and particularly intermarriages between these groups had led to the development of widespread multilingualism. While there had been mutual linguistic knowledge and occasional shifts between languages among Yukaghir, Eveny, and Chukchi for a long time, the Sakha and the Russian languages became dominant among the indigenous population in the twentieth century. However, the cultural influence of these newcomers has varied in different parts of the Kolyma basin.[1]

During the Soviet period, the rapid industrial development of Nizhnekolymsk Rayon transformed Chersky into a significant settlement in the Russian Arctic. Founded in 1931 as Nizhnye Kresty, the village soon grew into an urban-type settlement. In 1963, it was renamed Chersky after Polish geographer Jan Czerski, who had explored the region in the nineteenth century. By the 1960s, Russian and Sakha were both widely spoken in the town. Though Russian became the most important language of interethnic communication, over the last 70 years there was also a period of significant "Yakutization". The Sakha language spread through the boarding school system set up for the children of reindeer herders and hunters. As a result, for most Yukaghirs, Chukchis, and Eveny, the Sakha language became the main instructional language at school (Forsyth 1992). Nevertheless, in the so-called national villages where numerically-small indigenous peoples predominated, such as Andryushkino, the curriculum was entirely in Russian from the 1960s onward. Only in the 1980s were other languages like Yukaghir, Eveny and Sakha introduced in a modest amount (Vakhtin 2001). In Chersky, formal education was in Russian and those indigenous persons who settled in this regional center had fewer opportunities to speak or learn their own languages.

---

[1] According to Vakhtin et al. (2004), the Russian population has always been larger than that of the Sakha in the Lower Kolyma, and so the influence of Sakha culture has been significantly less present there than in the middle Kolyma.

In October 1992, shortly after the dissolution of the Soviet Union, Russian and Sakha were designated "state" languages (Zakon 1992). The languages of other indigenous groups received "official" status only in their "areas of settlement" (i.e. villages in which indigenous and numerically small peoples predominate). Some measures, such as reincorporating the languages into the school system as subjects of study, were put in place to encourage the development of Evenki, Eveny, Tundra Yukaghir, and Chukchi (Grenoble 2003), but as we discuss below, these languages are not visible in the local linguistic landscape. Since the end of the Soviet period, high rates of language attrition have been noted among young indigenous people in Nizhnekolymsk (Vakhtin 1991).

11.3.2 Chersky at the Present Time

The local nomadic population of Nizhnekolymsk Rayon was heavily influenced by Soviet policies aimed at promoting sedentary life (Kolesov 2003). As a consequence of collectivization in the 1930s, many small northern villages were first created and then consolidated into larger settlements from the late 1950s on (Yaglovsky 2003). As the center of Nizhnekolymsk Rayon, Chersky attracted many indigenous people, although the vast majority of the population was Russian. In the 1990s, the former Soviet military-industrial complex ground to a halt in the region, followed by the breakdown of scientific research in the Arctic, and most recently, the crises among civil aviation companies. Many researchers and scientists who came to the region have returned to the "mainland" of central Russia. This has been a general tendency for the Sakha Republic; the process is particularly conspicuous in Chersky, which has seen a population decrease from 11,176 in 1989 to 2,707 in 2010 (FSGS 2010). According to the unofficial data of the last census, there were 157 Eveny, 491 Sakha, 1597 Russians, 134 Chukchi, 128 Yukaghirs, 135 Ukrainians, and 53 Tatars in Chersky in 2010 (FSGS 2010). Thus, the ethnic groups belonging to the category of the small-numbered indigenous peoples of the North (e.g. *korennye malochislennye narody Severa*, such as Eveny, Chukchi, and Yukaghir) constitute 15.5 % of the population of Chersky, while in Nizhnekolymsk Rayon as a whole, 33.1 % of the population are representatives of these groups. More than half of the population in Chersky is Russian, with significant numbers of Sakha and Ukrainians.

11.4 Analysis of the Linguistic Landscape of Chersky

In our research, we analyzed 57 photographs from Chersky, which contained 101 linguistic landscape elements. This is essentially the entirety of the public linguistic landscape in this small settlement. Most of these items were photographed on the main streets, Tavrata and Burnasheva, along which such important and essential

Table 11.1 Linguistic landscape examples in the settlement of Chersky, analysed by category (from 95 photographs taken in April 2014)

Category	Number of examples
Political posters (slogans, signs with party logos, for example, 'United Russia,' election posters)	11
Memorial and commemorative posters	9
Building names	17
Street names	7
Advertisements	10
Informational signs	20
Communicative signs ('thank you for your purchase', etc.)	8
Graffiti	13
Total	95

sites are located, such as the village administration building, the school, the airport, the main-square, and grocery stores.

In our analysis, we have identified the following parameters as critical to our study: the use of languages of the peoples of Nizhnekolymsk Rayon; the sequence and location of the languages, and font size; and other non-linguistic (semiotic) aspects, such as the placement of signs and signage, colours, symbols, images. We considered the frequency and content of these examples under the following categories: political posters; memorial and commemorative posters; names of buildings; street names; advertising; informational signs; communicative signs ("Thank you for your purchase", *etc.*); and graffiti. These data are presented in Table 11.1.

11.4.1 Soviet Images as Indicative of Nostalgia for the "Golden Age" of Nizhnekolymsk Rayon?

Eight of the 95 examples we collected in the linguistic landscape bear a strong resemblance to the political posters and signs of the Soviet era, due to elements such as the prevalence of the colour red, banners celebrating the Red Army's victory over Germany during the Second World War (or Great Patriotic War, as it is popularly known in Russia) (Fig. 11.1), and rhetoric ("We are proud of our labour," see Fig. 11.2).

One of the key themes of political posters in Chersky is patriotism, which involves the creation of a new image of Russia as a homogeneous, uniform cultural space, expressed in rhetorical terms by a poster entitled "In a United Family" (Fig. 11.3). In Fig. 11.3, diversity and unity expressed by the central image that represents the future: a child in the arms of her mother who is dressed in Sakha traditional clothing, alongside a Russian man dressed in a Cossack army uniform. This image is similar to that of a well-known statue in Yakutsk commemorating the 1641 union

Fig. 11.1 This banner reads, in Russian, "The 9th of May. Happy holiday, dear veterans!" May 9th, or Victory Day, commemorates the Red Army's victory in the Second World War over Germany (in Russia, WWII is known as the Great Patriotic War)

Fig. 11.2 To the right of the polar bear beneath the colours of the Russian flag—the logo for the United Russia political party—the Russian text reads "We are proud of our labour"

of Semën Dezhnev, a Cossack, and his Sakha wife Abakayada Süchü (Fig. 11.4). The image also depicts them with their child and can be seen as symbolizing the unification of Russia and the Sakha Republic, while also underlining the particular degree of tolerance in the relations between the two nations.

However, here it is notable that none of the other indigenous groups are depicted: this is unity on the level of the major administrative unit (republic), between Sakha as a dominant indigenous group, and Russians, or the Russian state.

In Fig. 11.5, an image of a table laden with foods associated with different cultural groups is presented. The poster is entitled: "380 years since the date of entry (sic.) of the Republic of Sakha into structure of the Russian state." Ethnic cuisine

Fig. 11.3 Underneath the black text, "In a united family," in Russian, a woman and young child appear dressed in Sakha-style garments beside a man in Cossack army uniform

Fig. 11.4 Statue in Yakutsk of Cossack Semën Dezhnev and Abakayada Süchü, and their child

and representatives of all cultural groups are present, dressed in Russian, Sakha, Yukaghir, Eveny and Chukchi costumes.[2] Despite the ethnic diversity depicted through dress and food, the sole language of the posters is Russian. This unilinguality also carries a symbolic meaning, as during the Soviet period Russian was heavily promoted as the language of interethnic communication, which could bring together all cultures and groups. This image of plenty is perhaps also commemorating the Soviet period, or "golden age" of Chersky in the 1970s, a time seen as one of safety and abundance for the village and its inhabitants.

---

[2] In the Soviet historiography that deals with the colonial conquest of Siberia, the trope of the voluntary "entry" into the Russian state and friendship between peoples dominated. It was then customary to present colonial conquest as beneficial for the local peoples, who were depicted as less "developed" compared to the Russians (e.g. Bakhrushin and Tokarev 1953). Although there is slightly less explicit paternalism in the current representations, the Soviet language of voluntary entry and mutual benefit are strongly present today as well.

Fig. 11.5 The uppermost poster of people in traditional dress standing before a festive feast features Russian text reading "380 years since the entry of the Republic of Sakha (Yakutia) into the Russian state". Below it, we see a poster introducing the four local Communist party members underneath the image of a large ice-breaking ship. The bold white text reads, in Russian, "For the revival of the Arctic!"

11.4.2 Chersky as an Arctic Settlement

Chersky's image as an arctic village is reflected in the district's emblem, which features a white polar bear. While this image is not a strictly linguistic element, Elana Shohamy (2006) reminds us of the importance of not restricting language to words alone when analyzing the linguistic landscape, and also examining the ideologies reflected in non-verbal semiotic features. She argues that there are "multiple ways of representation that are not limited to words but rather include additional ways of expression consisting of a variety of creative devices [...] such as languaging through music, clothes, gesture, visuals, food [...]" (Shohamy 2006:21) Therefore, we analyze this polar bear as "speaking", or making a statement within the linguistic landscape about the region's connections to the Russian Federation. The image of the walking bear represents a strong animal that evokes feelings of fear and respect at the same time. This image underscores the geopolitical features of Russia; it is a northern country with significant territory in the Arctic. It is critical to note, though, that this symbolic bear has also been brought to the region by incomers; it is not a local indigenous icon. The polar bear is also an emblem of the "Polar Airlines" company, which is one of the carriers in Nizhnekolymsk Rayon; thus, the image of the bear is widely used in advertising signs around the Chersky airport. Overall, this image reflects the connection to the Arctic, as well as to Russia as a whole, and is replicated in public places, such as the primary and secondary schools, where the identities of the young students of Chersky are being formed (Fig. 11.6). While it is indeed a local animal, it is important to stress that the polar

Fig. 11.6 Below the polar
bear, the text in Russian
reads "Chersky
Comprehensive Primary
School"

bear image as a symbol of strength and power is not typically used within Chukchi, Even, or Yukaghir culture. Rather, this is again a remnant of the legacy of the development of the region by outsiders.

11.4.3   Images of the Indigenous Peoples of the North: Between Folklore and Exoticism

Images of the indigenous peoples of Chersky also play a central role in political advertising. Dressed in national costumes, the people in these posters are a representation of the 'multiethnic family of nations' of the North. This representation leans very much on the old Soviet model, which also carefully delineated the spheres in which ethnic identity could be expressed. Images of dancing Yukaghir or Chukchi people are heavily folklorized to produce a collective image of Northern peoples, and imply that the value of these people lies in their exoticism. This representation is stable and unchanging. Further notable is the fact that while the folklorized images are present, the indigenous languages of these peoples are not featured on any of our 44 examples. The languages by their very absence, also underscore an exotic and unchanging image of indigenous inhabitants. We suggest that their absence implies both exoticism and anachronism, inferring that there is no need to keep something which has no real value to modern life.

Thus, the scope and spread of the images is restricted to the sphere of folklore; this is furthered by the fact that the images of dancing and singing indigenous peoples are concentrated in the center of the building advertising the revival of local culture. According to Seitel (2002:6), the process of folklorization involves the "restyling the expressions [of culture] so that they become less complex aesthetically and semantically. They thus reify the notion of a dominant culture (the one whose knowledge informs and is developed by official administrative and educational institutions) [in which] folklore is not as complex or meaningful as the products of

Fig. 11.7 This sculptured image on the building corner depicts a figure with a drum. No text is present

high, elite, or official cultural processes." This folklorization of the culture of minority groups is a phenomenon characteristic of many northern Russian communities as well as other regions of the former Soviet Union (see Shay 2002; Foxall 2015), and Chersky is no exception. Part of the reason why indigenous languages do not appear on signs may also be the common stereotype that these minority languages are not written, even though written standards were indeed developed in the early Soviet period (Vakhtin 2001; Grenoble 2003) (Fig. 11.7).

11.4.4 The Social Role of the Indigenous Languages of Nizhnekolysmk Rayon

Our assessment revealed that the dominant language in the linguistic landscape is overwhelmingly Russian. Of the 95 items, 77 appear in Russian, with two being bilingual in both Russian and Sakha. Four examples include English, including graffiti on the walls and an airport reminder that Chersky is a border zone. Table 11.2 shows the frequency of use of the different languages.

In terms of bilingual examples, there are only two: one case of Russian-English and another of Russian-Sakha. As noted, English is found on an information sign at the airport, calling travelers' attention to the fact that they are entering the border zone (Fig. 11.8).

Sakha is a state language of the Sakha Republic alongside Russian and, at present, use of the language is high among the indigenous population of Nizhnekolymsk Rayon. While historically Sakha had been used as a *lingua franca* prior to the spread of Russian in this region, Yakutization did not become a strong trend until much more recently. As mentioned earlier, while most of the indigenous population in Nizhnekolymsk Rayon speaks some Sakha, the language only became widely distributed and implemented in the education system in the region during the 1970s.

Table 11.2 Frequency of use of the languages of ethnic groups in the linguistic landscape of Chersky (from 101 photographs taken in April 2014)

	Russian	Sakha	English	Chukchi	Yukaghir	Evenki	Even
Number of examples	77	6	12 (including 7 graffiti)	0	0	0	0

Fig. 11.8 The text in Russian at the top of the sign is similar to that of the English presented on the bottom: "Attention! Border zone! Entry (passage) only with passes and identity documents"

This contributed, alongside Russian, to the formation of a new Soviet-era cultural identity in which Sakha was elevated over the other indigenous languages of the territory.

One instance of Sakha language use that we observed was on a poster in the center of Chersky. It depicts an *alaas* (a shallow depression in the melted permafrost), used by Sakha as a place to establish homesteads, graze animals, and collect fish and berries. The poster also features proverbs in Sakha about the *alaas*, such as "*doidu surakhtaakh, alaas aattaakh*" (every place has its renown, each alaas—its name), which are also translated into Russian. The showcasing of this landform (which is a natural phenomenon characteristic of the central regions of the Sakha Republic but not of its arctic zone)[3] can be taken as a reminder of attempts to create a unified and symbolic cultural identity of the Sakha Republic (based on Sakha language and lifeways) that also erases the distinct and diverse cultural and geographical characteristics of Nizhnekolymsk Rayon.

11.4.5   Graffiti: A Contemporary Element of the Settlement

Graffiti noted by the authors consisted primarily of names and showcased the use of invective language in Russian. Examples are located on walls throughout Chersky, but are mainly present in the residential sector of the village. The authors are obviously young villagers, who seem to use whatever paint is available; the graffiti do not stand out with any intensity or brightness, but do add some colour to the living space of the village. Most graffiti are in the Latin script, if not in English (see Fig. 11.9).

---

[3] See Footnote 2 in Chap. 19 of this volume.

Fig. 11.9 Graffiti on the side of a building, with the English word "Ghost" in the center

The phenomenon of graffiti in Russia is often thought to have been imported from Western culture, but it also has Russian roots. Especially common in Russian graffiti (as in many cultures, the world over, since ancient times) is the phrase "*Zdes' byl ya*", or "I was here." None of the graffiti we photographed defaced or challenged other signage, but rather seemed to function as a way for youth to simply mark their presence in the settlement.

11.5 Conclusion

A study of the linguistic landscape of Chersky reveals that the languages of the indigenous peoples of Nizhnekolymsk Rayon are absent, despite the fact that a third of the population is comprised of Yukaghir, Chukchi and Eveny people. Their languages are not used on posters, signs or advertising, though folkloric images of indigenous northerners are present. Despite the popular idea that Nizhnekolymsk Rayon is a highly multilingual community, we see here that linguistic diversity has been reduced to only two languages—Russian and Sakha—a situation which reflects the hierarchy of languages in the Sakha Republic, and the status of Russian and Sakha as state languages. The use of Sakha highlights as well the process of Yakutization that has been ongoing over the past century both in education and public life, leading to many ethnically Yukaghir, Chukchi and Even individuals speaking Sakha but not always their own indigenous languages—which are still symbolically marked as "official" in Nizhnekolymsk Rayon. This official status is not corroborated by the actual status of these languages as used in the region, which speaks to threats to the stability of the cultures of these indigenous minorities.

The dominant language of the linguistic landscape is Russian, and the non-linguistic (semiotic) aspects of the characters, colours, and figures indicate a predominance of images that evoke the Soviet period. Many images hearken back to a harmonious multiethnic environment evoked through the Soviet rhetoric of internationalism, depicted by ethnic groups in Nizhnekolymsk Rayon in national costumes. The frequent use of the Russian language again highlights the significant socio-political role of Russian as a language of interethnic communication and the official language of memory, history and ideology.

References

Bakhrushin, S. V., & Tokarev, S. A. (Eds.). (1953). *Yakutiya v XVII veke* [Yakutia in the 17th century]. Yakutsk: Yakutskoe knizhnoe izdatelstvo.

Ben-Rafael, E., Shohamy, E., Amara, M. H., & Trumper-Hecht, N. (2006). Linguistic landscape as a symbolic construction of the public space: The case of Israel. *International Journal of Multilingualism, 3*(1), 52–66.

Forsyth, J. (1992). *A history of the peoples of Siberia, Russia's North Asian colony 1581–1990.* Cambridge: Cambridge University Press.

Foxall, A. (2015). *Ethnic relations in post-Soviet Russia: Russians and Non-Russians in the North Caucasus.* Oxford/New York: Routledge.

FSGS (Federalnaya Sluzhba Gosudarstvennoy Statistiki). (2010). *Tom 1. Chislennost' i razmeshcheniie naseleniya. Vserossiiskaya Perepis' Naselenie 2010g.* [Book 1. The number and location of the population. All-Russian Census 2010]. Electronic document. Retrieved 1 July 2015, from http://www.gks.ru/free_doc/new_site/perepis2010/croc/perepis_itogi1612.htm

Gorter, D. (Ed.). (2006). *Linguistic landscape: A new approach to multilingualism.* Clevedon: Multilingual Matters.

Grenoble, L. (2003). *Language policy in the Soviet Union.* Amsterdam: Kluwer Academic Publishers.

Hult, F. M. (2009). Language ecology and linguistic landscape analysis. In E. Shohamy & D. Gorter (Eds.), *Linguistic landscape: Expanding the scenery* (pp. 88–104). New York: Routledge.

Kolesov, M. I. (2003). Ustanovlenie Sovetskoy vlasti i obrazovanie ulusa [The establishment of Soviet power and the development of the county]. In E. P. Antonov (Ed.), *Nizhnekolymskiy Ulus: Istoriya, kul'tura, fol'klor* (pp. 99–115). Yakutsk: Bichik.

Landry, R., & Bourhis, R. Y. (1997). Linguistic landscape and ethnolinguistic vitality: An empirical study. *Journal of Language and Social Psychology, 16*(1), 23–49.

Pennycook, A. (2009). Linguistic landscapes and the transgressive semiotics of graffiti. In E. Shohamy & D. Gorter (Eds.), *Linguistic landscape: Expanding the scenery* (pp. 302–312). New York: Routledge.

Samoylova, G. V. (1993). *Zapolyarnaya tochka Gulaga* [A polar point of the Gulag]. Moscow: Vozvrashchenie.

Scollon, R., & Scollon, S. W. (2003). *Discourses in place: Language and the material world.* London: Routledge.

Seitel, P. (2002). Defining the scope of the term Intangible Cultural Heritage. Paper presented at the conference International Meeting of Experts on Intangible Cultural Heritage: Establishment of a Glossary, UNESCO Headquarters, Paris, France (pp. 1–14). Retrieved July 1, 2015, from www.unesco.org/culture/ich/doc/src/00270-EN.doc

Shay, A. (2002). *Choreographic politics: State folk dance companies, representation and power.* Middletown: Wesleyan University Press.

Shishlo [Chichlo], B. (2000). *Problem of identification in a polytechnic society: The case of the native people of the Kolyma region* (pp. 17–18). Paper presented at the Siberia and the Circumpolar North: Contributions by Anthropologists and NGO conference. Vienna: Institute of Ethnology, Cultural and Social Anthropology.

Shohamy, E. (2006). *Language policy: Hidden agendas and new approaches.* New York: Routledge.

Sloboda, M. (2009). State ideology and linguistic landscape: A comparative analysis of (post)communist Belarus, Czech Republic, and Slovakia. In E. Shohamy & D. Gorter (Eds.), *Linguistic landscape: Expanding the scenery* (pp. 173–188). New York: Routledge.

Vakhtin, N. (1991). *The Yukagir language in sociolinguistic perspective* (Monograph Series 2). Steszew: International Institute of Ethnolinguistic and Oriental Studies.

Vakhtin, N. (2001). *Yazyki narodov Severa v XX veke. Ocherki yazykovogo sdviga* [The languages of the peoples of Siberia in the 20th century. Essays on language shift]. St. Petersburg: Dmitriy Bulanin.

Vakhtin, N., Golovko, E., & Schweitzer, P. (2004) *Russkie starozhily Sibiri. Sotsial'niye i simvolicheskie askpekty samosoznaniya* [The Russian old settlers of Siberia. Social and symbolic aspects of self-identification]. Moscow: Novoe Izdatelstvo.

Yaglovsky, V. N. (2003). Sel'skoe khozyaystvo (1929–1985) [Rural economy (1929–1985)]. In E. P. Antonov (Ed.), *Nizhnekolymskiy Ulus: Istoriya, Kul'tura, Fol'klor* (pp. 116–124). Yakutsk: Bichik.

Zakon. (1992). *Zakon Respublika Sakha (Yakutia) ot 16 oktyabra 1992g. N. 1170-XII "O yazykakh Respublika Sakha (Yakutia)"*. [The Law of the Republic of Sakha (Yakutia) from 16 October 1992. N. 1170-XII. "On the languages of the Republic of Sakha (Yakutia)"]. Retrieved July 1, 2015, from http://www.sakha.gov.ru/node/41515

Chapter 12
Barriers to Sustainable Health Promotion and Injury Prevention in the Northwest Territories, Canada

Audrey R. Giles, Lauren A. Brooks-Cleator, and Catherine T.R. Glass

Abstract In this chapter, we examine barriers to providing sustainable health promotion and injury prevention programs in Northwest Territories (NWT), Canada. In the context of increasing healthcare costs in Canada, developing sustainable healthcare programs is especially relevant. Using a multiple case study methodology and a framework developed by Schell et al. (Implement Sci 8(15): 1–9, 2013), we evaluate the sustainability of two health promotion and injury prevention programs: Elders in Motion (EIM) and the NWT Aquatics Program. We apply Schell et al.'s (Implement Sci 8(15): 1–9, 2013) nine domains of capacity for sustainability to each program to illustrate the challenges of developing health programs in northern Canada. Our results suggest that largest barriers to sustainable health promotion and injury prevention programs in the NWT relate to program content and delivery, financial issues, and staffing. These findings can inform future health promotion and injury prevention strategies that are intended to improve health outcomes in northern Canada.

Keywords Health promotion • Injury prevention • Sustainability • Elders • Drowning

12.1 Introduction

In studies completed in northern Canada, the term "sustainability" is typically used to refer to resource development. Yet, sustainability in reference to public health is an area that is also deserving of attention. In this chapter, we examine the capacity for sustainability in programs that involve two aspects of public health, health promotion and injury prevention, in Canada's most populated territory, Northwest

A.R. Giles (✉) • L.A. Brooks-Cleator • C.T.R. Glass
University of Ottawa, Ottawa, ON, Canada
e-mail: agiles@uottawa.ca; lbroo049@uottawa.ca; cglas007@uottawa.ca

© Springer International Publishing Switzerland 2017   151
G. Fondahl, G.N. Wilson (eds.), *Northern Sustainabilities: Understanding and Addressing Change in the Circumpolar World*, Springer Polar Sciences, DOI 10.1007/978-3-319-46150-2_12

Territories (NWT). The World Health Organization (2015) defines health promotion as "the process of enabling people to increase control over, and to improve, their health. It moves beyond a focus on individual behaviour towards a wide range of social and environmental interventions" (para 1). Injury prevention on the other hand, includes activities to prevent, reduce, treat, and/or ameliorate injury-related death and disability (Hemenway et al. 2006). Sustainable health promotion and injury prevention efforts are especially necessary given rising healthcare costs that are occurring in Canada, and in light of increasing fiscal restraint and the retreat of the welfare state. Using a qualitative multiple case study and the framework developed by Schell et al. (2013), we assess the capacity for the sustainability of two NWT programs that involve health promotion and injury prevention components: Elders in Motion (EIM) and the NWT Aquatics Program.

12.1.1 Sustainable Public Health in the North

Sustainability within the realm of public health can be understood as a collection of contextual and organizational factors that allow for a program to be maintained over time (Schell et al. 2013). Scheirer and Dearing (2011) identified sustainability outcomes as a program's ability to maintain its activities, benefits to clients, organizational practices, community level partnerships, and salience of the program's core issue. In order to achieve these outcomes, Schell et al. (2013) developed a new framework of public health capacity for sustainability that was based on a comprehensive literature review and concept mapping process. This framework identifies nine domains of capacity for sustainability:

- **Political support:** the political environment that impacts program support, activities, and acceptance
- **Funding stability:** the ability for long term planning based on stable funding sources
- **Partnerships:** the relationships between the program and target communities
- **Organizational capacity:** the resources required to manage and run program activities
- **Program evaluation:** collecting data and evaluating program outcomes and activities
- **Program adaptation:** the ability to respond and modify program to ensure efficacy
- **Communications:** the dissemination of program activities and outcomes with decision-makers, stakeholders, and the public
- **Public health impact:** the program's impact on a target community's health perceptions, attitudes and behaviours
- **Strategic planning:** the process that dictates program content, delivery, and outcomes

Researchers, funders, and practitioners can apply the nine domains to small, community-level projects or larger programs. These domains are important to consider, as programs that are sustainable are more likely to produce healthier communities and long-lasting effects (Schediac-Rizkallah and Bone 1998).

The development of sustainable health promotion and injury prevention efforts is essential for decreasing health care spending in Canada, particularly in the North, where health care costs are the highest (Canadian Institute for Health Information 2013). A study by the Canadian Institute for Health Information (2013) that documented health expenditures from 1975 to 2013 found that NWT has the second highest (behind Nunavut) per capita cost at $10,686 or 9.0 % of the total health expenditure-to-territorial GDP ratio. As such, the creation of sustainable public health is key for the financial sustainability of the territory.

12.1.2 Elders in Motion

In 2009, the NWT Recreation and Parks Association (NWTRPA) partnered with the Canadian Centre for Activity and Aging (CCAA), which is based out of Western University in London, Ontario (CCAA n.d.), to establish Elders in Motion (EIM). To implement this program in NWT, community members rely on the NWTRPA for training, partial funding, and support with program development. Staff members from the NWTRPA train program leaders from NWT to deliver EIM in their communities, so that once implemented, the program can be maintained in the communities with little to no involvement from the NWTRPA (CCAA n.d.).

Typically, program leaders are recreation coordinators, community health representatives, Elder coordinators, and/or home support workers. EIM can be led in a recreation facility or home-based setting and aims to be inclusive of older adults with various ranges of independence and mobility. EIM consists of multiple activities and programs that are designed to help older adults lead active, healthy lives. A significant part of EIM is the Active Living Exercise Program that was adapted from the CCAA's Home Support Exercise Program for older adults in southern Canada, and which consists of 10 easy, but effective exercises for older adults (CCAA n.d). To adapt the Home Support Exercise Program for EIM, the NWTRPA modified the CCAA's program documents, materials, and audio/visual material to include northern people, images, voices, music, and themes. Other activities included in EIM are Balls and Balance Exercises, band exercises, Nordic walking, and walking bingo.

The NWTRPA's Active Communities Coordinator visits the communities involved with EIM throughout the year to help program leaders plan events; to give them ideas on motivating Elders to attend and on new activities to include; and to provide continuous support to the program leaders (NWTRPA n.d.). There is also an annual Training Gathering for EIM that includes training opportunities, best practices tips, Active Elder Awards, and discussions among participants (NWTRPA 2009). To help finance EIM program costs, program leaders can apply for a $2500 grant from the NWTRPA. EIM is supported financially by the Government of NWT

Departments of Health and Social Services, and Municipal and Community Affairs. EIM can be considered both a health promotion and injury prevention program because it helps Elders to become more physically active or maintain physical activity, which can play an important role in decreasing the risk of injuries.

12.1.3 NWT Aquatics Program

People have used NWT's bodies of water for recreation, transportation, and subsistence since time immemorial. The NWT Above Ground Pool Program (now NWT Aquatics Program) began in 1967 when the territory's first above ground pool was constructed in Fort Simpson. The program grew a great deal and soon seasonal swimming pools were constructed in municipal garages, arenas, and curling rinks across the North during summer months and seasonal waterfronts began operating in communities without pools. The NWT Aquatics Program quickly became a popular and prominent part of NWT's recreation programming. Originally a NWT government initiative, each summer, hundreds of southerners, typically college and university students, would apply for positions to run pools and waterfronts across the North (for further history, see Giles et al. 2007).

The program's initial goal was to have southern-based lifeguards and swimming instructors train local Aboriginal people to eventually run pools and waterfronts themselves, which would provide northern youth with the opportunity to learn and practice leadership skills. Shortly after the program's inception, drowning prevention was added as an additional goal (Giles et al. 2007). In its peak year of operation, 1997, the Program operated 41 aquatics programs in NWT (including what is now Nunavut): 21 pools, eight seasonal waterfronts, seven short-term waterfronts (where staff would visit the community to offer lessons rather than staying there), and five programs where people were bussed to neighbouring communities to access aquatics activities (Szabo 2002).

Despite its long duration in NWT, the program has struggled to produce northern aquatic leaders, and the number of aquatic facilities and programs has decreased dramatically since the program's peak years. Additionally, the drowning rate in NWT still remains six to ten times the national average in any given year (Canadian Red Cross 2013). The NWT Aquatics Program's focus on water, boat, and ice safety renders it an injury prevention program, while the non-structured activities offered at aquatic facilities provide NWT residents with opportunities for physical activity and thus health promotion.

12.2 Methodology

We used a multiple case study approach to understand issues related to sustainability in Elders in Motion and the NWT Aquatics Program. A case study is "both a process of inquiry about the case and the product of that inquiry" (Stake 2005:444).

A multiple case study approach is used to jointly examine a number of cases to understand a population, phenomenon, or general condition (Stake 2005). This approach is useful when the chosen cases will lead to better understanding about a larger collection of cases (Stake 2005); in this instance two public health programs in the North.

12.3   Methods

To investigate the capacity for sustainability of the programs, we employed a variety of qualitative research methods. For EIM, we used semi-structured interviews and archival research. For the NWT Aquatics Program, we used semi-structured interviews, focus groups, and archival research. The data that we gathered for this chapter were originally gathered for larger research projects by A. R. Giles (2006–2014) and L. A. Brooks-Cleator (2012–2014).

12.3.1   EIM

To examine EIM, Brooks-Cleator conducted nine semi-structured interviews: two with NWTRPA staff (Executive Director and Active Communities Coordinator), and seven with EIM program leaders from five different NWT communities, for a total of nine interviews. The program leaders included two Community Health Representatives, one Home Support Worker, and four Recreation/Health Promotion workers. The interview questions focused on the challenges and successes program leaders faced when implementing and sustaining EIM, the cultural relevancy of the program, the effectiveness of EIM, and their experiences with EIM training and activities. She also conducted archival research that included reviews of NWTRPA annual reports from 2009 to 2013, an evaluation of EIM that occurred in 2010, booklets about the Balls and Balance exercises and the Active Living Exercise Program, and training manuals for program leaders (Caseñas and Kalsbeek 2006).

12.3.2   NWT Aquatics Program

To examine the NWT Aquatics Program, Giles and her research assistants conducted semi-structured interviews with 120 residents from seven NWT and Nunavut communities (formerly part of NWT). Interview participants included past and current program staff, as well as community youth, adults, and Elders. Giles also conducted archival research with newspapers, territorial government documents, and municipal government documents. For both research projects, participants were given the option for their own names to be used or to remain anonymous.

12.3.3   Analysis

To analyze the data we used Braun and Clarke's (2006) approach to thematic analysis. Braun and Clarke (2006) described a six-step approach to thematic analysis. Giles analyzed the data related to the NWT Aquatics Program and Brooks-Cleator analyzed the data related to EIM. First, we familiarized ourselves with the transcribed interview data by (re)reading it and taking note of initial ideas. Second, we generated initial codes to systematically identify and organize all the data. We developed these codes by searching for ideas and quotes that related to the sustainability of the programs, such as funding, motivation of participants, community support, etc. Third, we arranged the codes into potential themes and reviewed the results to ensure that they fit in both their original context as well as the generated theme. Fifth, we concisely labelled the themes to reflect the message that we wanted the analysis to deliver. Finally, in the sixth step, we selected the most appropriate and compelling quotations from the data that supported the generated themes and that related to the research questions and literature reviewed, including Schell et al.'s (2013) nine domains of capacity for sustainability. To support the rigour of the study, the authors consulted each other on the results for this chapter and selected those with the greatest overlap.

12.4   Results and Discussion

After using Braun and Clarke's (2006) six stage approach to thematic analysis, we then identified three main themes in the data that were found to affect the sustainability of both EIM and the NWT Aquatics program: program content and delivery, financial issues, and staffing. All of these factors are especially relevant to the sustainability of programs in the NWT as they each relate to Schell et al.'s (2013) core domains that affect a program's capacity for sustainability. Below, we discuss how each theme pertains to each program and the key core domains that relate to each theme.

12.4.1   Program Content and Delivery

EIM Many of the program leaders discussed how the content of the EIM program is not relevant, in terms of the activities, language, etc., for all of the Elders in their communities. The leaders thought this irrelevance may have resulted in Elders being less likely to attend the program. One of the program leaders, Roslyn Firth, mentioned how "*the culture makes people a little bit reluctant to join in to something that is like a standardized series of exercises. It's a bit difficult to engage those people*" (personal communication, 19 June 2013). Another program leader noted, *"we had to really try hard to encourage most of them...A lot of them lead traditional lives too, so they are out on the land and hunting and fishing and whatever"* (Anonymous, personal communication, 19 June 2013).

With the delivery of the program, it seems that a significant factor for the sustainability of the program is dependent on where the program can be held. In most of the communities, the program can only be held if a recreation or community centre space is donated or if there is another common area for Elders to meet. One program leader discussed how it is difficult to sustain the program because the community centre for the older adults was closed down: *"actually, we used to have it here at the [seniors centre], the old folks home, but now they closed the centre down...That's one of the reasons why we quit too because we don't have any space to do it and everybody just lost interest"* (Anonymous, personal communication, 19 June 2013).

NWT Aquatics Program The NWT Aquatics Program's content is largely dictated by the curriculum produced by organizations that offer nation-wide water safety programs (such as the Canadian Red Cross Society and the Lifesaving Society of Canada) and by federal boating regulations produced by Transport Canada. The content of these programs and regulations are typically developed by and for those who live in the Canadian South, and taught by those who work in NWT on a seasonal basis.

We found that programs and regulations often do not reflect content that many northerners deem important, and those who offer the programs often have little or no knowledge of water, boat, and ice safety in a northern context or of traditional knowledge. For example, residents of Pangnirtung, Nunavut, noted that Transport Canada's pre-departure checklist does not include a harpoon, ammunition, or a knife, all of which they deem crucial for safety (Giles et al. 2013). Similarly, residents of Fort Simpson, NWT, described the importance of the Dene practice of making an offering of tobacco to the water prior to using it, which also is not included in water safety curricula. Community members in Taloyoak, Nunavut, detailed traditional Inuit knowledge that they have about resuscitation of a drowning victim that refers to the practice of not wiping away bubbles that form around a person's mouth, and noted that this practice is not included in water safety instructional materials (Giles et al. 2007). In all of the communities in which we conducted research, members noted that swimming lessons are often too similar to school in their structure and that this is a deterrent to local children's participation.

The issues associated with program content and delivery for both programs relate to three of Schell et al.'s (2013) core domains that affect a program's capacity for sustainability: organizational capacity, which relates to the resources required to deliver the program; program adaptation, which is how the program is adapted to ensure effectiveness; and partnerships, which involve the connections between the program and the community in which it is offered. NWT is unique with its 11 official languages, and because each community in NWT differs significantly in terms of its resources, infrastructure, size, geography, and demographics (Education, Culture, and Employment 2015; Municipal and Community Affairs n.d.). Not every community has the same organizational capacity or adaptability needed to run a program; the consequent variations in activity success rates can lead to challenges in territory-wide program sustainability. As a result, it is difficult to organize programs and adapt resources in ways that address diverse languages, cultures, and local capacity in the territory. To support greater sustainability, both EIM and the NWT Aquatics Program need to form stronger partnerships with communities and need to adapt to address each community's specific needs.

12.4.2   Financial Issues

EIM   The most common factor that was identified to be affecting the sustainability of EIM in NWT communities was financial constraints. A lack of finances to rent facilities, to transport Elders to facilities, and financial instability were all constraints highlighted by the majority of program leaders. One program leader discussed the importance of a program like EIM but she noted that "*the main [challenge] was just having to seek funding and that the funding was you know, just enough for a few classes. The challenge part was the financial part. The program itself was awesome*" (Anonymous, personal communication, 24 June 2013). The finances to sustain the program over the long term are just not available. One program leader mentioned that the community was just "*starting to get a little more Elders coming up in the spring, until we ran out of funding*" (Anonymous, personal communication, 24 June 2013). The sustainability of EIM is also significantly influenced by the funding for staff positions. The contract positions that are funded by grants end once the grant is finished, resulting in program termination. One program leader discussed his frustration with the unpredictability of funding: "*it's kind of always up in the air with the funding and stuff like that. They say it's a term position from now until March 31st and they say we'll have to see if we get our funding again*" (Anonymous, personal communication, 12 July 2013).

NWT Aquatics Program   Originally, the NWT Aquatics Program was delivered by the Government of NWT. In the late 1990s, the program was transferred to the NWT Recreation and Parks Association, which is a not for profit, non-governmental organization. At the same time, the government devolved increasing financial power to communities. As a result, rather than communities applying for funding specifically for aquatics programming, communities were instead given blocks of funding for which they determined their own priorities (Giles et al. 2007). While increased local control has tremendous benefits for communities, it also creates a situation whereby funding for aquatics programs enters into direct competition for funding with items like roads, water, and sewers.

Participants noted the prohibitive costs of running aquatics programs, particularly swimming pools, which are especially expensive to operate due to the high levels of maintenance. Further, due to a shortage of qualified northern residents, pools are typically staffed by seasonal workers from the South. Jessie, a former pool employee and Geoff, the NWTRPA's Executive Director noted that it would be more cost effective to hire northerners, as that would eliminate the customary practice of northern communities paying for a southern pool supervisor's flights and accommodations in the North (Jessie, personal communication, 27 June 2006; G. Ray, personal communication, 10 December 2014).

Receiving enough funding to run programs in a way that would meet residents' needs and not result in deficits to other community programs was found to be a significant challenge to sustaining both EIM and the NWT Aquatics Program. This relates directly to three of Schell et al.'s (2013) core domains: funding stability, which involves the stable funding required to make long-term plans; political support, which includes strong relationships between the program's umbrella organization and all levels of government; and partnerships within each community. To sustain the programs and to allow for long-term planning and program development, stable funding sources are necessary and require strong political support. Since aquatics programming is controlled by local community governments, partnerships between local governments could play an important role in reducing costs (for instance, through sharing the cost of bringing in an individual to train northerners). The same is true for EIM, but support would need to be coordinated at all levels of government.

12.4.3 Staffing

EIM Participants reported that not only was funding for staff to run EIM an issue; acquiring and maintaining knowledgeable staff was also a challenge for program sustainability. If the staff were not trained in delivering EIM due to a lack of knowledge or lack of training opportunities during short-term contracts, there would be no one to run the program. One program leader mentioned, *"the regular girl that did the Elder Day Program, she wasn't hired the second year and the girl that got it the second year didn't really do much exercises"* (Anonymous, personal communication, 12 July 2013). Another common issue across many of the communities in NWT is that there are usually only a few individuals who look after delivering recreation and physical activity programs in communities. There is not enough staffing capacity to maintain the programs, especially if they are not a priority in the communities. Roslyn identified this issues when she explained that the challenge of sustaining the EIM program:

> is having a staff person who can dedicate their time to delivering the program, 'cause you know we have all of the usual social issues that the communities in the North have...So the people who took the training, myself and the others, we already have full time jobs, so I think where everything broke down was, or one of the big reasons was, we were already working and so it's hard to add another responsibility into our day...There are so many ideas and great programs that come from the NWTRPA...but they don't always work because we don't have the capacity (R. Firth, personal communication, 19 June 2013).

Retaining staff and having adequate human resources available to begin with is a common challenge. One program leader mentioned that the community *"did have an Elders coordinator, but she's going back to school"* (Anonymous, personal communication, 12 July 2013). While attending school is important, it meant that there was no longer a trained individual available to run EIM, which resulted in the community program's cessation; this demonstrated how important key individuals are in the small, northern communities for sustaining health promotion programs.

NWT Aquatics Program In its early years and until the early 1990s, the NWT Aquatics Program had little trouble attracting southern seasonal staff to work for the program. As a government employee who used to be affiliated with the program stated:

> I think [it] was an allure, you know, the northern frontier…Canada's Arctic…and to a certain extent you were able to entice university students…to come up North, and, and to get free housing: come work in the land of the midnight sun and play in a pool all summer and we will pay you…and you don't have any expenses (Anonymous, personal communication, 11 February 2007).

Nevertheless, as wages for aquatics staff in southern Canada became more competitive and as travel to the Arctic became more common, it became harder to attract qualified southerners to work in the North. While perhaps the most straightforward solution would be to train northerners to run the pools and waterfronts, especially given their knowledge of northern cultures and waterways, such initiatives have proven to be extremely difficult to achieve. The certification process to become a lifeguard, swimming instructor, and pool operator is costly and time intensive. It is made all the more difficult by the fact that most pools in NWT are only three feet deep and the skills that lifeguards are required to learn and perfect require deep water. As a government employee noted:

> aquatics is a very responsible position when you are 17, 18, 19, 20, you know, the lives of people are in your hands, you're responsible for them. You can make the same money in pumping gas or mowing lawns or maybe do labour. Sometimes it's not a hard choice for people to do that (anonymous, personal communication, 18 June 2006).

Further, as a former government employee pointed out, *"it's really hard to retain any local people that you have recruited because of the huge job market in the NWT"* (Anonymous, personal communication, 11 February 2007).

While many residents we interviewed stated that they would like to see local people hired to work at the pool because they would be familiar with the community, its culture, and its residents, most felt that the most important issue was that the employee was adequately trained. As Mary, a resident of Taloyoak stated, *"[i]t doesn't bother me whether it is a person who comes from this community or if the person comes from the South. The most important thing is that the person knows what they're doing about safety instruction and the person is certified about the pool"* (Mary, personal communication, 25 June 2006).

Both programs demonstrate that even when the funding is available, if qualified staff are not hired and/or retained, it becomes challenging if not impossible for the programs to be sustained. Staffing is directly related to Schell et al.'s (2013) core domains of funding stability and partnerships, but also to organizational capacity. Recruiting and hiring appropriate staff are issues across many sectors, especially healthcare, in NWT (Health and Social Services 2014). Having stable funds to hire long-term staff not only attracts suitable applicants, but also increases the likelihood of staff retention and program sustainability. For both EIM and the NWT Aquatics Program, hiring local staff is important for improving the communities' economies, incorporating community culture, and developing skills in community members. However, this can only be done through the development of meaningful partnerships between the parent organization of the program and the communities in which the programs are offered. The programs need to be seen as important and worthwhile by community members in order to encourage local people to work with and sustain them. This also directly relates to the core domain of organizational capacity, as appropriate staff with knowledge of the community members and their traditions and cultures are those who are best situated to successfully deliver sustainable programs.

12.5 Conclusion

As we have demonstrated throughout this chapter, discussions of sustainability in a northern context must include public health, specifically health promotion and injury prevention programs. Given the diversity of NWT and the potential impact of these programs to improve the health of northern residents and address the rising costs of healthcare provision in the North, understanding how these programs can be sustained over the long-term is crucial. By examining EIM and the NWT Aquatics Program, we have shown that there are three main factors to consider for improving the sustainability of public health programs in the North: program content and delivery, financial stability, and staffing. We demonstrated through our case analysis and discussion of Schell et al.'s (2013) core domains that northern health and injury prevention programs are susceptible to various sustainability issues. While all of the domains are linked, EIM and the NWT Aquatics Program are specifically affected by organizational capacity, program adaptation, political support, funding stability, and partnerships. Analyzing these two programs allows us to consider how other public health programs in Canada's North can be established, advanced, and maintained to improve northerners' overall health, quality of life, and self-determination in community specific contexts.

References

Braun, V., & Clarke, V. (2006). Using thematic analysis in psychology. *Qualitative Research in Psychology, 3*(1), 77–101.
Canadian Centre for Activity and Aging [CCAA]. (n.d). *Elders in motion: Active living exercise program in the NWT*. London: Author.

Canadian Institute for Health Information. (2013). National health expenditure trends: 1975 to 2013. Retrieved from https://secure.cihi.ca/free_products/NHEXTrendsReport_EN.pdf

Canadian Red Coss. (2013). *Analytical report on Aboriginal open water fatalities: Promising practices for prevention.* Ottawa: Author.

Caseñas, C., & Kalsbeek, K. (2006, May). *Archival research tutorial.* Retrieved from http://www.library.ubc.ca/spcoll/Guides_UBC/Index.html

Education, Culture and Employment. (2015). *Official languages.* Retrieved from http://www.ece.gov.nt.ca/official-languages

Giles, A. R., Baker, A. C., & Rousell, D. D. (2007). Diving beneath the surface: The NWT Aquatics Program and implications for Aboriginal health. *Pimatisiwin, 5*(1), 25–49.

Giles, A. R., Strachan, S. M., Doucette, M. M., Stadig, G. S., & The Municipality of Pangnirtung. (2013). Using the vulnerability approach to understand aquatic-based risk and adaptation due to climate change in Pangnirtung, Nunavut. *Arctic, 66*(2), 207–217.

Health and Social Services. (2014). *Caring for our people: Improving the Northwest Territories health and social services system.* Yellowknife: Government of the NWT.

Hemenway, D., Aglipay, G. S., Helsing, K. L., & Raskob, G. E. (2006). Injury prevention and control research and training in accredited schools of public health: A CDC/ASPH assessment. *Public Health Reports, 121*(3), 349–351.

Municipal and Community Affairs. (n.d). *NWT communities.* Retrieved from http://www.maca.gov.nt.ca/?cmtylist=aklavik

Northwest Territories Recreation and Parks Association (NWTRPA). (n.d.). *Elders in motion.* Retrieved from http://nwtrpa.org/rpa/?page_id=55

NWTRPA. (2009). *Elders in motion: A fitness program for seniors.* Yellowknife: NWTRPA.

Schediac-Rizkallah, M. C., & Bone, L. R. (1998). Planning for the sustainability of community-based health programs: Conceptual frameworks and future directions for research, practice, and policy. *Health Education Research, 13*, 87–108.

Scheirer, M. A., & Dearing, J. M. (2011). An agenda for research on the sustainability of public health programs. *American Journal of Public Health, 101*, 2059–2067.

Schell, S. F., Luke, D. A., Schooley, M. W., Elliott, M. B., Herbers, S. H., Mueller, N. B., & Bunger, A. C. (2013). Public health program capacity for sustainability: A new framework. *Implementation Science, 8*(15), 1–9.

Stake, R. (2005). Qualitative case studies. In N. K. Denzin & Y. S. Lincoln (Eds.), *The SAGE handbook of qualitative research* (3rd ed., pp. 443–466). Thousand Oaks: Sage.

Szabo, C. (2002). *NWT aquatics program summary.* Yellowknife: Government of the Northwest Territories.

World Health Organization. (2015). *Health topics: Health promotion.* Retrieved from http://www.who.int/topics/health_promotion/en/

Chapter 13
Foreign Bodies in the Russian North: On the Physiological and Psychological Adaptation of Soviet Settlers and 'Oil Nomads' to the Oil-Rich Arctic

Rémy Rouillard

Abstract Based on more than a year of ethnographic field research in the Nenets Autonomous Okrug (NAO) in the Barents Sea region, as well as on a review of the literature existing in the field of Arctic medicine in Russia, this chapter focuses on the various discourses regarding adaptation to the Arctic of two groups who share the same ethnic background: the Russian settlers who arrived in the NAO to be involved in oil and gas prospecting, and the shift-workers – described as "oil nomads" in the chapter – who have been coming to the district since the mid-2000s from southern and central Russia to extract the oil discovered by their predecessors. This chapter examines the evolution of medical knowledge on adaptation to the Arctic from the Soviet period to the current day. I show that depictions of the bodies of settlers and oil nomads in both recent medical knowledge and popular discourses support the changes taking place in the organization of the workforce in the NAO, changes which appear to favor the oil nomads.

Keywords Russian Arctic • Oil industry • Medical sciences • Adaptation • Shift-work

13.1 Introduction

In the winter of 2009, I was invited by a group of oil workers employed by a company operating in the arctic region of the Nenets Autonomous Okrug[1] (NAO; Fig. 13.1) to go for *shashlyki* (barbeque) "in nature" outside of the okrug's capital,

---

[1] "Okrug" translates into "district" in English.

R. Rouillard (✉)
School of Psychoeducation, Université de Montréal, Montréal, QC, Canada
e-mail: remy.rouillard@mail.mcgill.ca

© Springer International Publishing Switzerland 2017
G. Fondahl, G.N. Wilson (eds.), *Northern Sustainabilities: Understanding and Addressing Change in the Circumpolar World*, Springer Polar Sciences,
DOI 10.1007/978-3-319-46150-2_13

163

Fig. 13.1 Nenets autonomous Okrug, Reference map (Ahlenius and UNEP/GRID-Arendal 2012)

Naryan-Mar.[2] As I was putting meat on the metal skewers, I made several small cuts on my fingers, which stubbornly refused to coagulate. Not understanding this, I showed my hands to one of the guests who had arrived a few months earlier from central Russia. "The level of oxygen is different here, you haven't been here long enough. It's like that for everyone who is not from here", he remarked. This was one of several events which led me to wonder about what arriving in an arctic region such as the NAO does to the bodies of those who are not from there, especially as the circumpolar world witnesses the growing presence of extractive industries and their workers.

In this chapter, I will examine medical knowledge as well as popular discourses commonly heard in the okrug regarding the health consequences of the arrival of outsiders to the Russian Arctic, discourses in which the notion of adaptation is central. More precisely, I will discuss the evolution of medical knowledge on the topic of adaptation to the Arctic from the Soviet period to the present day concerning two groups who most often share the same ethnic background, yet have come to the region in two different political, economic and ideological eras: the predominantly-Russian settlers who arrived in the NAO between the 1960s and 1980s to take part in oil and gas prospecting, and who now comprise the majority of the okrug's 48,000 inhabitants; and the "oil nomads", those 10,000 people who, since the mid-2000s,

---

[2] Naryan-Mar was founded in 1929 and means "Red Town" in the language of indigenous Nenets.

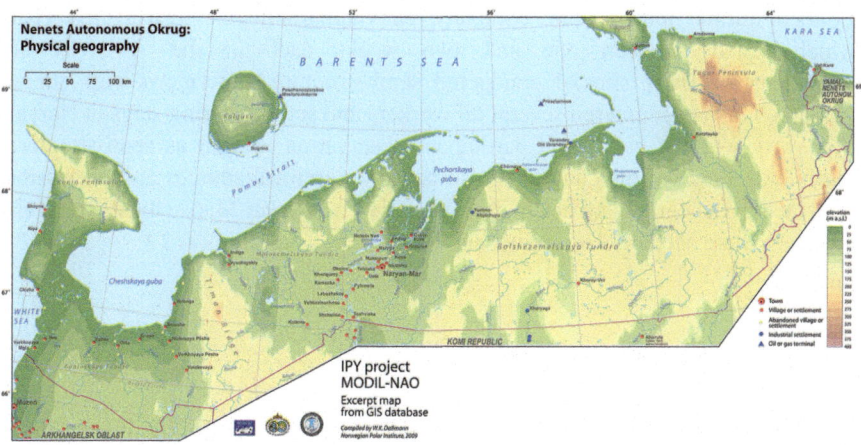

Fig. 13.2 Nenets autonomous Okrug: physical geography (Source: Dallmann 2009)

have arrived in the district as shift-workers from southern and central Russia to extract the oil discovered by their predecessors (Stammler and Peskov 2008).[3]

In the post-Soviet context, these discourses on adaptation bear on some of the issues associated with the colonial concept of "acclimatization", namely its problematic differentiation. Contrary to the colonial period, where differentiation was based on the various origins of the colonizers and the colonized – which served to justify their respective position in the political and socioeconomic hierarchies – in today's NAO, such differentiation is based on the ancestry of arrival to the okrug, in different political, economic and ideological contexts.

The data supporting this chapter stems from 14 months of ethnographic field research between 2008 and 2013 in two regions of the NAO where Nenets reindeer herders share the territory they migrate on with oil workers (Fig. 13.2). The first site is the island of Kolguev where nearly four hundred Nenets herders and villagers have as their "neighbours" two alternating shifts of 200 workers employed by Arktikmorneftegazrazvedka[4] (AMNGR). The other region is the Bolshezemelskaya Tundra ("Big Land Tundra") where seven herding brigades of the Kharp ("Northern Lights" in Nenets) agricultural cooperative migrate. It is also the site of one of the okrug's largest extraction sites, the Yuzhnoe-Khylchuyu[5] site where some 600 workers are based. This extraction site has been developed by both Lukoil and

---

[3] In this chapter I will not discuss the health issues nor the depiction in medical literature or popular discourses of the Nenets people, who are indigenous to the region and who represent 7000 of the okrug's residents. I discuss these matters and, more generally, the consequences of the Soviet and post-Soviet projects of oil-related developments on the Nenets way of life elsewhere (see Rouillard 2013).

[4] "Arctic and Sea Oil and Gas Exploration".

[5] "Southern-Khylchuyu," named after the river called by the Nenets "Khylchuyu".

ConocoPhillips' joint-venture company, Naryanmarneftegaz.[6] During my visits, which lasted between one and two weeks each at the AMNGR and Naryanmarfneftegaz sites, I conducted semi-structured interviews with the oil workers. I also encountered several oil workers and settlers during stays in Naryan-Mar, with whom I conducted participant observation as well as semi-structured interviews. Finally, this chapter is also based on archival sources and discussions with medical scientists and health care specialists in Naryan-Mar, at the Arkhangelsk Northern State Medical University, and in Moscow.

13.2   The Roots of Adaptation

The history of the concept of adaptation can be traced back to the notion of acclimatization born in the colonial period. Questions concerning the ability of humans to adapt to various natural environments and the effects of these environments on human constitutions first emerged in the eighteenth century, in the context of colonization (Osborne 2000). Different opinions circulated as to whether it was possible for human bodies to acclimatize to other climates, namely to those of the colonized countries, or if such changes were too drastic for bodies – specifically those of the colonizers – to be transplanted into other climates. Livingstone (1991:417) relates the words of British assistant surgeon A. Thomson, who, in the middle of the nineteenth century and following his experience in colonial India, suggested that "no length of seasoning[7] will diminish the deleterious influence of a tropical climate on the European constitution". In Victorian Britain, acclimatization came to be a part of medical discussions regarding disease and hygiene. In 1897, Dr. L. Westenra Sambon suggested that "the difficulties in the way of colonisation are not due to climate, but to parasitism" and that acclimatization "is, to a great extent, a mere question of hygiene" (cited in Livingstone 1999:97).

Lock and Nguyen (2010:150) remark that "[a]cclimatization was a powerful idea that linked climate, environment, constitution, and temperament through the trope of 'adaptation.'" The problematic aspect of the notion of acclimatization in the colonial context is that it implied differentiation: the bodies of those who were acclimatizing – the colonizers – were to be taken care of differently than those who were considered as well adapted, that is, the natives and the colonized.

I will now examine the roots of the widely used notion of adaptation in Soviet and post-Soviet medicine to explain the health challenges experienced by those who live and/or work in the Arctic, at a time where the concept has largely fallen out of

---

[6] Naryanmarneftegaz ("Naryan-Mar Oil and Gas" in Russian). In 2010, ConocoPhillips sold back its shares in Lukoil to the Russian company. As a result, the assets and production sites owned by Naryanmarneftegaz were absorbed by Lukoil.

[7] The notion of "seasoning" referred to the idea that Europeans should adapt their bodies gradually to the tropical regions they were colonizing in order to avoid falling victim to tropical fevers (Lock and Nguyen 2010:43).

favor in Western medical science and in the social sciences. In this discussion, my goal is not to engage in a critical manner with the knowledge produced by Russian medical scientists by trying to prove or disprove the findings they have produced, although I will point to findings stemming from Western medical sciences to indicate where they diverge from Soviet and post-Soviet medical sciences. Rather, I propose to reiterate the view of Lock and Nguyen (2010:108), who argue that "[...] biology is inevitably a snapshot, one situated in the time and space of a complex and shifting material reality, historically patterned by society, culture, economics, and politics."

13.3 The Soviet Conquest of the Arctic and the Body

As early as the late 1920s, the Soviet state had embarked on projects that demonstrated its capacity to conquer one of the harshest regions of the world: the Arctic. The Arctic thus became a mythical region over which the Soviet state and people would be able to demonstrate their mastery. Polar stations were established throughout the North, staffed by model citizens who had access to film collections and libraries so that they could replicate their "civilized" Soviet socialist way of life, even in the Arctic (McCannon 1998:46). Soviet leaders also possessed a longstanding and strong belief that the technologies developed by Soviet scientists would allow the state and its citizens to exploit nature rationally through conscious planning.

In the following decades, the North became a key source of natural resources for the Soviet state which had, until then, been extracted by prisoners in the infamous gulags. Following the death of Joseph Stalin, the new Soviet authorities under the leadership of Nikita Krushchev believed, instead, that the resources of the North should be extracted by workers who would be compensated for difficult working conditions (Thompson 2008). From then on, the state started relying on young, enthusiastic, and loyal workers who, in exchange, were to be given access to new privileges. These "northern benefits" (*severnye l'goty*) included faster access to an apartment, early retirement, longer vacations, and yearly trips to resorts in the "South", amongst other privileges. As Thompson (2008:44) remarks, "[i]f, earlier in the decade [in the 1950s], the North was still largely a destination for the regime's enemies, by the end a northern posting was considered a reward reserved for the Soviet labour elite."

Between 1955 and 1975, 800 new towns were created in the Russian North and Far East, most often for industrial purposes (Stammler and Eilmsteiner-Saxinger 2010). In the NAO, the last All-Soviet census conducted in 1989 revealed that 52.9 % of the okrug's inhabitants were not born in the region (Heleniak 2009a). Even today, Russia's northern regions are inhabited by 11.7 million people, a number that bears witness to the grandiosity of the Soviet development project in the North (Gudkov et al. 2012). The inhabitants of arctic and northern regions of Russia are still eligible for many of the northern benefits that have their roots in the Soviet period (Wengle and Rasell 2008).

Another means by which Soviet authorities expressed concern for northern settlers was the growing number of research projects focused on their health. With the increasing number of people arriving to the North, the study of adaptation became a key topic in Soviet medical research. In the mid-1950s, Aleksandr Bakulev, then President of the Academy of Medical Sciences, suggested that "the Academy has an obligation to the country to study the problem of acclimatization of humans to the North" (quoted in Danishevskii 1968:21). To illustrate the importance of this research agenda for the Soviet Union, by the late 1970s, 1500 researchers were studying the issue of human adaptation in 138 research or government institutions (Kaznatcheev and Strigin 1978).

One of the pioneers in the Soviet research on adaptation is physiologist G. Danishevskii. Based on his observations in Kolyma and in the Pechora regions, Danishevskii (1968) was convinced that the particular pathologies he observed among the inhabitants of the North were related to the process of acclimatization, and should therefore be treated accordingly. What he referred to as "meteorapathic reactions" (*meteorapaticheskie reaktsii*) were most commonly ones caused by environmental changes that disturbed a person's general condition, causing health problems such as general fatigability, somnolence and a loss of appetite, or various pains. Danishevskii's research in Russia's northwestern Pechora region nonetheless revealed that 54 % of healthy people investigated had experienced acclimatization-related problems during the first six months following their arrival to the area. The number decreased to 17 % for those who had been there between six and twelve months, and to 5 % after more than three years of living in the North.

Danishevskii (1968) remarked that in the process of adaptation, working conditions, the quality of housing, clothing, and most especially nutrition were decisive in a person's acclimatization. This belief in the possibility of improving the health of the settlers to the Arctic through better living and working infrastructures and services persisted through the following decade. In the late 1970s Kaznacheev and Strigin (1978:36) suggested that "without a doubt, on the basis of medical-biological data gathered and domestic norms, the working conditions and life in the Far North will guarantee the preservation of health to the same extent as in lower latitudes".

Kaznacheev (1986) also introduced a concept which is still commonly referred to today by medical researchers in Russia as "polar tension syndrome" (*sindrom polyarnogo napryazhenia*). As he remarks, this syndrome causes a vulnerability of the organism for those living in the Far North, a vulnerability which may manifest itself in infectious and inflammatory processes in the lungs and other vital organs, or in ischemic diseases of the heart (Kaznacheev 1986). This, according to him, concerned most especially the early stages of the adaptation cycle. However, Kaznacheev and Strigin (1978:36) also observed that working in the Arctic at certain periods of life may have a positive prophylactic impact on a person's life: "many of the processes involved in adaptation which are associated with tension, possible depletion and breakdowns (*polomki*), may provide an opportunity for training and perfecting if subject to rational and hygienic rectifications."

One of the ways in which the Soviet authorities monitored the health of their citizens, with the support of health care specialists, was through a widespread reliance on a system of southern resorts (*kurorty*). Stays in the South were said to have helped the new northerners to restore their physical energy in most cases. However, as Danishevskii (1968:368) remarks, "experience shows that the drastic change (*perestroika*) to the organism related to the significant climatic contrasts provokes among certain people a more or less strong and prolonged deterioration of their health". It was even observed that, upon experiencing such trouble acclimatizing to the southern climate, some of the relatively new "northerners" had been forced to go back to the North where their condition receded rapidly back to normal (Danishevskii 1968).

The medical research on the adaptation to the Arctic in the Soviet period thus generally conveyed the idea that, if monitored appropriately by health specialists, and if given access to appropriate living and working conditions, Soviet citizens could overcome the challenges associated with life in a rigorous environment and climate to such an extent that it could even become easier for them to live there than in their regions of origin.

Contrary to the view held by Soviet researchers, according to which living in the Arctic had produced important physiological changes, Western medical scientists studying the effects of the arctic climate and environment on the body tended to attribute less importance to physiological changes in the process of adaptation. For example, based on his research comparing the role played by nutrition for Inuit, American Indians and people of European descent in the North, Rodahl (1963: 112) concluded the following: "At any rate, any general physiological changes resulting from cold exposures may be small compared with the importance of factors, such as habituation, experience, training, fitness, etc."[8] Similarly, though later in time, Moran (1981:20), who studied the adaptation of indigenous peoples to arctic regions of the Western world, claimed that "[t]he bulk of the mechanisms for human adaptation to arctic areas are social and cultural rather than acclimatory, developmental or genetic".

13.4 The Roots of Shift-Work

Before examining the changes in the analysis of health problems of the workers involved in extractive industries in the North in the post-Soviet period, I briefly relate the roots of shift-work (*vakhtovyy metod* in Russian) in the NAO. The organization of the workforce along the principles of shift-work is not a new *modus operandi* for extractive industries. It can be traced back to the exploration and production operations of the oil reserves found in the Gulf of Mexico, starting in the 1950s.

---

[8] This text was published in *Medicine and Public Health in the Arctic and Antarctic* (1963), the report stemming from the first conference organized by the World Health Organization on the Arctic and Antarctic in 1962.

Since the 1970s shift-work or "fly in/fly out" was largely adopted by mining companies operating in remote regions of Canada and Australia (Storey 2010). One of the motivations leading to the adoption of shift-work was the state's reluctance to build new infrastructure to support more or less permanent communities in remote regions where the natural resources are often found. As Storey (2010:1162) remarks, […] "labour shortages and strong and rapid growth in demand for labour in the construction, mining and petroleum production sectors have further encouraged fly-in/fly-out as a solution to delivering labour to remote locations."

During the development of the Russian North, the Soviet authorities also relied on a form of shift-work referred to as "near shifts" (*blizhnye vakhty*) by Kvashnina and Krivoshekov (1998:7). With these types of shifts, the workers came from, or resided in, the same "natural-climatic zone" as their work environment, and arrived there by terrestrial or air transportation over relatively short distances. The oil industry in the NAO was organized in this way during the Soviet period: workers arrived from various regions of the Soviet Union and settled down in Naryan-Mar or in Murmansk, and from there would go on their shifts.

When the Soviet Union collapsed in 1991, northern regions such as the NAO experienced dramatic socioeconomic hardships. The two main economic sectors, the oil industry which employed a predominantly-Russian workforce and collective farms where Nenets reindeer herders worked, began to experience severe financial difficulties since the state reduced or stopped subsidizing their activities, often forcing them to cease operations. Struggling to find employment or sources of subsistence, 27.6 % of the population, most especially settlers, emigrated from the Okrug between 1989 and 2006 (Heleniak 2009b).

In the 1990s, a discourse which suggested that maintaining a large population in the Russian North and Siberia was too costly for post-socialist Russia gained influence among international and Russian analysts: this situation was referred to as the "Siberian curse" (Hill and Gaddy 2003). It was believed that the resources of the Russian North had to be extracted by a new, more flexible organization of labor that could limit costs and adjust more easily to the fluctuating prices of natural resources on the global market, thus requiring a high degree of mobility from the workers involved.

In the early 2000s, a drastic increase in the price of oil along with improvements in oil-extraction technologies made the NAO an attractive extraction site again – this time for both Russian and foreign oil companies. In 2006, the NAO had the most rapid increase in its oil production of any region in Russia, growing 61 % between 2003 and 2005 (Stammler and Peskov 2008). The main company responsible for the increase in oil-related projects was the private Russian company Lukoil.

The oil industry now relies on 10,000 workers who work as long-distance commuters, arriving at their work-sites from neighbouring Arkhangelsk or Murmansk Oblasts (Provinces), from other oil-producing regions such as Tatarstan, Bashkiria, and Perm Kray (Territory), and even from former Soviet republics such as Kazakhstan and Ukraine. The organization of this workforce resembles what Kvashnina and Krivoshekov (1998:7) refer to as "expedition type" (*ekspeditsionnye*) shifts. According to this labor organization, workers often come from distant regions of the country, from different climatic zones, and may travel over distances of two to three thousand kilometers or more to work for periods lasting most often between 2 and 4 weeks. Borrowing from Vaguet (2007:8), who conducted research among the workers of the oil and gas-rich regions of Western Siberia, I refer to these workers as "oil nomads" (*nomades du pétrole*) in contrast to their predecessors who settled in the NAO.

As the recent oil boom began in the NAO, many settlers began to feel anxious about the possibility of being replaced by shift-workers, even in economic sectors which were more or less directly associated with the oil industry. Acquaintances of mine, technicians and mechanics, mentioned rumours concerning the potential purchase of the Naryan-Mar Airline's helicopter fleet by Moscow-based YuTAir. One rumour suggested that if YuTAir bought the airline, it would bring its own workers and send employees living in Naryan-Mar to work in the South in order to avoid paying the Soviet-inherited northern benefits.[9] A somewhat similar story was related by one of the owners of a popular restaurant among transiting oil workers where food from the Arctic is served year-round. I asked the owner, who had come from Moldova, why all their employees were shift-workers from Moldova, the Caucasus, or other regions of Russia. "You know, the level of oxygen is different here," he told me. "Those who settled here, they have had to adapt to this. As a result, they are unable to work quickly, they do things more slowly. I am telling you from experience."

13.5   Adapting to the Post-Soviet North

With the growth of shift-work in the recent decades, more research has been conducted among shift-workers in Russia and elsewhere. In Canada and Australia, the research has focused predominantly on the political and socioeconomic consequences of fly in/fly out on the resource-rich regions in which they work and on their home regions, as well as on the challenges and opportunities associated with shift-work for the workers themselves and for their families (e.g., Carson 2011; Ferguson 2011; McIntosh 2012; Misan and Rudnik 2015; Storey 2010). Research focusing on similar aspects of shift-work in Russia has also been conducted (e.g., Stammler and Eilmsteiner-Saxinger 2010). However, most of the research conducted by Russian scientists studying shift-work is unique due to the centrality of the health-related consequences of adaptation to the arctic environment and climate.

---

[9] The Russian law "On government guarantees and compensation for persons working and residing in the regions of the Far North and equivalent areas" provides a range of benefits to the residents of regions situated in the Far North, such as the NAO (Federal Law RF 1993 [2014]). They are eligible to housing benefits, such as a subsidy to acquire a house or apartment, early retirement, longer vacations, and free transportation to any region of Russia every second year.

The most commonly used definition of adaptation today derives from the work of Anokhin (1969, 1975), whose approach came to be termed "the theory of functional systems". Inspired by his work, psychologist Simonova (2011:26) summarizes the recent conception of adaptation as a "return of the parameters of the adapting system to the previous level or as their stabilization to the new level". Another key characteristic of research on adaptation which has its roots in the late Soviet period is the idea of stages of adaptation, as identified by Avtsyn et al. (1985). The first stage, that of "adaptive tension" (*stadya adaptativnogo napryazhenya*), lasts from the first 2– 6 months; the second stage is the "stabilisation of functions" (*stabilizatsii funktsii*), from the sixth or eighth month up to 2 or 3 years; the third phase, that of "adaptativeness" (*adaptirovannosti*), begins in the third or fourth year of life in the North. Gudkov and Popova (2011:1) suggest that the "success or failure of the whole process of adaptation to the conditions of the North is determined in a decisive way by the character and outcome of the adaptive reactions of the migrant during the first and most difficult stage: that of adaptive tension". Considering the significance of this stage, I relate some of the aspects characterizing it.

During the initial phase of adaptation, individuals arriving in the North are said to experience changes affecting the cardio-respiratory system as well as cerebral activity. Sarychev (2006:62) suggests that the cardio-respiratory system is the first to be involved in the process of adaptation to unfavorable climatic conditions. According to his research, it is the central nervous system that initially stimulates changes to the heart rate. Another important element of change is the reduction of Vital Lung Capacity (*zhiznennaya emkost' legkikh*), most likely due to an increase in blood flow to the lungs. Interestingly, Gudkov et al. (2012) identified the higher blood supply to the lungs as a characteristic of northerners, in contrast to inhabitants of central regions of Russia. During the first two months following arrival in the North, there tends to be increased pressure on the newcomer's cardio-respiratory system, leading to a reduction in its efficiency and thus requiring a greater energy-expenditure for each vascular and respiratory cycle (Gudkov and Popova 2011). During the fourth and fifth months, the system begins to adapt to the new climatic and environmental conditions. Russian settlers tend to have lower losses of energy due to more efficient gas exchange systems as well as hypometabolism, thought to be developed through their time spent in the North (Grishin and Ustyuzhaninova 2010).

The intensity of the first phase of adaptation is apparently so strong that it prompts changes in the brain, affecting memory and emotions as well as activity in both cerebral hemispheres. Krivoshschekov et al. (2009) suggest that adaptive tension experienced in extreme conditions, such as those of the Arctic, induces what is termed a functional asymmetry of the brain (*funktsional'naya asimmetriya mozga*) as one response mechanism. They even report the existence of a "polar metabolic type"; that is, a type which has already adapted to the North, characterized by levels of insulin and cortisol that are evidence of differing degrees of activity in the brain (Krivoshshekov et al. 2009:207).

Depicted this way, adapting to the North seems to be a highly demanding process, not only physically, but also mentally. Simonova (2010:89) suggests that in cases when a person requires treatment, "negative changes experienced which regularly happen in the organism and psyche of the person are compensated for much faster, and the working capacity re-established if the person experiences recovery [...] in his familiar middle latitudes where there is no negative influence of the 'northern tension.'" Some researchers argue that the length of shifts should be limited. Having conducted medical research with oil workers on the island of Kolguev, where shifts last 52 days, Sarychev (2006) recommends to reduce shifts to a period lasting between 30 and 35 days in order to reduce the "physiological costs" of the adjustment reactions.[10]

According to Kvashnina and Krivoschekov (1998:14), the scientific and medical research conducted on adaptation of the shift-workers should have three main goals:

- the increase of efficiency of work
- the preservation of human health
- an all-encompassing (*vsestoronnyy*) development of the person (*lichnost'*).

The question of efficiency and productivity is inextricably linked to that of profits. One of the ways in which both the oil companies and state administrations generate savings is by reducing the amount of infrastructure and number of installations, and by reducing the number of potential recipients of northern benefits (see footnote 9). In relation to the second goal mentioned by Kvashnina and Krivoschekov – the preservation of human health – the health costs appear to be quite high for those individuals who have to adapt to the region. It is recommended that they adapt gradually, that they work in shifts lasting periods not exceeding about one month, and that they then go home for the equivalent period of time. Thus, the question of adaptation can be compared to that of "acclimatization" or the "seasoning" of colonial days.

In relation to the third goal of medical research on adaptation – the development of the person – the workers I have encountered generally agree to work in the Arctic because they believe that by doing so they are helping the country to flourish, and/or because they see this as an opportunity for higher salaries and the possibility of career advancement within the industry. They could choose to find a job elsewhere, in another region where capital is flowing, as investments are made in other oil fields. Borrowing loosely from Deleuze and Guattari (1987), one could potentially see the oil nomads as "deterritorialized biologies", available to follow different flows of capital if they are more advantageous in terms of salaries or more promising in terms of career advancement.

I mention above that after a number of years living in the Arctic, one mechanism of adaptation described by Russian medical scientists is that of hypometabolism and the emergence of a "polar metabolic type". One wonders how it would be for those

---

[10] I happened to stay at the main extraction site on Kolguev – that of Arktikmorneftegazrazvedka – in the winter of 2009, that is, at the lowest point of the global financial crisis. Due to the difficult financial situation, the company had asked its workers to remain on the island for 75–80 days instead of 52 in order to reduce the costs of transportation. This indicates that even companies which have served as the sites for research promoting shorter shifts may favor reducing the costs of their operations, at the expense of the health-related consequences on their workers.

people who have settled in the NAO to resettle back in the South. There now seems to be some degree of direct pressure from the job market for them to do so. In his research on the Russian settlers in Chukotka, Thompson (2008) heard settlers express the belief that resettling in the South after having adapted to the North could be fatal. However, as he relates it, the "actual causes of mortality, as some northerners reflected, might rather be linked to the psychological stress of displacement and the practical consequences of losing family and community support" (Thompson 2008:232). During a discussion, a local doctor mentioned that Russian settlers, especially those who are older, would have great difficulties re-adapting to the South physically. One could thus say that the Russian settlers seem to have reterritorialized in the North, both biologically and socially, and feel they can hardly deterritorialize their bodies to another climate.

13.6   Conclusion

In Russia, the conception of adaptation has been the key lens through which to explain the health issues of those arriving in the Arctic to extract natural resources, in both the Soviet and post-Soviet periods. Even though recent research has relied on concepts developed in the Soviet period, the consequences of adaptation to the Arctic on a person's health are described differently. In the Soviet period, it was believed that the provision of decent living and working conditions would allow for settlers to adapt to the Arctic to such an extent that moving back to central or southern regions could put their health at risk. Since the 2000s, Russia has been relying largely on its oil and gas industry to support its economy, thus requiring a large number of workers in these regions. The descriptions of the oil nomads' bodies in recent medical science are associated with recommendations to limit their stays in the Arctic in order to avoid putting their health at risk. One might wonder to what extent these recommendations could be seen as a fear that oil nomads will adapt too well to the Arctic, adding to the burden of the state and companies to provide them with benefits and build necessary infrastructure. It thus appears that the issue associated with the concept of acclimatization in colonial days – that of creating differentiation – is still present today in the NAO, this time based not on the geographical and climatic origins of people, but on the socioeconomic and ideological context in which they arrived in the North.

References

Ahlenius, H., & UNEP/GRID-Arendal. (2012). *Nenets Autonomous Okrug, Reference Map.* Retrieved December 19, 2012, from http://www.grida.no/graphicslib/detail/nenets-autonomous-okrug-reference-map_feab#

Anokhin, P. K. (1969). *Sistemnaya organizatsiya fiziologicheskikh funktsiy* [Systemic organization of physiological functions]. Moscow: Meditsina.

Anokhin, P. K. (1975). *Ocherki po fiziologii funktsional'nykh sistem* [Essays on the physiology of functional systems]. Moscow: Meditsina.

Avtsyn, A. P., Marachev, A. G., & Milovanov, A. P. (1985). *Patologiya cheloveka na Severe* [Human pathology in the North]. Moscow: Meditsina.

Carson, D. (2011). Skilled labour migration flows to Australia's Northern territory 2001–2006: Beyond periphery? *Australian Journal of Labour Economics, 14*(1), 15–33.

Dallmann, W. (2009). *Nenets Autonomous Okrug: Physical geography.* Norwegian Polar Institute. Retrieved December 15, 2012, from http://ipy-nenets.npolar.no/illustrations/NAO_01_phys-geo_En_reduced.pdf

Danishevskii, G. M. (1968). *Patologiya cheloveka i profilaktika zabolevaniy na Severe* [Human pathology and profilactics of disease in the North]. Moscow: Meditsina.

Deleuze, G., & Guattari, F. (1987). *A thousand plateaus: Capitalism and schizophrenia.* Minneapolis: University of Minnesota Press.

Federal Law RF. (1993; amended in 2014). *O gosudarstvennykh garantiyakh i kompensatsiyakh dlya lits, rabotayushchikh i prozhivayushchikh v rayonakh Kraynego Severa i priravnennykh k nim mestnostyakh* [On government guarantees and compensation for persons working and residing in the regions of the Far North and equivalent areas].

Ferguson, N. (2011). From coal pits to tar sands: Labour migration between an Atlantic Canadian region and the Athabasca oil sands. *Just Labour: A Canadian Journal of Work and Society, 17 & 18*, 106–118.

Gudkov, A. B., & Popova, O. N. (2011). Reaktsiya dykhatel'noy sistemy pri migratsii na Evropeiskiy sever [Reaction of the respiratory system during migration to the European North]. In G. N. Degteva (Ed.), *Problemy zdravookhraneniya i sotsial'nogo razvitiya arkticheskoy zony Rossii* (pp. 433–451). Moscow: Paulsen.

Grishin, O. V., & Ustyuzhaninova, N. B. (2010). Gipometabolizm v uslovyakh deystviya nizkikh temperatur [Hypometabolism in conditions of low temperatures]. *Byulleten' Sibirskogo otdeleniya Rossiyskoy akademii meditsinskikh nauk, 30*(3), 12–17.

Gudkov, A. B., Popova, O. N., & Nebuchennykh A. A. (2012). *Novosely na evropeyskom severe: fiziologo-gigienicheskie aspekty* [New settlers in the European North: Physiological and hygienic aspects]. Arkhangelsk: Northern State Medical University.

Heleniak, T. (2009a). The role of attachment to place in migration decisions of the population of the Russian North. *Polar Geography, 32*(1–2), 31–60.

Heleniak, T. (2009b). Growth poles and ghost towns in the Russian Far North. In E. Wilson Rowe (Ed.), *Russia and the North* (pp. 129–163). Ottawa: Ottawa University Press.

Hill, F., & Gaddy, C. G. (2003). *The Siberian curse: How communist planners left Russia out in the cold.* Washington, DC: Brookings Institution Press.

Kaznacheev, V. P. (1986). *Klinicheskie aspekty polyarnoy meditsiny* [Clinical aspects of polar medicine]. Moscow: Meditsina.

Kaznatcheev, V. P., & Strigin, V. M. (1978). *Problema adaptatsii cheloveka. Nekotorye itogi i perspektivy issledovaniy* [The problem of adaptation of humans: Some conclusions and perspectives in research]. Novosibirsk: Siberian Branch of the Academy of Medical Sciences.

Krivoshshekov, S. G., Mel'nikov, V. N., & Novitskaya, S. Y. (2009). Funktsional'nye asimmetrii u cheloveka v usloviyakh severa [Functional asymmetry in humans in northern conditions]. In A. L. Maksimov (Ed.), *Materialy vserosiyskoy nauchno-prakticheskoy konferentsii s mezhdunarodnym uchastiem, Posvyashchennoy III Mezhdunarodnomu Polyarnomu Godu* (pp. 206–208). Arkhangelsk: Northern State Medical University.

Kvashnina, S. I., & Krivoshekov, S. G. (1998). *Okhrana zdorov'ya rabotnikov vakhtovogo truda na kraynem severe* [Preservation of shift-workers' health in the far North]. Ukhta: Ukhta Industrial Institute.

Livingstone, D. (1991). The moral discourse of climate: Historical considerations on race, place and virtue. *Journal of Historical Geography, 17*(4), 413–434.

Livingstone, D. (1999). Tropical climate and moral hygiene: The anatomy of a Victorian debate. *The British Journal for the History of Science, 32*(1), 93–110.

Lock, M., & Nguyen, V.-K. (2010). *An anthropology of biomedicine*. Chichester: Wiley-Blackwell.

McCannon, J. (1998). *Red Arctic: Polar exploration and the myth of the North in the Soviet Union, 1932–1939*. Oxford: Oxford University Press.

McIntosh, A. (2012). Ten truths about Australia's rush to mine and the mining workforce. *Australian Geographer, 43*(4), 331–337.

Misan, G. M., & Rudnik, E. (2015). The pros and cons of long distance commuting: Comments from South Australian mining and resource workers. *Journal of Economic and Social Policy, 17*(1), 1–37.

Moran, E. F. (1981). Human adaptation to arctic zones. *Annual Review of Anthropology, 10*(1), 1–25.

Osborne, M. A. (2000). Acclimatizing the world: A history of the paradigmatic colonial. *Osiris, 2*(15), 135–151.

Rodahl, K. (1963). Nutritional requirements in the polar regions. In *Medicine and public health in the Arctic and Antarctic: Selected papers from a conference* (pp. 97–115). Geneva: World Health Organization.

Rouillard, R. (2013). *Nomads in a petro-empire: Nenets reindeer herders and Russian oil workers in an era of flexible capitalism*. Unpublished Ph.D dissertation, McGill University.

Sarychev, A. S. (2006). Metody otsenki stepeni adaptirovannosti organizma neftyanikov k ekstremal'nym usloviyam truda v zapolyar'e [Methods of evaluation of the level of adaptivity of oil workers' organism in extreme polar working conditions]. *Ekologiya cheloveka, 8*, 62–64.

Simonova, N. N. (2010). *Psykhologiya vakhtovogo truda na Severe* [The psychology of shift-work in the North]. Arkhangelsk: Pomor State University.

Simonova, N. N. (2011). *Psikhologicheskii analiz professional'noy deyatel'nosti spetsiyalistov neftedobyvayushchego kompleksa (na primere vakhtovogo truda v usloviyakh kraynego severa)* [Psychological analysis of the professional activities of specialists in the oil industry (based on the example of shift-work in the Far North]. Dissertation summary. Moscow: Moscow State University.

Stammler, F., & Eilmsteiner-Saxinger, G. (2010). Introduction: The northern industrial city as a place of life and of research. In F. Stammler & G. Eilmsteiner-Saxinger (Eds.), *Biography, shift-labour and socialisation in a northern industrial city – The Far North: Particularities of labour and human socialisation* (pp. 9–16). Rovaniemi: Arctic Centre, University of Lapland.

Stammler, F., & Peskov, V. (2008). Building a 'culture of dialogue' among stakeholders in North-west Russian oil extraction. *Europe-Asia Studies, 60*(5), 831–849.

Storey, K. (2010). Fly-in/fly-out: Implications for community sustainability. *Sustainability, 2*(5), 1161–1181.

Thompson, N. (2008). *Settlers on the edge: Identity and modernization on Russia's Arctic Frontier*. Vancouver: UBC Press.

Vaguet, Y. (2007). *Les hydrocarbures, les villes et les hommes dans le nord-ouest Sibérien* [Fossil fuels, cities and men in northwestern Siberia]. Actes du Festival International de géographie De Saint-Dié Des Vosges. Retrieved May 20, 2011, from http://archives-fig-st-die.cndp.fr/actes/actes_2007/vaguet/hydrocarbures_siberiens.pdf

Wengle, S., & Rasell, M. (2008). The monetisation of *l'goty*: Changing patterns of welfare politics and provision in Russia. *Europe-Asia Studies, 60*(5), 739–756.

Chapter 14
Rights and Responsibilities: Sustainability and Stakeholder Relations in the Russian Oil and Gas Sector

Emma Wilson

Abstract Oil and gas industry expansion into arctic waters is a high-profile concern for global sustainability, although factors such as the price of oil and the prospects for gas marketization have slowed the rate of expansion in recent years. Less attention is currently paid to some of the major onshore oil and gas developments in the Russian North, where production is ongoing, often since Soviet times, and in many cases with significant impacts on fragile arctic ecosystems. Such projects also have considerable (positive and negative) social impacts for Russia's indigenous and local populations, many of whom depend on the land and waters of oil-bearing regions for their livelihoods. Thus, the sustainability of Russia's arctic oil and gas industry is of critical importance whether or not the activities are taking place offshore in the glare of international scrutiny. Usinsk *Rayon* (county) in Russia's Komi Republic grabbed international headlines in 1994 with one of the world's biggest oil spill disasters, but today is much less well known internationally, despite continuing to suffer regular spills. This article explores the current situation in Usinsk Rayon, comparing global sustainability standards and companies' understandings of their own responsibilities to society with the ways that 'corporate social responsibility' in practice is perceived locally. The paper concludes that there is a need for local people to redefine the 'social licence' that determines relations between communities, government and industry, so as to break the expectation that environmental pollution can be traded off for social benefits.

Keywords Corporate responsibility • Community engagement • Indigenous peoples • Oil and gas • Komi Republic

E. Wilson (✉)
Scott Polar Research Institute, Cambridge, UK
e-mail: emma.wilson@ecwenergy.com

© Springer International Publishing Switzerland 2017   177
G. Fondahl, G.N. Wilson (eds.), *Northern Sustainabilities: Understanding and Addressing Change in the Circumpolar World*, Springer Polar Sciences,
DOI 10.1007/978-3-319-46150-2_14

14.1 Introduction

Despite uncertainty around oil prices and gas markets, the potential of the oil and gas industry to expand into arctic waters remains highly significant to global sustainability debates. Protests against such expansion can make headlines in this era of concern over climate change and fossil fuel dependency. Less international attention is currently paid to some of the major onshore oil and gas developments in the Russian North, where production has been ongoing, often since Soviet times, and in many cases with significant impacts on fragile arctic ecosystems. In the north of Russia's Komi Republic, to the west of the Ural Mountains, the oil industry has been operating on a large scale since the 1960s. Annual oil production in Komi Republic increased from nine million tons in 2000 to 13.7 million in 2012 (Staalesen 2014). In 1994, Usinsk *Rayon* (county) suffered the third largest oil spill in history (eight times the size of Exxon Valdez), with over 100,000 tons of oil spilled across the tundra and rivers (Jernelov 2010). Karjalainen and Habeck (2004) observed that the 1994 spill was a landmark event in local environmental consciousness. The global reaction to the disaster had given local residents 'the feeling that all of a sudden their concerns were receiving attention in many other parts of the world' (Karjalainen and Habeck 2004:177). Yet the region has enjoyed comparatively little international scrutiny since, although the environmental situation remains poor. Despite largely successful efforts to clean up the aftermath of the 1994 spill, much of the old pipeline infrastructure is still in place and in need of replacement, so spills remain a daily hazard. This is particularly serious for the rural communities of mostly indigenous Komi, whose livelihoods are based on a combination of fishing, small-scale agriculture, reindeer herding, civil service employment and oil industry service jobs (see Chap. 19 by Crate in this volume).

This article explores the interaction between local communities and the oil industry. It considers international sustainability standards and companies' understandings of their own responsibilities to society, with a focus on public reporting and stakeholder engagement. The article explores how 'corporate social responsibility' is perceived locally and how relations between industry, government and communities are negotiated and played out. The article draws on fieldwork carried out in November 2013, March 2014, and June 2015. This included 32 formal and informal interviews in villages of Usinsk Rayon and neighbouring Izhma Rayon, in the rayon capitals, Usinsk and Izhma, and in Syktyvkar and Moscow. These interviews were supplemented by participant observation, including a meeting of 14 villagers in a local library (Novik Bezh village) to discuss the research questions; observation of a meeting in Usinsk to approve environmental permits; and informal discussions with fellow hotel residents – including oil workers and researchers – in Usinsk and Izhma. The field material was supplemented by analysis of publicly available materials from company, government and civil society websites, academic journals and local newspapers.

14.2 Sustainability, Corporate Social Responsibility and Stakeholder Relations

Sustainability means different things to different people. The diversity of definitions of sustainability in oil company policy and public reporting is a case in point. For example, Statoil, which is a partner in the Kharyaga onshore oil project located to the north of Usinsk in Nenets Autonomous *Okrug* (District), states in its 2014 *Sustainability Report* that it "aims to be recognised as the most carbon-efficient oil and gas producer, committed to creating lasting value for communities" (Statoil 2014: 3). In its integrated *Sustainable Growth* report, Total, the operator of the Kharyaga project, states its aim as being to deliver "clean, safe and competitive energy for as many people as possible" (Total 2014: 8). The company's specific commitment to society and the environment is "to respect the environment, protect human health, ensure product and facility safety, and promote social and economic development in our host countries" (Total 2015). In its 2014 *Annual Report*, Lukoil, the main company operating (as Lukoil-Komi in Usinsk Rayon), boasts "over 20 years of sustainable growth", using the term "sustainable" to indicate the durability of the business itself (Lukoil 2014). In its 2011–2012 *Sustainability Report*, Lukoil uses the terms "responsible business practice" and "social responsibility" to refer to environmental and social sustainability (Lukoil 2012).

The origin of the modern notion of corporate social responsibility (CSR) is attributed to Howard Bowen and his 1953 book, *Social responsibilities of the businessman*. Rooted in the US business context of the time, it defined CSR as the aligning of company policies and decisions to the objectives and values of society, while also emphasising that government and others also have a key role to play in ensuring responsible industrial practice (Bowen 1953). Academic and organisational thinking around CSR evolved throughout the 1960s to the 1980s, incorporating more analysis of practice and consideration of related concepts such as corporate citizenship (Carroll 1999). Milton Friedman's 1970 essay, *The social responsibility of business is to increase its profits*, set the foundation for the 'shareholder model' of business, which maintains that the primary responsibility of business is towards its shareholders (Friedman 1970). Stakeholder theory subsequently emerged to challenge the intellectual foundation of the shareholder model, focusing on the relationship between the business itself and the groups or individuals who could affect it or be affected by it, and underpinned by the moral argument that companies should share their value not only with shareholders, but with all their normative stakeholders (Hasnas 2013; Freeman 1984). In company reporting, the tension between these models has yet to be resolved. For example, in its 2012 *Sustainability Report*, Lukoil states that "the Company's key goal for the next decade will be to increase shareholder value" (Lukoil 2012: 6). However, Rosneft, which also operates in Komi Republic, states in its 2013 *Sustainability Report* that: "Stakeholder engagement is the foundation of the company's sustainable development" (Rosneft 2013: 38).

Over the years, CSR has been challenged as being misleading as a term (Morrison 2014) or ineffective as a practice (Blowfield and Frynas 2005). Corporate responsibility has been re-framed by business in other ways, such as 'shared value' (Porter and Kramer 2011) and 'social licence to operate' – a term used by businesses to refer to the broad acceptance of their activities within the local community and wider society (Thomson and Boutilier 2012; Morrison 2014). A 'social licence' may exist in various forms, from informal relations to more formal 'impact-benefit agreements', and its negotiation is based on rights and responsibilities defined by international standards of good practice, national and sub-national laws, and local social norms and expectations (Black 2013; Novikova and Wilson 2015). The notion of the stakeholder remains central to these concepts, and there is increasing interest among researchers, non-governmental organisations and policymakers in how 'meaningful' stakeholder engagement processes are – *i.e.* to what extent they enable trust to be built between government, industry and communities, and result in positive outcomes for local society (BIC 2014; OECD 2015).

A major recent influence on perceptions and understanding of corporate responsibility was the UN Human Rights Council's endorsement of the *UN Guiding Principles on Business and Human Rights* or UNGPs (United Nations 2011), which were supported by Russia. The UNGP calls for governments to protect human rights, for business to respect those rights, and for adequate forms of remedy in cases of human rights violations. This emphasis on the relative responsibilities of government and industry re-focuses attention on the roles and responsibilities of governments in ensuring the sustainability of industrial operations. The legal framework and regulatory institutions in a country and/or region of operations may or may not be adequate to support the implementation of international standards. The Russian Federation law framing environmental and social issues in an industrial context plays out differently in different Russian regions, with some regional legislation supporting company-community relations and protecting local rights more than others (Wilson and Stammler 2006; Sirina et al. 2008; Martynova and Novikova 2012; Novikova 2014). For example, the 1999 Russian Federation law "On guaranteeing the rights of small-numbered indigenous peoples of the North" allows peoples who meet the necessary criteria to "possess and use their lands, free of charge, in places of traditional habitation and economic activities in the pursuit of traditional economic activities" through establishing an *obschina* (roughly translatable as 'commune') and receiving a land allotment for the obshchina.. Technically, this could allow the land user to negotiate with an oil industry operation over use of that land. However, the Komi people are not recognised by Russian Federation law as 'small-numbered indigenous peoples of the North' (*malochislennie korennyie narody Severa*). To be recognized as such, a people needs to have a population of less than 50,000, while the Komi number 300,000. The northern Komi (the *Izhma* Komi or *Izvatas*) number just 40,000 and see themselves as distinct from the southern Komi; for instance, they practice reindeer herding, which the southern Komi do not. They have been trying to become legally recognised as a small-numbered indigenous people of the North for many years (Donahoe et al. 2008; Kim et al. 2015). Without being granted that status they cannot benefit from the right to set up tax-free *obschinas* to establish their legal rights to lands that they use in a traditional manner. Nor can the Komi call for a process of free, prior, and informed consent (FPIC) prior to industrial operations taking place on their lands. FPIC is an indigenous peoples' right, established in the 1989 ILO Convention 169 on Indigenous and Tribal Peoples and the 2007 UN Declaration on the Rights of Indigenous Peoples (see, e.g., Buxton and Wilson 2013). Russia has not ratified ILO 169, and abstained from voting on the adoption of the UN Declaration. The only tool the indigenous Komi have to leverage influence with the oil industry is the mandatory public consultation required as part of an environmental expert review of an environmental impact assessment prior to starting an industrial project. The Russian Federation law "On environmental expert review" (1995) and the "Environmental impact assessment regulations for proposed business and other activities" (2000) require public consultation prior to commencement of an industrial project. However, this does not apply to all industrial activities, such as seismic testing, for instance, which can damage local hunting or berry-picking grounds.

In light of inconsistencies in the legal framework in Russia, the so-called 'voluntary' sustainability standards – introduced via certification schemes, or project finance conditionalities, or through the implementation of corporate policies (which might be introduced by a multinational joint venture partner, for instance) – are becoming increasingly significant in shaping company practice in Russia (Novikova and Wilson 2015). Most practitioners see the benchmark for sustainable industry operations as being the Environmental and Social Performance Standards of the International Finance Corporation (IFC), the private sector arm of the World Bank. The IFC Performance Standards are applied to the companies to which it lends money. Other international financial institutions, such as the European Bank for Reconstruction and Development (EBRD 2014) or the Asia Development Bank (ADB 2009) have their own performance standards, while the IFC standards have also been adopted by the 80 Equator Principles Financial Institutions (Equator Principles 2013). The IFC Performance Standard 7 (PS7) requires companies to seek the free, prior and informed consent of indigenous peoples in cases of resettlement from their lands, and to draw up an indigenous people's development plan (IFC 2012). Yet while there are examples from elsewhere in Russia where project finance conditionalities have had a considerable effect on the way that companies engage with indigenous communities (e.g., EBRD involvement in financing the Sakhalin-2 project in the Russian Far East), Lukoil-Komi is not using project finance from international financial institutions. Nonetheless, the parent company, Lukoil, is registered on the London and Frankfurt stock exchanges; it reports to its shareholders and investors on environmental and social issues, and it cares about its global image. Key questions, then, are to what extent the company's reporting reflects the reality on the ground in its regions of operations, and whether its voluntary policies and initiatives are effective at ensuring the sustainability of its local operations.

14.3   Usinsk Rayon

Usinsk Rayon is located in the far north of Komi Republic. The villages of this rayon lie on the Kolva and Usa rivers, which flow into the great Pechora River that winds its way northwards and westwards through Komi Republic, then through Nenets Autonomous Okrug to the Barents Sea and the Arctic Ocean. Traditionally the Pechora River has supported fishing and commerce, and the lands around the Pechora and its tributaries were important as agricultural land for Soviet collective and state farms. Villagers, who are mostly ethnic Komi, yearn back to the days of the collective farm system. In recent decades agriculture has declined in the oil-producing north of the Republic, due to the lack of competitiveness and the loss of government support for market protection and local enterprise development. While some people have managed to set up their own small trading businesses, or work in the public sector, the oil industry is now one of a relatively small range of employment options available to local people. Even local environmental activists might also work for the industry.

By contrast with the villages, the rayon's capital Usinsk appears much better off, with malls and supermarkets, cafes, and expensive clothes shops. Lukoil-Komi's modern head office building towers over the Usinsk administration building in the centre of town. Usinsk's citizens – who include many incomers from Azerbaijan, Uzbekistan, Bashkiriya, and Ukraine who came for oil jobs – seem to be happy with the benefits brought by the oil industry, though some complain about the cost of housing. There is a distinct difference between the village communities – mostly ethnic Komi – and the mixed, largely incomer (and possibly temporary) population of the city, where most of the decisions are made. An exception is the village of Kolva, which is close enough to Usinsk for people to commute there, and whose population is more mixed than those of the other villages in the rayon. Yet the contrast between the rural and urban lifestyles is considerable, in particular due to the villagers' close relationship to the land and the Kolva River.

Lukoil-Komi is the dominant – but by no means the only – oil and gas company operating in Usinsk Rayon. Others include Severnaya Neft' (Rosneft'), Rusvietpetro (a Russian-Vietnamese joint venture), and Kolvaneft' (Nobel Oil). In an interview with me, Lukoil's Moscow representatives came across as well versed in CSR theory, based on the notion of three pillars: economic, social, and environmental. Lukoil has signed up to the Social Charter of Russian Business, has adopted a corporate *Social Code* (Lukoil 2002), and is one of only three oil company members of the Global Compact in Russia. It sees itself (and is perceived externally) as progressive on social welfare, sustainability reporting, and environmental management (Hudina 2010). Lukoil publishes a *Sustainability Report* once every two years, following the global sustainability reporting standard, the Global Reporting Initiative. The report includes an interesting feedback mechanism, based on regional stakeholder dialogues (see, e.g., Lukoil 2012). To date, the company has not held a dialogue in Komi Republic but, in any case, the purpose of the dialogues is to comment on the *Sustainability Report*, rather than address and resolve specific local issues or engage with host communities.

Lukoil's Moscow representatives told me that engagement with local Komi Republic NGOs is the responsibility of Lukoil-Komi. A statement in Lukoil's 2011–2012 *Sustainability Report* provides some insight into the company's perception of local-level corporate responsibility, and the relative responsibilities of government and business: "We consider it to be our duty to pay our taxes in a timely manner, which helps the state to solve social challenges in the regions of our company's presence" (Lukoil 2012:2). The company also negotiates an annual agreement with the government of Komi Republic to provide support to the region (Poussenkova 2014), and with the host rayon administrations, including Usinsk, as well as with *Komi Voityr*, the official republic-wide association of Komi people.

Lukoil-Komi faces particular sustainability challenges in Komi Republic. In 1999, Lukoil bought up the old state enterprise Komineft and, since then, has been trying to clean up the effects of the 1994 spill and replace the aging network of pipelines. Indeed, in its 2006 *Sustainability Report*, the company reported that it had eliminated the inherited consequences of the 1994 spill and the status of 'ecological disaster zone' had been removed from several parts of the rayon (Lukoil 2006). Yet the region continues to suffer from spills, largely related to the age of the existing infrastructure, which is being replaced at a slow pace and, according to local respondents, due to the less than responsible practices associated with that replacement (corrupt deals and use of poor quality pipe were mentioned more than once). In May 2013 another large spill of between 20 and 200 tonnes was discovered. It actually happened in November 2012 but was kept quiet until the oil came down the Kolva River with the ice during the spring thaw. Rusvietpetro eventually took responsibility for the spill (Staalesen 2014).

The Kolva River is particularly vulnerable to oil and gas industry operations. From its source in Nenets Autonomous Okrug, it passes about 1 km from the Kharyaga field (where the Russian company Lukoil, Total (France), and Statoil (Norway) operate), then past two further major fields in Usinsk Rayon – Vozeiskoye and Usinskoye – before flowing into the Usa River which in turn flows into the Pechora River. The main Usinsk-Kharyaga highway, which is the major road transport route for equipment, chemicals, and personnel heading to the oil sites, crosses the meandering Kolva several times on its way north. Herders from the reindeer herding enterprise '*Severnyi*', based in the village of Mutnyi Materik, pointed out that the reindeer of two of their reindeer brigades use the forest land close to the Usinsk-Kharyaga highway on their migration routes. Thus it is a zone of extremely high ecological and social risk for all the companies operating along this corridor.

I visited four villages located in this corridor or farther along the shores of the Usa and Pechora Rivers – Kolva, Novik Bezh, Ust' Usa and Mutnyi Materik. Local people reported how they suffered directly from the 2013 spill (and indeed had suffered from the 1994 spill, too). Like Karjalainen and Habeck (2004), I observed that environmental concern often referred back to the experience of 1994, even 20 years on. Villagers were divided among those who do not want the oil industry to be present at all in their region and those who accept its presence, but do not tolerate the sloppy practices and what they see as a lack of respect shown to them by the industry over the past decades. A local entrepreneur stated: 'We don't oppose the oil industry as such, we just wish they would operate more cleanly. We know it is possible elsewhere in the world, why not here?' Local residents complained about the negligence of oil operators who allow spills to happen. They also faulted the slow pace of official emergency response efforts in 2013, which meant that they had to do their own cleanup with shovels and plastic bags. The residents also complained of delays in receiving any official information following the 2013 spill. They felt that rayon officials and regulators, companies and the local media had failed to take the incident seriously; indeed they believed that individuals from the companies and from government organs had suppressed the information. Given the lack of official information, the NGO Save Pechora Committee and their *Ecological Bulletin*, the association of northern Komi *Izvatas*, the Izhma Komi newspaper *Veskhyd Syornyi* and the *7 × 7* internet journal and blogging site played a key role in disseminating information about the 2013 spill. The Save Pechora Committee and *Izvatas* have volunteers in all the villages around the Pechora Basin, and those who do not have a reliable internet connection can be contacted by phone. Sometimes international attention is brought to bear through Greenpeace and the online journal *The Barents Observer*. For example, Greenpeace was instrumental in providing information to the public prosecutor which led to Lukoil-Komi being fined 614 million roubles ($18.5 million US) for oil spill damage in September 2013, though Lukoil appealed the fine, which was then dropped (Ottery 2014).

There is a distinct lack of trust towards the officials based in the Usinsk Rayon Administration and regulatory agencies, who local people feel do not provide them enough information or defend their interests sufficiently well. Partly this is due to a simple lack of communication. The villagers claimed that the rayon officials and regulators ignore the photos and other evidence that they provide. To some extent this reflects the conclusions of Karjalainen and Habeck (2004:179) who observed that that while local concerns were expressed by referring to 'personal experiences and observations', the knowledge of officials failed to 'link up with local people's reality'. However, the one official who was willing to talk to me in Usinsk argued that the villagers' evidence is in fact used, but in negotiations with the company around social investment payments to the region, thus implying that more pollution can result in more social payments. This tendency is a cause for concern among environmentalists, along with the fact that payments from environmental fines currently go straight into the rayon budget, rather than into a separate environmental fund (which used to be the case). One environmentalist, for instance, criticised the rayon officials in Usinsk for 'closing their eyes to the environmental issues and focusing on the social benefits'.

In 2014, between my second and third visits to Komi Republic, there was a remarkable turn of events. In Izhma Rayon, which neighbours Usinsk Rayon to the west, Lukoil-Komi's failure to carry out the legally required public consultation prior to starting work on four new wells close to the village of Krasnobor led to protest action. Local activists called for Lukoil-Komi to be banned from operating in Izhma Rayon until a number of grievances had been resolved, from adequate consultation to pipeline maintenance, and they demanded more social support for local communities (Greenpeace 2014). At the time, an environmentalist criticised the protest: 'They are being conservative in arguing for more money for social projects. They should demand "no more spills" and proper consultation and respect for local communities.' Nonetheless, the campaign against Lukoil-Komi – including its strong environmental message – gained global resonance. The protests were picked up by Greenpeace and the international 350.org climate campaign, and – in an echo of 1994 – people from around the world sent messages of support to the *Izhma* Komi. The wave of protest quickly spread from Izhma Rayon to Usinsk Rayon, with protest meetings held in Novik Bezh relating to the siting of a drilling tower in a marshy area prone to floods. However, the oil industry itself and the scale of its problems in Usinsk Rayon are much more extensive and deep rooted than in Izhma Rayon. Activists do not have the leverage to halt well-established ongoing operations or force pipelines to be replaced in Usinsk Rayon. As a result of the Izhma protest, *Izvatas* has now secured a direct agreement with Lukoil-Komi, which includes support for the training of local young people in a range of professional skills, as well as support for the running costs of the association. This will also be of benefit to the Komi living in Usinsk Rayon. Nonetheless, not all the protestors' demands have been met (notably the pipeline repair). It remains to be seen whether the *Izhma* Komi can maintain the balance between negotiation of benefits and a continued hard line on environmental and social responsibility following the signing of this agreement.

14.4   Conclusions

Despite Lukoil's reputation as a socially responsible company, the practical experience of its regional branch, Lukoil-Komi, in community engagement has failed to build trust with local residents outside of Usinsk. This could be a major risk for Lukoil-Komi, which has been suffering an above-average amount of negative press since early 2014. The Usinsk Rayon government also bears responsibility for making community engagement meaningful. At a basic level, providing information to local communities about spills when they happen and consulting on proposed construction and exploration activities before they start would increase local people's trust in both government and industry. A respectful relationship established through timely dialogue (whether required by law or not), could have reduced or avoided damage to resources and lessened the intensity of the conflict in Izhma Rayon. Yet the pattern currently being played out is familiar: negligence followed by protest leading to the signing of a benefit-sharing agreement. This is against a backdrop of environmental fines being used to top up rayon budgets. This research suggests the need for local rayon officials and civil society groups alike to redefine the oil industry's 'social licence to operate' in northern Komi Republic, starting with a demand for greater environmental responsibility ('no more spills'), greater respect for local needs and livelihoods, and more meaningful engagement between government, industry and local communities.

Acknowledgements I would like to thank my research partner Dr. Kirill Istomin of the Komi Science Centre, who facilitated and collaborated on the field work for this paper that was carried out in Komi Republic. I also thank Valentina Semyashkina of Save Pechora Committee who provided contacts for local people in Usinsk Rayon, and all the people who provided interviews and otherwise helped us during our field work. The project *Sustainability and Petroleum Extraction*, funded by the Norwegian Research Council from January 2013 to December 2015, is a collaboration between the Norwegian Institute of Foreign Affairs (NUPI), University College London (UCL), Menon (Oslo), Poyry (Oslo), and the Institute of World Economy and International Relations (IMEMO, Moscow) (see http://csroil.org/). The interview questionnaires for this field work were developed in collaboration with partners on this project, Julia Loe of Menon and Elana Wilson Rowe of NUPI, who have conducted field research for the same project in Murmansk Oblast' and Nenets Autonomous Okrug respectively. The third field visit was supported by the project *Indigenous peoples and resource extraction in the Arctic: Evaluating ethical guidelines*, led by the Arran Lule Sami Centre, Norway and funded by the Norwegian Ministry of Foreign Affairs from 2013 to 2016.

References

ADB. (2009). *Safeguard policy statement*. Manila: Asian Development Bank. Retrieved July 10, 2015, from http://www.adb.org/sites/default/files/institutional-document/32056/safeguard-policy-statement-june2009.pdf

BIC. (2014). *Environmental and social management (ESAM): Proposal for strengthening the World Bank's safeguards*. Washington: Bank Information Centre. Retrieved July 10, 2015, from http://www.bicusa.org/issues/safeguards/environmental-and-social-assessment/

Black, L. (2013). *The social licence to operate: Your management framework for complex times*. Oxford: Do Sustainability.

Blowfield, M., & Frynas, G. J. (2005). Setting new agendas: Critical perspectives on corporate social responsibility in the developing world. *International Affairs, 81*(3), 499–513.

Bowen, H. (1953). *Social responsibilities of the businessman*. New York: Harper and Row.

Buxton, A., & Wilson, E. (2013). *FPIC and the extractive industries: A guide to implementing the spirit of free, prior and informed consent in industrial projects*. London: IIED. Retrieved July 7, 2015, from http://pubs.iied.org/16530IIED.html

Carroll, A. B. (1999). Corporate social responsibility: Evolution of a definitional construct. *Business and Society, 38*(3), 268–295.

Donahoe, B., Habeck, J. O., Halemba, A., & Santha, I. (2008). Size and place in the construction of indigeneity in the Russian Federation. *Current Anthropology, 49*(6), 993–1020.

EBRD. (2014). *Environmental and social policy*. London: European Bank for Reconstruction and Development. Retrieved July 10, 2015, from http://www.ebrd.com/who-we-are/our-values/environmental-and-social-policy/performance-requirements.html

Equator Principles. (2013). *The Equator Principles-III*. London: The Equator Principles. http://www.equator-principles.com/index.php/ep3

Freeman, R. E. (1984). *Strategic management: A stakeholder approach*. Boston: Pitman.

Friedman, M. (1970, September 13). The social responsibility of business is to increase its profits [Electronic version]. *The New York Times Magazine*, p. 32. Retrieved July 7, 2015, from http://www.colorado.edu/studentgroups/libertarians/issues/friedman-soc-resp-business.html

Greenpeace. (2014). *Indigenous people of Komi decide to kick Lukoil out of their lands*. Auckland: Greenpeace. Retrieved July 10, 2015, from http://m.greenpeace.org/new-zealand/en/high/blog/indigenous-people-of-komi-decide-to-kick-luko/blog/48808/

Hasnas, J. (2013). Whither stakeholder theory?: A guide for the perplexed revisited. *Journal of Business Ethics, 112*(1), 47–57.

Hudina, A. (2010). *The role of the UN global compact in promoting corporate social responsibility in Russia*. Masters dissertation, Central European University.

IFC. (2012). *Environmental and social performance standards and guidance Notes* (2012 ed.). Washington, DC.: International Finance Corporation. Retrieved July 7, 2015, from http://www.ifc.org/wps/wcm/connect/topics_ext_content/ifc_external_corporate_site/ifc+sustainability/our+approach/risk+management/performance+standards/performance+standards+-+2012

Jernelov, A. (2010). The threats from oil spills: Now, then, and in the future. *AMBIO, 39*, 353–366.

Karjalainen, T. P., & Habeck, J. O. (2004). When 'the environment' comes to visit: Local environmental knowledge in the Far North of Russia. *Environmental Values, 13*(2), 167–186.

Kim, H. J., Shabaev, Y. P., & Istomin, K. V. (2015). Lokal'naya gruppa v poiske identichnosti [A local group in search of an identity]. *Sotsiologicheskie issledovaniya, 8*, 94–101.

Lukoil. (2002). *Social code*. Moscow: Lukoil. Retrieved July 10, 2015, from http://www.lukoil.com/static_6_5id_262_.html

Lukoil. (2006). *Sustainability report 2005–2006*. Moscow: Lukoil. Retrieved July 10, 2015, from http://www.lukoil.com/materials/doc/reports/Social/Report_eng_2006.pdf

Lukoil. (2012). *Sustainability report 2011–2012*. Moscow: Lukoil. Retrieved July 10, 2015, from http://www.lukoil.com/materials/doc/reports/Social/Lukoil_OD_eng.pdf

Lukoil. (2014). *Annual report 2014*. Moscow: Lukoil. Retrieved July 10, 2015, from http://www.lukoil.com/materials/doc/AGSM_2015/LUKOIL_AR_eng_2014.pdf

Martynova, E. P., & Novikova, N. I. (2012). *Tazovskie Nentsy v usloviyakh neftegazovogo osvoeniya*. (Taz Nentsy in the context of oil and gas development.). Moscow: A.G. Yakovlev Publishing House.

Morrison, J. (2014). *The social licence: How to keep your organisation legitimate*. Basingstoke: Palgrave Macmillan.

Novikova, N. I. (2014). *Okhotniki i neftyaniki: Issledovanie po yuridicheskoy antropologii*. (Hunters and oil workers: Research in legal anthropology). Moscow: Nauka.

Novikova, N., & Wilson, E. (2015). Korporativnaya sotsial'naya otvetstvennost': transformatsiya ponyatiya na zapade i znachimost' dlya korennykh narodov Rossii [Corporate social responsibility: Evolution of the concept in the West and the implications for indigenous peoples of Russia]. *Ural'skiy istoricheskiy vestnik, 2*(47), 108–117.

OECD. (2015). *Due diligence guidance for meaningful stakeholder engagement in the extractives sector: Draft for comment*. Paris: OECD. Retrieved July 10, 2015, from http://www.oecd.org/daf/inv/mne/OECD-Guidance-Extractives-Sector-Stakeholder-Engagement.pdf

Ottery, C. (2014). *Oil swamps in Usinsk: Investigation finds Russian oil spills six times the size of deepwater horizon*. Greenpeace Energy Desk (online). Retrieved July 10, 2015, from http://energydesk.greenpeace.org/2014/08/11/oil-swamps-usinsk-investigation-finds-russian-oil-spills-six-times-size-deepwater-horizon/

Porter, M., & Kramer, M. (2011). Creating shared value. *Harvard Business Review, 89*(1/2), 62–77.

Poussenkova, N. (2014). Neftegazovoe osvoenie Arktiki: Chtoby v NAO bylo kak v OAE. (Oil and gas extraction in the Arctic: if only it could be like the United Arab Emirates in the Nenets Autonomous Region). *Ekologicheskiy vestnik Rossii, 11*, 25–31.

Rosneft. (2013). *Sustainability report 2013*. Moscow: Rosneft. Retrieved July 10, 2015, from http://www.rosneft.com/attach/0/10/92/RN_SR_2014_ENG_WEB.pdf

Sirina, A. A., Yarlykapov, A. A., & Funk, D. A. (2008) Special edition of Etnograficheskoe obozrenie (Ethnographic review) on oil, ecology and culture. *Etnograficheskoe obozrenie, 3*.

Staalesen, A. (2014, January 10). Quietly flows the Kolva: An arctic story. *Barents Observer*, Retrieved July 10, 2015, from http://barentsobserver.com/en/nature/2014/01/quietly-flows-kolva-arctic-oil-story-10-01-0

Statoil. (2014). *Sustainability report 2014*. Stavanger: Statoil. Retrieved July 10, 2015, from http://www.statoil.com/no/InvestorCentre/AnnualReport/AnnualReport2014/Documents/DownloadCentreFiles/01_KeyDownloads/Sustainability_report_2014.pdf

Thomson, I., & Boutilier, R. (2012). *What is the social license?* Retrieved July 10, 2015, from http://www.socialicense.com/

Total. (2014). Sustainable growth report 2014. Paris: Total. Retrieved July 10, 2015, from http://www.total.com/sites/default/files/atoms/files/rapport_croissance_durable_va.pdf?xtmc=sustainable%20growth%20report&xtnp=1&xtcr=2

Total. (2015). *Society and environment: Operating sustainably and responsibly every day*. Web content. Retrieved July 10, 2015, from http://www.total.com/en/society-environment-operating-sustainably-and-responsibly-every-day

United Nations. (2011). *UN guiding principles on business and human rights*. Geneva/New York: United Nations. Retrieved July 10, 2015, from http://business-humanrights.org/en/un-guiding-principles

Wilson, E., & Stammler, F. (Eds.). (2006). Special issue of *Sibirica: the Interdisciplinary Journal of Siberian Studies, 5*(2).

Chapter 15
When Municipalities Met Goliat on the Coast of Finnmark: Collaborative Dynamics Between Local Authorities and an International Oil and Gas Company

Toril Ringholm

Abstract A meeting between a multinational oil and gas company and small municipal organisations appears an uneven match. The company has years of experience entering new localities, from all over the world, whereas for the local authorities this will most often be a once in a lifetime experience. Norwegian municipalities traditionally have a central position in local development. Collaboration is a crucial asset, and the question is to what degree this experience works in a new setting. How do the local authorities act in ensuring that the area benefits from development, and how does the company respond to this? What collaborative arenas are established and how do they work? What means are used to encourage collaboration, and what actions do they activate? In what way is power and trust employed in collaboration? The analysis in this chapter is based on a four year research project examining the interaction between four municipalities and a multinational resource company during the development of the Goliat off-shore oil-field near the western part of Finnmark in Northern Norway. The study reveals a process where both parties make use of power, and where trust is developed. This is also described as a process of mutual learning. The data contains repeated interviews with key actors, observations of meetings and mapping of outcomes.

Keywords Municipality • Oil-company • Collaboration • Local development • Trust • Power

T. Ringholm (✉)
Lillehammer University College, Lillehammer, Norway
e-mail: toril.ringholm@hil.no

© Springer International Publishing Switzerland 2017
G. Fondahl, G.N. Wilson (eds.), *Northern Sustainabilities: Understanding and Addressing Change in the Circumpolar World*, Springer Polar Sciences,
DOI 10.1007/978-3-319-46150-2_15

15.1   Introduction

The oil and gas industry holds a prominent position in the Norwegian economy, and has done so for decades. The northern part of Norway, though, has only recently gained experience with having geographical proximity to oil and gas installations. The local authorities, local industry and other local actors are, in general, inexperienced in dealing with the big, multinational companies that operate the production licenses. This situation, however, is changing. New technology and promising findings have made the northernmost part of Norway increasingly relevant for the oil and gas industry. Stavanger, in the south-western part of the country, has developed into a growth-pole, based on petroleum-activity, and there certainly exists a desire for the northern part of the country to experience a similar development.

The "institutional infrastructure" in the northernmost part of Norway however, is not the best to make such dreams come true. Small businesses and small municipalities[1] characterise the area. Granted, the municipalities do hold a prominent position in the development of local business, welfare and culture. There are historical reasons for this, as the municipalities have carried out local development in a broad spectrum of areas since the 1830s (see Bukve 2001; Finstad and Aarsæther 2003; Teigen et al. 2013). In the present day, the important position of municipalities in Norway also makes them a relevant actor for the licence holder to interact with, in order to carry out corporate social responsibility on the local and regional level. Collaboration is an institutionalised procedure for the municipalities in local development. Mayors, in particular, have been the driving force, and their orientation towards local industrial and societal development seems to have become stronger over the years (Rose and Ståhlberg 2005; Sandberg and Ståhlberg 2001).

That said, licenses for drilling are granted by the state, and the local authorities in the affected area have no decision-making authority on this matter. They, therefore, need to make their voices heard in other arenas, like the consultation rounds, through the party system and in the media. The license normally comes with terms. The terms are ways of ensuring the operators' commitment to social responsibility and of allocating resources to strengthen the development of the local communities and the region.

In this paper, I examine the relationship between the four municipalities located near the Goliat oil-field in the Barents Sea and Eni, a multinational, Italian-based company that operates the field, with a specific focus on the question of how the interaction between the local authorities and the license holder unfolds. The municipalities in this area are Hasvik, Måsøy, Nordkapp and Hammerfest; they are most likely to be affected by an oil spill from the Goliat off-shore installation and the term "influence-area" is often used for this particular group of municipalities. For all of these municipalities, with the exception of Hammerfest, the petroleum industry is entirely new. The question is how the local authorities act to ensure that the area benefits from the resource development, and how the operator responds to local demands.

---

[1] Small in terms of population, but often large in terms of geographical area.

The interaction between local authorities and the petroleum industry is a new field of research. Thus far, Norwegian research has not addressed this particular aspect of the impact of the petroleum industry. Nordic municipalities, and Norwegian ones in particular, have a long tradition of being the driving force of local development, in collaboration with business actors and others. Is their experience also relevant when faced with this new actor, of a calibre hitherto unseen in this region?

15.2   Collaboration, Trust and Power

Norwegian municipalities engage in all kinds of local development: industrial and business development; infrastructure; public service; democracy; culture; and leisure activities. The first general mapping of their efforts in this area showed that even very small municipalities, with a population of less than 2000, engaged in up to ten projects simultaneously (Ringholm et al. 2009). Mostly the projects are carried out in collaboration with other municipalities, regional and state authorities, local businesses and voluntary organisations (Teigen 2000; Ringholm et al. 2009). Hence, municipalities have decades of experience with network governance as a tool for local and regional development (Bukve 2001; Andersen and Røiseland 2008; Hall et al. 2009). Norwegian legislation allows municipalities to take on what responsibilities they desire, as long as these responsibilities are not delegated to another institution.

Definitions of network governance emphasise horizontality, interdependency, stability, autonomy of the actors, and negotiation as the way solutions are achieved (Rhodes 1997; Torfing 2007). This does not rule out hierarchical structures (Scharpf 1997), but the significance of such structures would need to be downplayed for the sake of an optimal solution. When confronted with a big, multinational company that gets its drilling licence from the state, the small municipality is the junior partner. What interdependency exists between the two? Is there a chance that a process of network governance can emerge between them? This depends on whether they have a common interest in particular policy solutions and to what degree each of them has resources that the other needs in order to implement the policy.

Hence, the following research questions: Are collaborative arenas established in order to develop common solutions? What resources do the two parties possess and how are these activated in order to reach a desired policy? How does communication and possible collaboration between the municipalities and the company develop over a period of time? Is there a mutual understanding of roles, challenges and solutions developed?

These questions explore the relationship between power and trust in the interaction between a multinational company and municipalities. The way resources are activated show "the positioned practice" of power (Flyvbjerg 1991; Reed 2001). Rather than merely mapping the means of power, this perspective of power emphasises the importance of studying how the means are actually used in policy making interaction. Trust is regarded by several scholars, however, as equally important as power in identifying how resources are distributed between organisations (Bachmann 2001). Trust is often described a "lubricant" for collaboration (Coleman 1988), and as an essential ingredient of the decision making process. The knowledge of how trust and power relations develop and interact in the relationship between the company and the municipality will deepen our understanding of the creation of ripple effects in the form of local community development.

15.3 Methods

The data for this paper consist of three types: Interviews; observation of meetings; and mapping of competence enhancing and local development projects financed by the operator, Eni Norway.

The data were collected within the framework of a trailing research project (Schwebs 2003; Olsen and Lindøe 2004; Stensaker 2013) over a period of nearly four years, from early 2010 to the second half of 2013. This means that we have gathered data and carried out analyses since the start of the development, in order to assess the dynamics of the impact that the project has on local and regional development and the possible ripple effects to emerge from it. A pilot study was carried out in 2009–2010 (Eikeland et al. 2010) and interview data from the pilot study are included in the analysis.

15.3.1 Interviews

The interviewees fell into three categories: municipal representatives; representatives from Eni Norway; and leaders of the projects that Eni Norway has funded. Most of the interviewees were interviewed several times. Table 15.1 gives an overview of the different informants and the year they were interviewed.

The municipalities were mostly represented by their mayors and project leaders. Interviews with Eni Norway included the regional director, the manager of competence-building projects and local development projects, and two administrative officers at the company's Hammerfest office.

Table 15.1 Overview of informants

Type of informant	2009[a]	2010	2011	2012	2013	2014	Total # interviews	Total # interviewees
Mayor of municipality	4	4		4	4		16	4
Municipal administrator			2		1		3	3
Eni representative		4		2	2	2	10	4
Project coordinator		1			7	7	16	15
Sum	4	9	2	6	12	6	29	19

[a]Interviews from the pilot-study

In addition to interviews, the researchers attended three meetings involving Cooperation-group Goliat (see Sect. 4.1). The two researchers who were present at each meeting made detailed notes from the presentations and discussions. Eni Norway also provided data on the funds it allocated to finance and co-finance projects for local development, culture and capacity-building. The project leaders confirmed the amounts provided by the company.

15.4   The Process of Shaping Cooperative Relations

Over the life of the research project, we observed a process of mutual learning in which arenas were established for collaboration and problem solving, actions were taken, power relations were in use and trust was developed. The following sections will elaborate each of these areas in more detail.

15.4.1   Arenas for Collaboration and Problem Solving

As soon as the decision for developing the Goliat oil field was made, the mayor of Hammerfest took the initiative to establish "Cooperation-group Goliat" (CGG), as a forum for the exchange of mutual information between Eni Norway and the local authorities (Ringholm and Nilsen 2011, 2014). The initiative was inspired by Hammerfest's previous experiences with a similar forum between Statoil, the license holder for the Snøhvit gas-field, established a few years earlier on Melkøya island near Hammerfest.

At the outset of the process, Eni Norway was reluctant to participate, not being used to this form of collaboration with local authorities. The main reason that the company eventually decided to participate was that the arena provided an opportunity to keep the local authorities informed about the oil-field development and about the oil-spill protection system that the company was establishing. Regular meetings would make it possible to answer questions on the spot and avoid misunderstandings. The CGG mainly became a forum for the mutual exchange of information; it is important to note that the idea for one larger capacity-building project, the "virtual classroom" in Hasvik and Måsøy municipality, was also launched there.

The CGG also became an arena for reconciliation, at least with regard to one difficult conflict. In 2009–2010, there arose a particularly strong and lasting disagreement between Hasvik municipality and Eni Norway about what the municipality could expect from the company in terms of general local development and oil-industry installations. Both Hasvik and Måsøy municipalities held up what they had interpreted as promises from the operator in the months preceding the drilling decision, a time when Eni Norway was meeting with different kinds of actors to gain support for a drilling decision. The company's response to Hasvik's and

Måsøy's contention was that these were not promises, but rather possibilities that had not been prioritised at such an early stage.

During the first year, it became clear that the ripple effects in the form of employment and local purchases would be marginal. The company would build storing facilities for oil-spill protection equipment in both Hasvik and Måsøy, but these would bring few, if any, jobs. This was far from what the mayors had interpreted from the initial meetings. The situation became especially tense, causing the local council of Hasvik to terminate a meeting with Eni, in June 2010, on the grounds that there was nothing to discuss. The controversy received quite a lot of attention in the local and regional media, and prompted Eni's top management from Italy to come to Norway for a meeting with the local council. After this meeting, bilateral relations between Hasvik and Eni took a positive turn. One informant from the municipality commented:

> After this meeting we had a more concrete dialogue between us and Eni in the sense that they have behaved more tidily with regard to what they have said and what plans they have. [...]We feel that the dialogue has become more concretely directed at what Eni can offer Hasvik and what Goliat can mean for our municipality.[2]

One outcome of this meeting was financial support from Eni Norway for a slow-food[3] centre in Hasvik. While the ongoing controversy caused a halt in the CGG meetings for a few months, the fact that such an arena existed eventually urged all the members back into meeting, and gradually the dialogue improved.

Other meetings also took place, though not with the same regularity as the CGG. Eni Norway also met with each of the municipalities separately in bilateral negotiations about particular development projects that the municipalities wanted the company to support. The four municipalities did not inform each other about the content of these meetings, but it was common knowledge that they took place. There was clearly competition between them for the limited number of onshore installations that came with the development, and for other local development projects. Hammerfest, where Eni Norway's regional office is located, had more of these meetings, because of the need to determine the impact that locating the company's regional office there would have on local housing and public services.

15.4.2   Means and Actions

As the decision in favour of opening the Goliat field appeared inevitable, local actors were determined to get as much out of it as possible (Eikeland et al. 2010). One way was to influence the licence terms. This was done mainly by making use of political contacts with the central government. These contacts were good. The Minister of Fisheries at the time came from Finnmark County, where the four

---

[2] Translation by the author.

[3] The ambition is to achieve certification from the slow-food movement for local food.

municipalities are located. She also held a high position in the Social Democratic Party, the leading party in the coalition government. Moreover, the former mayor of Hasvik municipality was a member of the National Assembly. The mayor of Hammerfest was a veteran of the Social Democratic Party, and carried a great deal of influence within party circles. The interviews reveal that these connections helped local political actors to influence the license terms (St. Prp. nr 64 2008–2009) that were made. As one mayor noted:

> We have political contacts with central decision making actors and we are willing to use them. That way we are in a position to make noise about unreasonable decisions. In that connection we have control of the opinion and parts of the political apparatus, especially the Social Democratic Party.[4]

This quote reveals a solid self-consciousness regarding the impact of the local actors, and even if slightly exaggerated, the bulk of the interviews supported the notion that local actors had good connections to decision-making institutions.

Eventually it turned out that although the main part of the terms concerned energy supply and environmental issues, in particular the oil-spill protection system, they also turned out to be unusually detailed[5] in terms of outlining the company's social responsibility and the expected benefits for the region. It is unusual for such terms to outline particular benefits for specific municipalities, but in this case the four municipalities are mentioned in the terms. For example, the terms demand the company to establish a regional office in Hammerfest, a contract regime that allows local and regional businesses to compete for projects associated with the development,[6] particular capacity-building programs and support for industrial development in Finnmark County as a whole and in the four municipalities; in particular in the fishery and tourist industries, the two most important industries in these municipalities.

The oil-spill protection system was a major issue in several of the CGG meetings, as oil spills can cause severe damage to the fisheries. Eni Norway made a significant effort both with regard to actually developing the protection system and to presenting and explaining the steps in its development to the municipalities. Both parties participated in the process of identifying local development projects. There was not always agreement about the definition of a "good project". Eni Norway needed to ensure that the project served the purpose of fulfilling the licence terms and the defined areas of the company's corporate social responsibility policy, and that the company should not be suspected of corruption. The project also needed to be viable and stand a good chance of achieving the planned outcomes, which was naturally in the interest of the municipalities.

---

[4] Translation by the author. The quote in Norwegian is to be found in Ringholm and Nilsen 2011: 21–22.

[5] This is, compared to other drilling licences.

[6] In essence this means that the tenders for contracts should be set up in as small units as possible and not in large clusters of technology and/or other services. This would at least allow small businesses to compete even if it would not guarantee them a contract.

15.4.3 Trust and Power

The collaboration process demonstrated both the use of power and the process of trust building in the relationships between the municipal authorities and the oil company. Specifically, power relations are evident in:

- The municipalities' use of their political network to define the licence terms;
- Eni Norway's control over what development purposes it wanted to support;
- The possibility of the municipalities' opinion of the company reaching the media and influencing the reputation of the company.

This can somewhat crudely be summed up as follows. While the company holds the power over the desired goods – funding of local development projects and an oil spill protection system, the municipalities' power is based on: (1) their ability to influence the reputation of the company by the media and through political channels: and (2) their knowledge of what projects will encourage local development, and how to realise these projects. Reputation is an important asset for the company, as it affects its relationship with both central authorities and local actors. Cultivating a reputation of being a supporter of local development can help a company gain access to new oil fields and, in general, build trust with government and communities, a view that is confirmed in the interviews with Eni representatives. For the municipalities, almost any kind of local development is an asset in the perpetual struggle of keeping and recruiting businesses and inhabitants. In summary, the relationship between the municipalities and Eni Norway can be described as an exchange relation, as both parties possess resources that are of interest to the other.

There is certainly power involved in the cooperation process, but to leave it at that would present us with a rather one-dimensional impression of the cooperation that took place. Cooperation also requires the building of trust. Trust comes in many shapes. A typology of trust is offered by Mari Sako (1998), who describes it as a continuum, and distinguishes between three types. *Contractual trust* is the trust that the contract partner will fulfil the obligations described in the contract. *Goodwill trust* describes a situation where one is confident that the partner will put mutual benefit before its own profit. *Competence-based trust* lies between these two, and describes the trust that the partner is capable and willing to fulfil what has been agreed upon.

The general impression from the interviews, both with the municipal and company informants, is that the relations between the two parties developed into a competence-based trust over the four years covered by the study. The starting point was different for the four municipalities. Whereas distrust characterised the relationship between the two smallest municipalities, Hasvik and Måsøy, and Eni Norway, this was not the case for Hammerfest and Nordkapp. In latter two municipalities' relations to Eni Norway, the starting point was more of a contract-based trust. The "contract" in this case refers to the licence terms. This trust, however, developed into competence-based trust over the course of the negotiations. Two developments in particular supported the transition from contract-based to competence-based trust: (1) the municipalities' experience of Eni Norway taking the oil spill protection seriously; and (2) the recognition by the company of the municpalities' important role in local development. The trust, in other words, went both ways.

The dialogue with the municipalities gained importance as the company learned that the municipalities were often a key actor with regard to finding local development projects that could actually help it meet the licence terms. Therefore, effort was put into filling leading positions in the company with people who possessed communication skills appropriate for the particular setting.

For the municipal actors, and especially for those in the small municipalities, the trust building process followed certain "stages". Firstly, there were the *high expectations* in the run up to the license decision. This was followed by *disappointment*, when municipal actors realised that their expectations would not be met. After this was a phase of *reconciliation* as they found a new basis for the dialogue which identified what was realistic to expect from the Goliat development.

On the whole, this was also a process of learning. In the interviews carried out in 2014, all the mayors agreed that what they had learned from this process would be an asset in similar situations in the future; the knowledge of both how to communicate with a multi-national company and how to establish a trust-building institutional framework for such dialogue. This quote, from one of the mayors, expresses the general impression from interviews with both the municipal informants and those from the company:

> I would describe it in the way that Eni Norway and we have worked us closer to each other, in the way that we understand more of how the other thinks and the framework that each of us works within.[7]

15.5   What Local Development?

Table 15.2 presents an overview of the projects and events that Eni Norway funded or co-funded in the four municipalities. As many of these projects are still in an early phase, and the implementation of some of them have been delayed, it is impossible at this stage to calculate what the ripple effects will be with regard to employment or other economic indicators.

In the years from 2008 to 2012, Eni Norway allocated nine million Norwegian kroner (NOK) for local development projects in the municipalities. The municipalities were involved in securing these funds. Moreover the company also sponsored cultural events such as concerts and festivals, to a level of close to 900,000 NOK. The degree that the municipal actors were involved in securing this funding is not known, as these events are basically run by local voluntary organisations.

There are not many education or research institutions located in this region; this is most likely the reason why only a small proportion of the funding for educational

---

[7] Translation by the author. The quote in Norwegian can be viewed in Ringholm and Nilsen 2014:32.

Table 15.2 Local development projects in the influence municipalities with funding from Eni Norway 2008–2012

Project	Municipality	Norwegian kroner
"Live Here" (recruiting initiative)	Hasvik	100.000
Course, knitting factory	Hasvik	75.000
Certification of Slow Food competence	Hasvik	450.000
Products from knitting factory	Hasvik	130.000
Seabird counting	Hasvik	400.000
Seabird centre	Nordkapp	2.499.000[a]
"Newton-room" – for educating school-children in physics	Hammerfest	504.450
EnergyCampus North	Hammerfest	598.500
Course oil-spill protection	Måsøy	453.150
Arctic Slow Food Centre	Hasvik	2.496.600[a]
Digital classrooms for secondary an higher education purposes	Hasvik/Måsøy	2.000.000
Course, welding	Måsøy	196.650
Slow food festival	Hasvik	196.650
Ingøy-days (Local festival)	Måsøy	50.000
Sørøya-days (Fishing festival)	Hasvik	442.500
Nordkapp Filmfestival	Nordkapp	100.000
North Cape Simulators for oil spill protection education	Nordkapp	250.000
Design of multiservice vessel	Hasvik	750.000
Sum		**11.692.500**

Source: Ringholm and Nilsen (2014)
[a]Projects that were not implemented by June 2014

and research and development purposes stayed in these municipalities. Around 35 million NOK for research and development and education was allocated to institutions and businesses in northern Norway. Around 4.5 million NOK went to such purposes in the four municipalities.[8] In addition to the projects listed in Table 15.2, some of the secondary schools in the county (among them, the one in Hammerfest) have cooperation agreements with Eni Norway. In one such agreement, funding is provided for students to make field visits to the company, and to support chemistry activities and other relevant arrangements in the school.

All these projects were initiated by the local politicians in the municipalities and, in particular, by the mayors. They also fit well into the license terms for the Goliat field. However, so would many other projects. When we asked Eni Norway why it chose to support these particular projects, the answers pointed to two different rationales. Firstly, the terms stated that the support was supposed to go to the region that was taking the bulk of the risk with regard to oil spills. Secondly, the company

---

[8] These sums are estimated from material in Ringholm et al. (2015).

claimed that there were not that many good projects to choose between and those that got funded were the ones most likely to produce the desired outcomes.

15.6   An On-going Process

The experiences of the Goliat-development process show how Eni Norway worked in a local context with institutionalised practices for local development. Collaboration is the *modus operandi* of local development in Norway. Therefore, the company was invited, and actually expected, to join in such collaboration. Although this was not its usual approach, the company adjusted to the situation in various ways and, as a result, both parties have benefitted. The trust between the municipalities and the company increased and important local development projects received financing.

It would be naive, though, to overestimate the importance of trust. The trust was mixed with power: the local actors had an impact on the company's reputation, and the company had the power to decide both which oil-related installations should be built and what local development projects to support.

The mix of trust and power is not a stable alloy. In this analysis two aspects in particular stand out as important for the future intrinsic value of trust and power in these situations: the realisation of the projects and the functioning of the oil-spill protection system. Some of the funded projects have so far not been realised according to plan; the most significant ones being those that involve the most monetary funding and possible employment in the smallest of the affected municipalities. If they are not realised, a matter that relies on the municipal effort put into the projects, it is possible that Eni's confidence in the municipality will weaken. On the other hand, if the oil-spill protection system should fail, despite the effort put into it, this can cause severe damage to the competence-based trust that the municipalities have in the company.

This paper has shown that the collaborative tradition and political capacity of local actors can have a clear impact on how companies exercise their corporate social responsibility in northern and arctic settings. The role and power of the local authorities varies among the arctic countries, and among different industries. The Norwegian case demonstrates that a great diversity of cooperation models may evolve from collaboration between industry and municipalities, and a deeper knowledge of their institutional framing and actions within it could represent a source for learning and development in other parts of the Arctic.

References

Andersen, O. J., & Røiseland, A. (Eds.). (2008). *Partnerskap, problemløsning og politikk* [Partnership, problem solving and politics]. Bergen: Fagbokforlaget.

Bachmann, R. (2001). Trust, power and control in trans-organizational relations. *Organization Studies, 22*(2), 337–365.

Bukve, O. (2001). *Lokale utviklingsnettverk* [Local development networks] (Report No. 76). Bergen: University of Bergen, Institute of Administrative and Organizational Sciences.

Coleman, J. S. (1988). Social capital in the creation of human capital. *American Journal of Sociology, 94*(Suppl), 95–120.

Eikeland, S., Karlstad, S. Nilsen, T., & Ringholm, T. (2010). *Regionale forventninger om ringvirkninger ved utbygging og drift av Goliat* [Regional expectations of ripple effects from the development of the Goliat project] (Report 2010:5). Alta: Norut Alta.

Finstad, N., & Aarsæther, N. (Eds.). (2003). *Utviklingskommunen* [Municipal Development]. Oslo: Kommuneforlaget.

Flyvbjerg, B. (1991). *Rationalitet og magt. Det konkretes videnskab* [Rationality and power. The Science of the Concrete]. Copenhagen: Akademisk forlag.

Hall, P., Kettunen, P., Löfgren, K., & Ringholm, T. (2009). Is there a Nordic approach to questions of democracy in studies of network governance? *Local Government Studies, 35*(5), 515–538.

Olsen, O. E., & Lindøe, P. (2004). Trailing research based evaluation; phases and roles. *Evaluation and Program Planning, 27*, 371–380.

Reed, M. I. (2001). Organization, trust and control: A realist analysis. *Organization Studies, 22*(2), 201–228.

Rhodes, R. A. W. (1997). *Understanding governance*. Buckingham: Open University Press.

Ringholm, T., & Nilsen, T. (2011). *Finnmarkskart i endring? Samhandling mellom nabokommuner og Eni Norges Goliatprosjekt* [Changing maps of Finnmark? Interaction between neighboring municipalities and Eni Norway's Goliat project] (Report 2011:5). Alta: Norut Alta.

Ringholm, T., & Nilsen, T. (2014). *Lokal samfunnsutvikling i kjølvannet av Goliatprosjektet. Møter mellom influenskommunene og Eni Norge under utbygging av Goliatfeltet* [Local community development in the wake of the Goliat project. Meetings between the influence municipalities and Eni Norway during the development of the Goliat project] (Report 2014:7). Alta: Norut Alta.

Ringholm, T., Aarsæther, N., Nygaard, V., & Selle, P. (2009). *Kommunen som samfunnsutvikler* [The municipality as a societal developer] (Report 8/2009). Tromsø: Norut Tromsø.

Ringholm, T., Bye, G., & Nilsen, T. (2015). *Skaper Goliat-utbyggingen kompetanseutvikling?* [Does the Goliat development contribute to capacity building?] (Norut-rapport 2015:4). Tromsø: Norut

Rose, L., & Ståhlberg, K. (2005). The Nordic countries: still the 'promised land'? In B. Denters & L. Rose (Eds.), *Comparing local governance – Trends and developments* (pp. 83–99). New York: Palgrave Macmillan.

Sako, M. (1998). Does trust improve business performance? In C. Lane & R. Bachmann (Eds.), *Trust within and between organizations* (pp. 88–117). Oxford: Oxford University Press.

Sandberg, S., & Ståhlberg, K. (2001). Utvecklingspolitiska inställningar hos nordiska region- och lokalpolitiker [Development-policy attitudes among Nordic regional- and local politicians]. In H. Baldersheim, S. Sandberg, K. Ståhlberg, & M. Øgård (Eds.), *Norden i Regionernas Europa* (Nord 2001, 18, pp. 89–113). Copenhagen: Nordic Council of Ministers.

Scharpf, F. W. (1997). *Games real actors play: Actor centred institustionalism in policy research*. Oxford: West View Point.

Schwebs, R. (2003). *Følgeforskning som innfallsvinkel for organisasjonslæring i offentlig virksomhet* [Trailing research as a gateway to organisational learning in public institutions]. Copenhagen: Copenhagen Business School.

St. Prp. nr 64 (2008–2009). *Utbygging og drift av Goliatfeltet* [White book on the development and extraction of the Goliat-field].

Stensaker, J. G. (2013). Methods for tracking and trailing change. *Research in Organisational Change and Development, 1*, 149–174.

Teigen, H. (2000). Bank- og bygdeutvikling: plan og marknad frå sjølbergingssamfunnet til den globale kapitalismen [Bank- and rural development: Plan and market from the self-containing society to the global capitalism]. In H. Teigen (Ed.), *Bygdeutvikling: Historiske spor og framtidige vegval* (pp. 183–222). Trondheim: Tapir Akademisk Forlag.

Teigen, H., Ringholm, T., & Aarsæther, N. (2013). Innovatør frå alders tid [Innovator from times immemorial]. In T. Ringholm, H. Teigen, & N. Aarsæther (Eds.), Innovative kommuner [Innovative municipalities]. Oslo: Cappelen Damm Høyskoleforlaget.

Torfing, J. (2007). Introduction: Democratic network governance. In M. Marcussen & J. Torfing (Eds.), *Democratic network governance in Europe* (pp. 1–22). New York: Palgrave McMillan.

Chapter 16
Human Capital and Sustainable Development in the Arctic: Towards Intellectual and Empirical Framing

Andrey N. Petrov

Abstract One of the aspects of sustainable development in the Arctic is shifting the region's reliance from non-renewable resources and the public sector to economies increasingly based on knowledge and innovation. Knowledge, creative, and cultural economies represent economic sectors heavily embedded into internal community capacities and intangible competitive advantages. This chapter extends the discussion of the relevance of formal and informal education and knowledge to sustainable development in the Arctic. It discusses patterns and trends in post-secondary educational attainment and attendance in the last decade and provides an assessment of human capital and knowledge production in the Arctic. It also identifies and discusses persistent and emerging human formal education gaps in the Arctic (spatial, gender, Indigenous/non-Indigenous, formal/informal and entrepreneurial). The chapter offers conceptual and qualitative links between human capital accumulation, the knowledge economy and sustainable regional development.

Keywords Human capital • Economy • Education • Sustainable development • Arctic

16.1 Introduction

Although natural resources are typically conceived as the main driver of the arctic economy, the true treasure of the Arctic is its people, or, in other terms, its *human capital*. Since pre-historic times arctic residents have developed skills and knowledge that enabled them to live in harsh conditions. At the same time, the ability of arctic societies to benefit from standardized codified knowledge and formal education so far has been rather limited. The first and second *Arctic Human Development Report* (AHDR 2004, 2014) highlighted a shift from viewing knowledge as a

A.N. Petrov (✉)
University of Northern Iowa, Cedar Falls, IA, USA
e-mail: andrey.petrov@uni.edu

© Springer International Publishing Switzerland 2017   203
G. Fondahl, G.N. Wilson (eds.), *Northern Sustainabilities: Understanding and Addressing Change in the Circumpolar World*, Springer Polar Sciences,
DOI 10.1007/978-3-319-46150-2_16

standardized commodity to seeing it as a distributed resource for decision making, local adaptations, and increased use of technology to access knowledge. The attainment of education (whether formal or not) is an investment in human capital. The outcome of this investment is knowledge production and transfer that ensures the livelihoods and prosperity of arctic communities. This chapter builds upon the author's contribution to the second (AHDR 2014), extending the discussion of the relevance of formal and informal education and knowledge to human wellbeing and sustainable development in the Arctic.

The following analysis and discussion pertain to human capital largely described in terms of formal education (received in formalized settings at educational institutions). Certainly, economic development in the Arctic also benefits from informal education and knowledge transfer, whether it is a part of traditional culture and activities, or related to skills and knowledge earned from living in communities and on the land or sea (AHDR 2014). Here my analysis mostly focuses on formal education because informal skills attainment is difficult to measure, and the data are insufficient.

16.2 Human Capital in Sustainable Development in the Arctic: General Considerations

16.2.1 Defining Human Capital

Human capital can be defined as the stock of knowledge and skills embodied in a human population that has economic value (OECD 2009). Most generally, it refers to formal and tacit knowledge and skills, which can be deployed for general economic returns. Human capital incorporates three components: general skills (literacy), specific skills (related to particulate technologies and operations), and technical and scientific knowledge (mastery of specific bodies of knowledge at an advanced level). There is a broad agreement in the literature that human capital is closely related to both individual and aggregated labor outcomes: higher individual wages and enhanced employability on the one hand and greater productivity and accelerated technological progress on the other (De la Fuente and Ciccone 2002).

All three components of human capital are closely tied to schooling. General skills are typically acquired at the primary education level. Specific skills are acquired during secondary education and advanced scientific and technical knowledge is obtained through higher education. However, formal education is not a sole source of human capital. In fact, recent studies suggest that in the arctic human capital is less related to formal levels of schooling than it is in the south (Petrov and Cavin 2013). This difference is very important and is attributable to the human capital associated with skills and knowledge, not obtained through attending educational institutions.

Human capital is necessary to maintain a stable or expanding economic base and ensure human well-being. Education is an integral part of human development and well-being as identified both by the *UN Human Development Reports* (UN Development Program 2014) and the Arctic Social Indicators project (ASI 2010). These studies emphasize the role of education in ensuring economic well-being, fate control (empowerment), and cultural continuity, especially if standard schooling practices are intertwined with local and traditional contexts. *Arctic Social Indicators* (2010) named education as one of the six key domains of human development in the Arctic.

16.2.2 Human Capital and Sustainable Economic Development

Human capital is a crucial factor of regional economic growth and development and a key attribute of a modern post-industrial economy. First, human capital is the most important ingredient in the "knowledge sector," which includes technologically advanced industries and services (e.g. information technology, high tech manufacturing, financial services, etc.), the most intensively growing elements of modern economy that define the overall success and competitiveness of regional economic systems in a globalizing world. Second, knowledge underpins 'old' industries, including the whole array of primary (e.g., extractive industries, agriculture), secondary (manufacturing), and tertiary (services) sectors. In other words, the modern economy is an economy heavily based on knowledge (Bell 1973).

Studies show that in the European Union, for each additional year of schooling, an individual gains 6.5 % in wages. The same is true in North America (De la Fuente and Ciccone 2002; Psacharopoulos and Patrinos 2004). This figure is even higher for Indigenous residents: in Saskatchewan, Canada, the lifetime earnings of an Indigenous male were found to increase by 38 % and for female by 59 % if a university degree was completed (Howe 2011). Higher education levels are also associated with greater labor force participation and lower unemployment. Schooling has a role in reducing poverty and providing the means for economically disadvantaged groups to improve their standard of living (ASI 2010). In Canada, studies show that the completion of university education is the most financially rewarding in terms of improving the earnings of Aboriginal Canadians, followed by the completion of non-university postsecondary education (Hossain and Lamb 2012). In addition to benefitting an individual, investment in education also increases overall productivity by as much as 5 % per additional year of schooling (De la Fuente and Ciccone 2002) and, thus, improves the competitiveness of national and regional economies. Other positive outcomes of education are felt both in households and society at large, such as the increased ability of educated women to manage their lives and financial situations through reproductive control (Oxaal 1997).

16.2.3 Human Capital and Sustainable Economic Development in the Arctic

While the reliance of arctic economies on the resource and public sectors is persistent, globalization has brought new opportunities and challenges for arctic communities in the new knowledge-driven global economy. Northern regions can become "learning" regions (Morgan 1997) that adopt and adapt innovations while developing their own body of economically relevant knowledge and skills based on local experiences and traditions. In this respect, a development strategy based on enabling local human capacities to advance economic development is appealing. However, the Arctic faces formidable challenges to becoming such a region: internally-generated growth is inhibited by limited local capacities (institutional, financial and infrastructural) and, most importantly, by the shortage of human capital (Petrov 2011).

There is a conceptual argument about why human capital is central to ensuring sustainable economic development in the Arctic and other peripheral regions around the world (see Petrov 2014). It is typical for peripheral regions, which rely heavily on resources or the public sector to: develop a culture of dependency that discourages local entrepreneurship and innovativeness (Polèse et al. 2002; Suorsa 2009); be disconnected from communities and networks of practice (Lagendijk and Lorentzen 2007); be forced to follow rigid techno-economic trajectories (Clark et al. 2001). All these structural conditions impede the acquisition of the knowledge and capacities necessary to achieve sustainable development in the face of the knowledge-based, ultracompetitive, globalized economy of the present (and future).

Human agency is a key transformative factor of the economy: agents of transformation are critical and necessary components of development. These agents can be political institutions, firms, or non-governmental organizations. However, in the end, the agents of change are individuals and their groups who 'write' the economic history of the region. In other words, human capital, and particularly *creative capital* (a stock of creative abilities and knowledge that has economic value), effectively define the region's future in a modern (and future) economy.

The importance of human and creative capital in transforming remote regions has been demonstrated in a number of studies (Aarsæther 2004; Copus and Skuras 2006; Jauhiainen and Suorsa 2008; Hall and Donald 2009; Petrov 2007; Petrov and Cavin 2013). Some researchers have even observed that less favorable business and social environments amplify the importance of creativity and human capital (Aarsæther 2004; North and Smallbone 2000). Evidence from northern success stories suggests that the economic returns of human capital tend to be more connected to local economies. This connectedness is partially determined by the nature of the knowledge-based economy in general, but also by the tight relationship between human capital and other forms of societal capital in the periphery (Petrov 2011). In addition, communities can capitalize on Indigenous knowledge and tradition, and facilitate institution building and the formation of civic society.

16.2.4 Human Capital and Fate Control

The link between education and empowerment is another important element of achieving sustainability in arctic communities. One aspect is the empowerment of individuals to pursue education (whether western or traditional) and develop an ability to make informed decisions about various aspects of their lives. These individuals may be empowered to fulfill important functions in the community. *Arctic Social Indicators I* (2010) offered a new way to conceptualize empowerment through the notion of fate control (i.e. the ability of individuals and communities to define their own destiny). Education is an integral part of fate control. Stronger fate control may strengthen communities, making them more resilient and adaptable. It may, however, eventually result in unexpected and unwanted out-migration patterns, and thus the loss of the new human capital gained, for example, through the expansion of educational opportunities (Rasmussen 2011). Another possible consequence of formal education is the erosion of cultural identity and loss of contact with nature (ASI II 2014; Battiste 2000).

16.2.5 Sustaining Human Capital

A common problem of non-metropolitan, peripheral regions is the 'flight' of human capital. With an increased level of education the ability (and desire) of local residents to find new job or educational opportunities elsewhere also increases. Many arctic residents, particularly women, move away from the Arctic to pursue or use their education. At the same time, many arctic regions are attracting human capital from the south as skilled professionals take advantage of high earnings in certain industries (mining, oil, etc.). Unfortunately, however, most of them stay in the Arctic only for a limited time. Departing educated Indigenous northerners and returning 'newcomers' create a 'brain drain' from the Arctic (Handland 2004; Petrov and Vlasova 2010). In many regions this coincides with a 'brain turnover' (intensive in- and out-migration of human capital) and 'brain waves' (surges and dips of human capital associated with the boom-and-bust economic cycle) (Heleniak 2010). The study of human capital mobility may illuminate the ways in which human capital can be retained in place and/or attracted (back) to the Arctic. With the growing access to education for arctic residents in their regions and elsewhere, the issue of retention becomes even more critical (Rasmussen 2011).

16.3 Human Capital and Post-secondary Education in the Arctic: Patterns, Trends and Gaps

16.3.1 Higher Education Opportunities in the Arctic

Access to post-secondary education is a key component in harnessing human capital. Today, there are more educated people in the Arctic than a decade ago, and options for post-secondary training in the regions have increased. In recent decades, arctic regions have seen an increase in the number and capacity of post-secondary institutions. Colleges and universities exist in all major arctic jurisdictions, although there are more in the Russian and Nordic sectors (Fig. 16.1). Umeå University, the University of Alaska system, Lulea University, University of Oulu, University of Iceland, University of Tromsø, and Murmansk State Technical University are the leading arctic institutions in terms of enrollment.[1] All of these universities are located in larger cities, although some have branch campuses in rural areas. Remote regions are served to a much lesser degree. In most arctic colleges and universities, female students constitute the majority of enrollment. Higher education institutions serving indigenous populations exist in various jurisdictions, e.g., Sámi University College (Norway), Taimyr College (Russia), Nunavut Arctic College (Canada), and Ilisagvik College (Alaska). In addition to national systems of post-secondary education, an international and inter-institutional higher education framework has been built by the University of the Arctic.

However, regional differences within the Arctic persist: post-secondary educational opportunities are still more extensive in northern Europe and Russia than in northern Canada and Alaska. In addition, the ability of Indigenous people to access institutions of higher learning albeit improving, is still problematic (Rasmussen 2011). Many arctic residents are still compelled to pursue education elsewhere, and their return to their home region after completing their education is far from guaranteed.

16.3.2 The Geography of Arctic Human Capital

Educational attainment (a level of education achieved by a person or a group) is a traditional proxy measure of human capital. The differences in post-secondary educational policies and the legacies of national systems make it difficult to compare jurisdictions. However, we can see emerging patterns that supersede national boundaries and historical legacies. High levels of education are often observed in

---

[1] For this particular assessment we only included colleges and universities in cities located in the Canadian territories, Alaska, Greenland, northern Norway, Sweden and Finland, as well as in Murmansk Oblast', and Nenets, Yamal-Nenets, Taimyr and Chukchi autonomous okrugs of Russia. We also considered only institutions with publically available enrollment data.

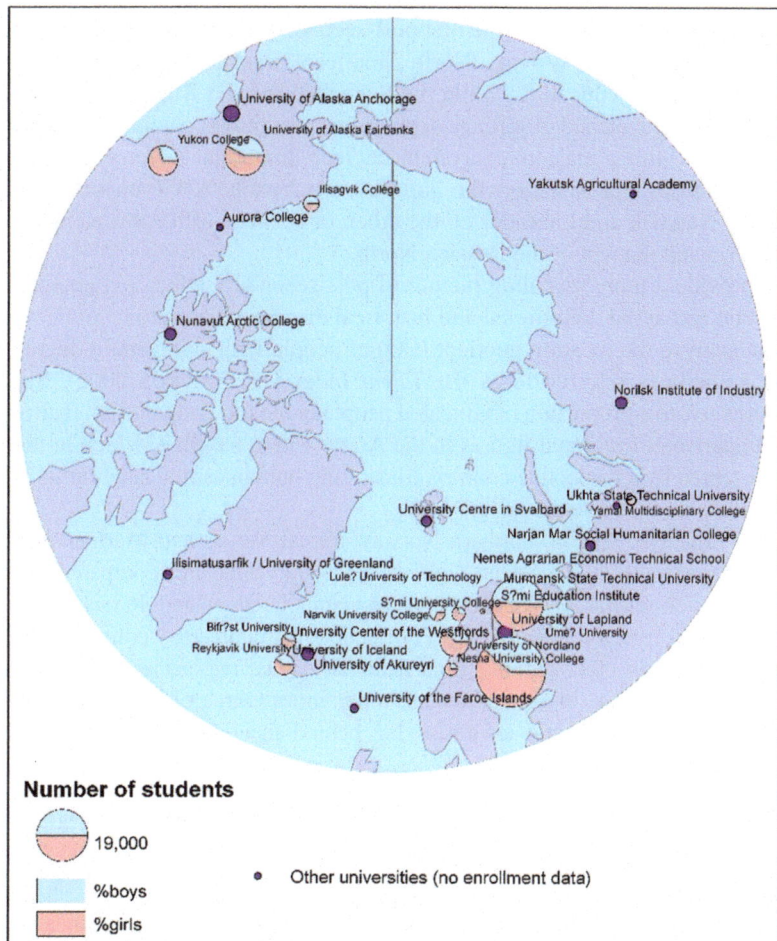

Fig. 16.1 Selected arctic colleges and universities: Enrollment and gender balance (*Source*: institutional websites)

booming resource regions, such as the Yamal-Nenets Autonomous *Okrug* (District) in Russia, and in more diversified regional economies, such as Yukon Territory in Canada. Higher levels of educational attainment are also seen in the more urbanized parts of the Arctic, e.g., Murmansk *Oblast'* (region) of Russia, northern Scandinavia, and Iceland. Studies indicate that the bulk of educated professionals are "newcomers" or non-Indigenous residents (although there are some exceptions) mostly working in the resource sectors and public administration (Rasmussen 2011; Voswinkel 2012). Human capital is also heavily concentrated in urban settlements, such as Tromsø in northern Norway, Rovaniemi in Finland, Murmansk, Salekhard, Magadan and Noril'sk in northern Russia, Whitehorse and Yellowknife in Canada, and Anchorage, Fairbanks and Juneau in Alaska.

Tertiary education (beyond the first post-secondary degree) is skewed by the differences in educational systems. North American jurisdictions have high levels of post-baccalaureate education, while the Scandinavian and Russian North demonstrate a lower prevalence of tertiary degrees because post-secondary education frequently ends with a masters-equivalent degree.[2] Important differences are found within countries: for instance, the gap between Yukon, NWT on one hand and Nunavut, Nunavik and Labrador on the other, or between affluent regions of western Siberia and the rest of the Russian North.

To provide a more revealing picture of post-secondary education attainment in the Arctic and avoid definitional and historical discrepancies among arctic regions, we can analyze the location quotient (LQ) of people with a university degree (Fig. 16.2), sometimes referred to as the *Talent Index (TI)*[3] (Florida 2002). Since TI measures *relative* proportion of educated people in the total population, it allows for the comparison of regional figures in the Arctic with a baseline, which in this case is represented by a respective country, illustrating human capital accumulation relative to the country's average (Fig. 16.2).

More urbanized Yukon, northern Norway, Yamal-Nenets and Murmansk regions have higher proportion of residents with post-secondary education compared to the rest of Russia, as do several Alaskan boroughs and Nuuk in Greenland. In fact, some arctic regions demonstrate a relatively high TI of people with university education. High TI is also observed in other regional (and national) capital regions as well as booming resource frontiers (e.g., in Yamal-Nenets Autonomous Okrug of Russia). At the same time, the majority of arctic territories lag behind their southern metropoles with

---

[2] In Russia and Scandinavia undergraduate university degrees historically require more time than in North America and may lead to roughly the same qualification as North American masters degree. However, according to the definition, these degrees are not considered tertiary since they represent a first post-secondary degree for students.

[3] Location quotient (LQ) refers to a relative proportion of adults in a certain occupation or education group compared to a baseline (national level). It is an advantageous measure, because it compares all regions (communities) with a single common denominator (their country's baseline), whether a national benchmark or some other chosen indicator. Most measures are computed for the labour force of 15 years and over.

$$LQ_i = \frac{\lambda_n}{\lambda_C},$$

where LQ_i is a location quotient of i (occupation, education, etc.), λ_n is the share of population having the measured characteristic i in region n and λ_C is the share of population having the same characteristic in the reference region (nation).

The following indices are utilized in this study:

Talent Index (TI) – is a LQ of adult population who have a university degree.
Bohemian Index (BI) – is a LQ of people with artistic and creative occupations.
Leadership Index (LI) – a LQ of people with leadership and managerial occupations.
Entrepreneurship Index (EI) – a LQ of people with business occupations.
Applied Science Index (ASI) – a LQ of people with occupations in applied and natural science, computer science and engineering (not used in this study due to data constraints).

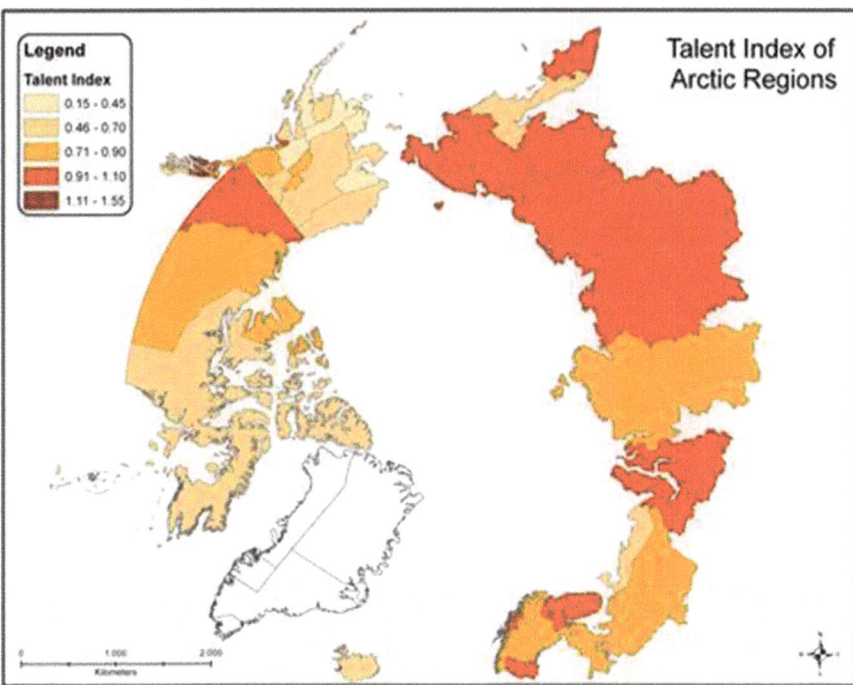

Fig. 16.2 Talent Index: Location quotient of people with university degree in arctic regions (*Source*: compiled by author from U.S. Census Bureau, Statistics Canada, Statistics Norway, Statistics Greenland, Rosstat, Eurostat; various years)

respect to formal education levels. This gap is especially evident in areas with relatively high Indigenous populations, such as Nunavut, rural Alaska, and the Nenets, Koryak and Taimyr Okrugs of Russia. Small remote urban communities, such as Iqaluit, Dudinka, Tura, and Susuman also demonstrate low levels of educational attainment (Fig. 16.3). A slightly higher, but still relatively low Talent Index is observed in the 'old' industrial cities of the Russian North: e.g. Noril'sk, Apatity, Olenegorsk, Monchegorsk, and Vorkuta.

16.3.3 Beyond Regions: A Municipal-Level Analysis

If human capital metrics are well documented at the regional level, data constraints limit our ability to measure human capital at the municipal level. For larger settlements, however, the educational attainment data required for computing *Talent Index* are mostly accessible. At the city-level, analysis includes arctic cities selected based on population (generally exceeding 20,000) and "regional importance" (all regional capitals, if available, are included).

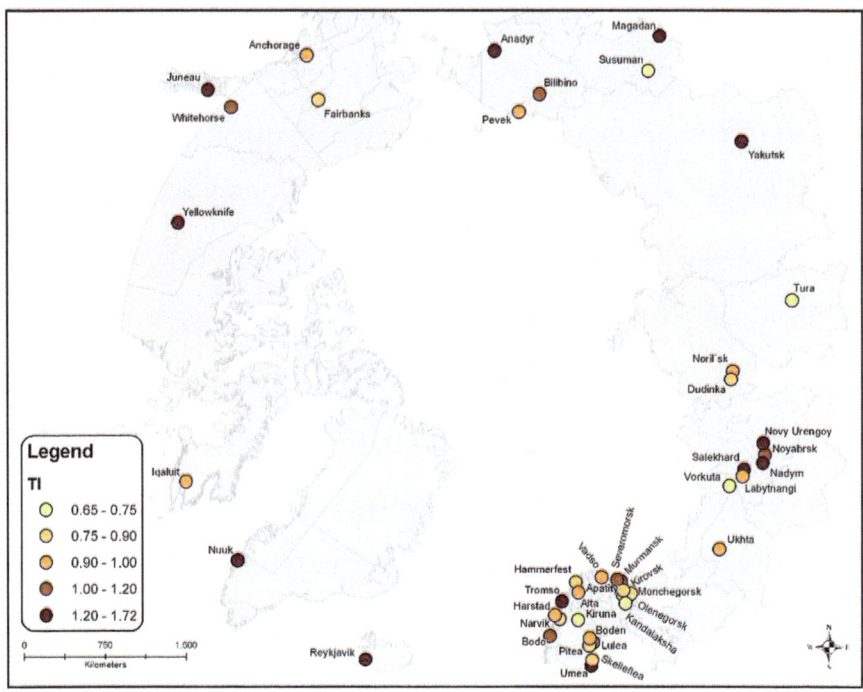

Fig. 16.3 Talent Index (TI) in selected arctic cities (*Source*: compiled by author from U.S. Census Bureau, Statistics Canada, Statistics Norway, Statistics Greenland, Rosstat, Eurostat; various years)

As seen in Fig. 16.3, arctic cities demonstrate varying degrees of 'talent' concentration in comparison to the national level. Some are certainly 'creative hot spots': for example, Anadyr's TI (1.72) is comparable to Moscow's (1.79). Very high TIs are also recorded in other regional (and national) capitals both in Russia and across the Arctic, including in Salekhard, Yakutsk, Umeå, Magadan, Juneau, Yellowknife, Tromsø, and Reykjavik. Another large, highly educated labor force cluster is observed in the Yamal-Nenets Autonomous Okrug, where Salekhard, Novy Urengoy and Nadym have TIs above 1.4. This may reflect the influx of educated labor migrants in the last decade as TIs in these cities exhibited substantial growth between 2002 and 2010.

These observations should not disguise a wide gap in educational attainment (and thus TI) between urban and rural areas in the Arctic. For example, in the most urbanized part of the Arctic, the Russian North, the percent of formally educated individuals with a post-secondary degree varies from region to region, from 5 to 20 %. Many other northern cities and towns, however, have low levels of educational attainment, especially compared to the South.

16.3.4   Gaps and Trends in Arctic Human Capital

16.3.4.1   A Double Spatial Gap

A simple geographic analysis demonstrates *two spatial gaps* in arctic post-secondary education: the gap between the Arctic and southern regions and the gap between urban/industrial arctic regions and the rest of the Arctic. Generally, a pattern of low post-secondary education levels in the North is prevalent. However, such a generalization fails to reflect the variability and diversity of arctic regions. Recent research demonstrates substantial levels of creativity based on non-codified informal knowledge, which might not conform the conventional notion of human capital (Aarsæther 2004; Petrov 2007; Petrov and Cavin 2013). As argued above, creative human capital is critical for economic development and socio-economic transformation in the Arctic, as it often becomes the engine of the economic reinvention and revitalization of a region.

16.3.4.2   A Gender Gap

Considerable changes in gender patterns of education took place during the last 10–15 years. Although the gender gap in education still persists in many regions, women tend to be increasingly more educated as a group compared to men. By the 1990s, women had become the majority group in relation to higher education in several countries, and by the late 1990s, this occurred throughout most regions in the Arctic. Alaska is one region where the feminization of human capital over the last decade has led to an increase in women with post-secondary education (52.3 % as compared to 47.7 % among men). Alaskan women are also more highly educated: 29.8 % of female Alaskans have post-secondary education as opposed to 25.2 % of males. At the same time, men lead among adult residents with post-baccalaureate degrees (especially in professional and doctoral degrees, where the male to female ratio is 1.5:1) (U.S. Census Bureau 2012).

Data from Northwest Territories (NWT) show that while there are still more men with post-secondary education, women experienced steady gains in the last ten years and by 2009 constituted 46 % of post-secondary educated population over the age of 15 (NWTBS 2009). In Yukon, as many women as men have post-secondary certificates (YBS 2008). In recent decades women also gained considerable ground in some male-dominated fields such as engineering, applied sciences, mathematics, computer sciences and physical sciences. In the Russian Arctic, women dominate among residents with higher (post-secondary) education. The gap between men and women is fairly substantial: for example, 24.5 % of adult female residents of Murmansk Oblast' have completed a post-secondary degree, compared to 20.8 % of male residents. In Yamal-Nenets Autonomous Okrug, a region with quite a different population structure and settlement history, the gender gap is even more pronounced: post-secondary education is attained by 31.1 % of women and 22.8 % of men.

16.3.4.3   An Indigenous/Non-indigenous Gap

The levels of engagement in post-secondary education among Indigenous and non-Indigenous people illustrate another 'education gap.' Closing this gap is not only beneficial for Indigenous individuals in terms of lifetime earnings (Howe 2011; Hull 2005), but will likely to inject considerable revenues into the economy (Sharpe et al. 2007). The gap is dramatic in some regions, and less evident (but still present) in others. NWT is a case in point as a region with a drastic education differential: only 4.9 % of NWT Aboriginal residents held a university degree in 2009 compared to 32.3 % of non-Aboriginal residents (NWTBS 2009). Although Indigenous people in NWT made formidable gains in the last decades (only 1.8 % had university degree in 1999), the educational gap is still wide (NWTBS 2009).

In Russia, the discrepancy between Indigenous and non-Indigenous populations can be approximated using statistics for urban (predominantly non-Indigenous) and rural (heavily Indigenous) areas. The gap in percent of formally educated individuals with a post-secondary degree varies from region to region between 5 and 20 %. For example, only 14.3 % of rural residents in Yamal-Nenets Okrug have attained the post-secondary level compared to 29.2 % of urban dwellers. Similarly, in Chukotka 7.6 % of adults living in rural areas reported having higher education in contrast to 28.4 % of adults living in urban settings. Only 5.9 % of rural men (mostly Indigenous) in Chukotka had a post-secondary degree, reflecting both the urban/rural and Indigenous/non-Indigenous education gaps (Rosstat 2012).

16.3.4.4   A Formal/Informal Education Gap

Recent studies revealed that human capital in the Arctic is not always associated with the western educational paradigm (e.g., Petrov 2007). This is especially true for cultural capital, a form of human capital based on the economic engagement of cultural activities and practices. Another strong economic endowment associated with informal knowledge transfer is the social economy, which is based on long-lived traditions of cooperation, sharing, and community-building to sustain prosperity. This knowledge is useful in the growing social economy of the Arctic (AHDR 2014).

The gap between formal and informal education in the Arctic is problematic for a region's sustainable development. Arctic communities have informal, highly adaptive traditional knowledge systems (Cruikshank 2005). These systems include a unique combination of human, social and civic capital that provides opportunities for enhancing material well-being while simultaneously ensuring cultural vitality and sustaining traditional livelihoods and the natural environment. The minimal role of traditional knowledge in formal education and the lack of recognition given to it as a component of human capital of the arctic regions continues to limit sustainable development options in northern communities. Integrating traditional knowledge transfer into the formal educational system, incorporating informal education in human capital metrics and establishing an improved understanding of its role in economic developments are all required to close the gap.

16.3.5 Beyond Education: Arctic's Creative Capital and the 'Entrepreneurial Gap'

The centrality of human capital in regional reinvention in the periphery appears to be especially important in respect to breaking with resource path-dependency and facilitating a sustainable economic future for remote regions. Community-level research conducted in peripheral areas, although mostly outside the Arctic, squarely points to a pivotal role of creativity (spanning beyond education, experience or technical expertise or any other "traditional" attributes of human capital) in local economic success. For example, a study of local innovation in northern Scandinavia stressed "the importance of key local actors in innovative processes that take place in remote regions" (Aarsæther 2004: 244). The authors concluded, "almost every innovation has had a clear core agent to manage the process. Very often this agent, initiator and 'engine' of the process has been a local person, who has committed him/herself to the development of a new idea" (Aarsæther 2004:244). Similar evidence has been cited in other marginal regions (e.g., Stöhr 2000; Petrov 2011), where entrepreneurs and inventors have been credited with revitalizing economies in their communities. All these studies suggest that human capital in general, and *creative capital* in particular, are important and organic ingredients of local development in the periphery.

Figure 16.4 presents *Leadership Index (LI), Entrepreneurial Index (EI)* and *Bohemia Index (BI)* maps for the circumpolar region. Similarly to TI, these indices are calculated as location quotients of residents in leadership, entrepreneurial and artistic occupations respectively. Since these are occupational measures, the indices not only reflect formal education, but occupational skills, and thus are more inclusive of various forms of human capital. To some extent they capture skills based on traditional knowledge (mostly in terms of artisanship and leadership); however, admittedly, the Census occupation categories do not always adequately encapsulate high-skilled traditional activities.

Taken together with Fig. 16.2 (TI), it is evident that there are regions that have high TI, LI and BI (and they are not necessarily the same ones). As mentioned earlier, Yukon, certain parts of Russia (e.g., Murmansk and Yamal-Nenets), and northern Scandinavia demonstrate levels of TI near or exceeding respective national averages. Yamal-Nenets Autonomous Okrug and Kamchatka Oblast' were ranked 9th and 10th among top Russian regions in 2002. At the same time, some arctic regions register a remarkably high LI; a pattern observed in other studies, (e.g., Petrov and Cavin 2013). The highest indices are associated either with larger urban and administrative centers or with very remote and sparsely populated regions. The geographic distribution of BI largely reflects the prevalence of an Indigenous

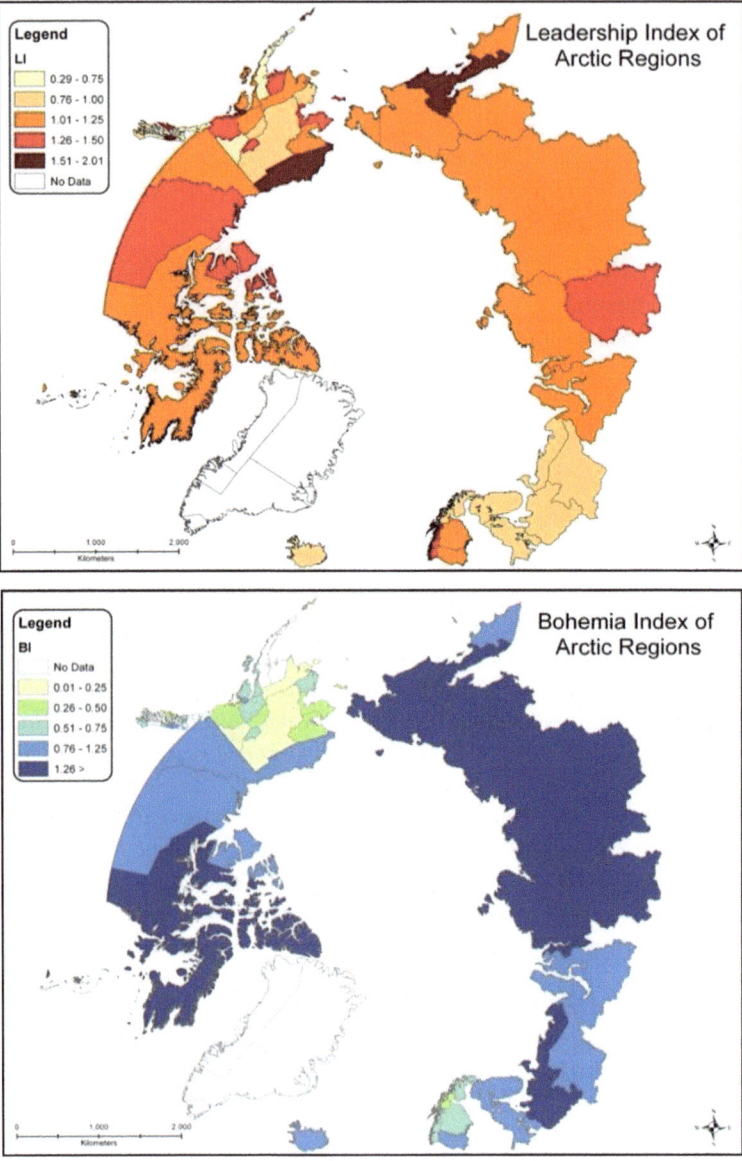

Fig. 16.4 Bohemia (BI), Leadership Index (LI), and Entrepreneurial Index (EI) Index in the arctic regions (*Source*: compiled by author from U.S. Census Bureau, Statistics Canada, Statistics Norway, Statistics Greenland, Rosstat, Eurostat; various years)

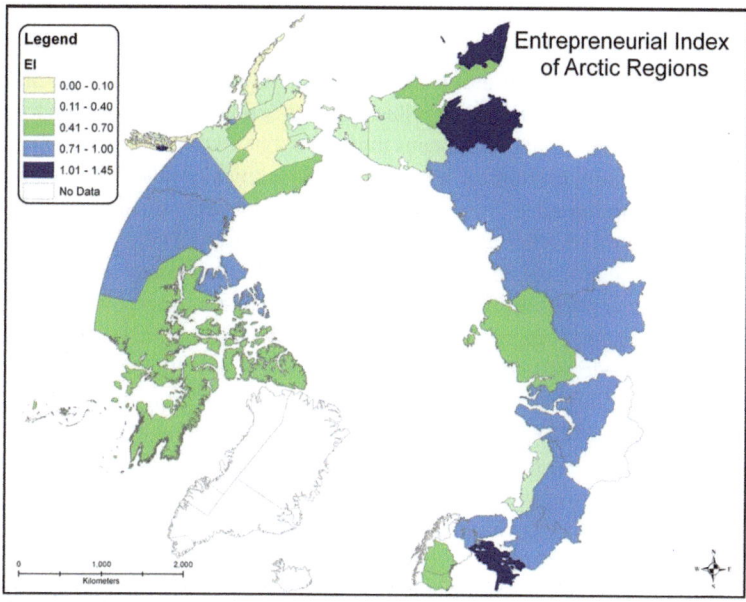

Fig. 16.4 (continued)

population. Most arctic regions exceed national baselines in relative proportion of residents with occupations in arts and culture suggesting a presence of cultural capital and a considerable potential of an arctic cultural economy. In Russia, the Taimyr, Koryak, Chukotka Autonomous Okrugs and Republic of Sakha (Yakutia) ranked among top 10 regions in terms of BI in 2002.

In contrast, the Entrepreneurship Index in the Arctic is generally low. The entrepreneurial class in most of the circumpolar regions is rather weak, and demonstrates a *gap* versus other forms of creative capital. This fact presents an additional challenge for sustainable economic development, because the lack of entrepreneurial activity becomes an impediment for engaging other types of human capital in the production of economic wealth. Other studies also point to the "peripheral disconnect" or lack of synergy between the entrepreneurial capital and other forms of human creativity (Petrov and Cavin 2013). Investing in all forms of *local* human capital is one of the elements of a successful economic development strategy in the northern frontier.

16.4   Discussion and Concluding Remarks

This brief analysis demonstrates that although the educational opportunities and educational attainment of arctic residents has increased, there are still serious shortages of human capital in circumpolar regions. We are able to observe a variety of 'gaps' (some of them persistent, and others emerging) among arctic regions and populations. The spatial gaps represent the difference in educational attainment (and opportunities) between arctic and southern regions, and between urban and rural Arctic places. The gender gap reflects the tendency of arctic women to achieve higher levels of educational attainment than men, but also the continuing male predominance in post-baccalaureate education and technical fields of knowledge. The Indigenous/non-Indigenous gap illustrates a persistent discrepancy between Indigenous people and other northerners in respect to formal education. The lack of integration between formal and informal education is also a continuing problem that marginalizes arctic populations. Finally, the entrepreneurial gap points to a shortage of local entrepreneurs compared to other types of creative capital (e.g., cultural). All of these gaps impede sustainable social and economic development in the Arctic.

The internal differences illuminate a few successful regions (and mostly urban communities that serve as administrative and resource centers) against the backdrop of regions with limited human capital, brain drain and brain turnover. However, the "formal education-only" view of human capital does not adequately present the full picture of human capital. Other forms of human creativity that correspond with knowledge and skills within or beyond those attained through formalized education must be considered. This consideration makes the palette of human capital more complete and changes the overall understanding of the geography of human capital in the Arctic. This is most important in respect to cultural capital that is least tied to formal education, but may not be less vital for sustainable community development.

Engaging human capital and the knowledge economy must be an integral part of a larger *sustainable development strategy* for the Arctic. Addressing human capital gaps provides a new opportunity for northern communities to achieve the diversification of their economic base, break away from the boom-bust cycles, reduce dependency on external economic and political actors and ultimately improve the quality of life for local residents. As we start to quantify and understand human capital in the Arctic, it is evident that some arctic regions have considerable concentrations of highly educated workers and creative professionals. These individuals are predominantly located in administrative and economic centers, such as Yellowknife, Juneau, Salekhard and Anadyr. Some of these cities rival their much larger southern counterparts such as Moscow, St. Petersburg and Krasnoyarsk. On the other hand, remote communities still have very limited human capital. While unavoidable, the leakage of students and highly educated professionals from the North is a problem that can be partially alleviated if cities, regions and national actors introduce meaningful efforts to attract and retain human capital in the North. Without such actions even the most highly ranked arctic regions and communities may lose their potential in times of economic difficulties (as has been demonstrated in the Russian and Canadian Norths in the 1990s). Investment in education is important, but it should come hand-in-hand with policies targeting human capital that are based on a better understanding of its needs and behavior in the Arctic. This pertains both to professionals and to local artisans and crafters (the pillars of Arctic cultural economy). Although not all arctic regions can strongly benefit from building a knowledge economy, it is certainly a key ingredient necessary for achieving sustainable development in northern communities that will also allow for the partial reconciliation of traditional economies and cultures with modern capitalism. In other words, by sustaining creativity cities will be able to create sustainability.

Acknowledgement This research was partially supported by NSF PLR# 1338850 and 1360365.

References

Aarsæther, N. (Ed.). (2004). *Innovations in the Nordic periphery.* Stockholm: Nordregio.

AHDR. (2004). *Arctic human development report.* Akureyri: Stefansson Arctic Institute.

AHDR. (2014). Arctic human development report II: Regional processes and global linkages. (J. Larsen & G. Fondahl, Eds.). Copenhagen: Nordic Council of Ministers.

ASI I. (2010). Arctic social indicators – A follow-up to the *Arctic Human Development Report* (J. N. Larsen, P. Schweitzer, & G. Fondahl, Eds.). Copenhagen: Nordic Council of Ministers.

ASI II. (2014). Arctic social indicators. Implementation. (J. N. Larsen, P. Schweitzer, & A. Petrov Eds.). Copenhagen: Nordic Council of Ministers.

Battiste, M. (2000). Maintaining Aboriginal identity, language, and culture in modern society. In M. Battiste (Ed.), *Reclaiming indigenous voice and vision* (pp. 192–208). Vancouver: UBC Press.

Bell, D. (1973). *The coming of post-industrial society.* New York: Basic Books.

Clark, P., Tracey, P., & Lawton Smith, H. (2001). Agents, endowments, and path-dependence: A model of multi-jurisdictional regional development. *Geographische Zeitschrift, 89,* 166–181.

Copus, A., & Skuras, D. (2006). Accessibility, innovative milleux and the innovative activity of businesses in the EU peripheral and lagging areas. In T. N. Vaz, E. J. Morgan, & P. Nijkamp (Eds.), *The New European rurality: Strategies for small firms* (pp. 29–40). Aldershot: Ashgate.

Cruikshank, J. (2005). *Do glaciers listen? Local knowledge, colonial encounters and social imagination.* Vancouver/Seattle: UBC Press/University of Washington Press.

De la Fuente, À. & Ciccone, A. (2002). *Human capital in a global and knowledge based economy.* Final report. European Commission: Employment and Social Affairs.

Florida, R. (2002). The economic geography of talent. *Annals of the Association of American Geographers, 94*(2), 743–755.

Hall, H. & Donald, B. (2009). *Innovation and creativity on the periphery: Challenges and opportunities in Northern Ontario* (Ontario in the creative age Working Paper Series). Toronto: Martin Prosperity Institute: REF: 2009-WPONT-002.

Handland, J. (2004). Alaska's "brain drain": Myth or reality? *Monthly Lab Review, 127,* 9.

Heleniak, T. (2010). Migration and population change in the Russian Far North during the 1990s. In C. Southcott & L. Huskey (Eds.), *Migration in the circumpolar North: Issues and contexts* (pp. 57–91). Edmonton: Canadian Circumpolar Institute Press.

Hossain, B., & Lamb, L. (2012). The impact of human and social capital on aboriginal employment income in Canada. *Economic Papers: A Journal of Applied Economics and Policy, 31*(4), 440–450.

Howe, E. (2011). Bridging the Aboriginal education gap in Saskatchewan. *Gabriel Dumont Institute of Native Studies and Applied Research, 6.*

Hull, J. (2005). *Post-secondary education and about market outcomes Canada, 2001.* Winnipeg: Prologica Research Inc.

Jauhiainen, J. S., & Suorsa, K. (2008). Triple helix in the periphery: The case of multipolis in northern Finland. *Cambridge Journal of Regions, Economy and Society, 1*(2), 285–301.

Lagendijk, A., & Lorentzen, A. (2007). Proximity, knowledge and innovation in peripheral regions. On the intersection between geographical and organizational proximity. *European Planning Studies, 15*(4), 457–466.

Morgan, K. (1997). The learning region: Institutions, innovation and regional renewal. *Regional Studies, 31*, 491–503.

North, D., & Smallbone, D. (2000). The innovativeness and growth of rural SMEs during the 1990s. *Regional Studies, 34*(2), 145–157.

NWTBS, Northwest Territories Bureau of Statistics. (2009). *Education: Highest level of schooling.* Retrieved September 2, 2014, from http://www.statsnwt.ca/education/highest-level/

OECD. (2009). Human capital and its measurement. In: D-B Kwon (Ed.), *The 3rd OECD World Forum on "Statistics, Knowledge and Policy": Charting progress, building visions, improving life* (pp. 1–15). OECD World Forum.

Oxaal, Z. (1997). *Education and poverty: A gender analysis.* Sussex: Institute of Development Studies at the University of Sussex.

Petrov, A. (2007). A look beyond metropolis: Exploring creative class in the Canadian periphery. *Canadian Journal of Regional Science, 30*(3), 359–386.

Petrov, A. (2011). Beyond spillovers: Interrogating innovation and creativity in the peripheries. In H. Bathelt, M. Feldman, & D. F. Kogler (Eds.), *Beyond territory: Dynamic geographies of innovation and knowledge creation* (pp. 168–190). New York: Routledge.

Petrov, A. (2014). Creative Arctic: Towards measuring Arctic's creative capital. In L. Heininen (Ed.), *Arctic yearbook* (pp. 149–166). Akureyri: Northern Research Forum.

Petrov, A., & Cavin, P. (2013). Creative Alaska: Creative capital and economic development opportunities in Alaska. *Polar Record, 49*(4), 348–361.

Petrov, A., & Vlasova, T. (2010). Migration and socio-economic well-being in the Russian North: Interrelations, regional differentiation, recent trends and emerging issues. In L. Huskey & C. Southcott (Eds.), *Migration in the circumpolar North: New concepts and patterns.* Edmonton: CCI Press.

Polèse, M., Shearmur, R., Desjardins, P. M., & Johnson, M. (2002). *The periphery in the knowledge economy: The spatial dynamics of the Canadian economy and the future of non-metropolitan regions in Quebec and the Atlantic Provinces.* Montreal: INRS – Urbanisation, Culture et Societe.

Psacharopoulos, G., & Patrinos, H. A. (2004). Returns to investment in education: A further update. *Education Economics, 12*(2), 111–134.

Rasmussen, R. O. (2011). *Megatrends.* Copenhagen: Nordic Council of Ministers.

Rosstat. (2012). *Ekonomicheskie i Sotsialnye Pokazateli Rayonov Kraynego Severa i Priravnennykh k Nim Mestnostei 2000–2011* [Economic and social indicators in the territories of the far north and equated areas 2000–2011]. Moscow: Rosstat.

Sharpe, A., Arsenault, J., & Lapointe, S. (2007). *The potential contribution of Aboriginal Canadians to labour force, employment, productivity and output growth in Canada, 2001–2017.* Ottawa: Centre for the Study of Living Standards. Retrieved from http://www.csls.ca/reports/csls2007-04.PDF

Stöhr, W. B. (2000). Local initiatives in peripheral areas: An intercultural comparison between two case studies in Brazil and Austria. In *Developing frontier cities* (pp. 233–254). Dordrecht: Springer.

Suorsa, K. (2009). *Innovation systems and innovation policy in a periphery.* Nordia Geographic Publications 38(4). Oulu: University of Oulu.

United Nations (UN) Development Program. (2014). *Human development report.* Retrieved from http://hdr.undp.org/en/2014-report

U.S. Census Bureau. (2012). *American community survey.* Retrieved from http://www.census.gov/acs/www/

Voswinkel, S. (2012). *Survey of Yukon's knowledge sector.* Whitehorse: Yukon Research Centre.

YBS (Yukon Bureau of Statistics). (2008). Education. Information sheet #C06-09. Retrieved September 2, 2014, from http://www.eco.gov.yk.ca/stats/pdf/2006censuseducation.pdf

Part III
Advancing Sustainability

Chapter 17
From Lone Wolves to Relational Reindeer: Revealing Anthropological Myths and Methods in the Arctic

Stacy Rasmus and Olga Ulturgasheva

A Malemut shaman from Kotzebue sound near Selawik lake told me that a great chief lives in the moon who is visited now and then by shamans who always go to him two at a time, as one man is ashamed to go alone... (Nelson 1900: 515)

Abstract This chapter examines how a new methodological approach, *peer observation of research*, was used as part of a comparative, ethnographic study of social resilience in Alaska and Siberia. The approach evolved through the collaboration of two indigenous researchers working in two different regions of the Arctic. Breaking from what has become a standard auto-ethnographic or self-reflexive enterprise in anthropology, our study aimed to document the collaborative ethnographic interaction from multiple perspectives and positions. We present two fieldwork episodes demonstrating the process and potential utility of a peer observation method for social researchers working in collaboration with indigenous communities and people in the Arctic. Peer observation of research reveals: (1) the ways in which our methods and models of collaborative research are relational and negotiated within an indigenous community and cultural context and (2) the degree to which our own indigenous kinship and association influences our ethnographic outcomes in a fieldwork setting leading to productive points of orientation and disorientation.

Keywords Collaborative ethnography • Indigenous research methodologies • Arctic indigenous engagement in research • Peer observation of research • Arctic social science

S. Rasmus (✉)
Center for Alaska Native Health Research, University of Alaska Fairbanks,
Fairbanks, AK, USA
e-mail: smrasmus@alaska.edu

O. Ulturgasheva
Department of Social Anthropology, University of Manchester, Manchester, UK
e-mail: olga.ulturgasheva@manchester.ac.uk

© Springer International Publishing Switzerland 2017
G. Fondahl, G.N. Wilson (eds.), *Northern Sustainabilities: Understanding and Addressing Change in the Circumpolar World*, Springer Polar Sciences,
DOI 10.1007/978-3-319-46150-2_17

223

17.1 Introduction

Contrary to popular belief, no one is every truly alone in the arctic and activities seldom go unseen or unobserved whether it be by another human being, or by an animal or spirit. There is a Yup'ik saying that it is always good where there are two, two hunters or two Elders speaking together, as each will keep the other true. The quote above is a from an oral tradition recounted in an early ethnography of Yup'ik[1] groups living on the Bering Sea Coast and speaks to this cultural value in traveling and presenting oneself with another of similar status and background. When the story claims it would shame a man to go alone to a new place and meet a chief, the implication is that one man alone will have trouble establishing his claims and proving himself to be who he says he is. We extend the local value and metaphor in this story and in broader Yup'ik tradition from two shamans, two hunters or two Elders and apply it to ourselves as two anthropologists.

The starting point for this study was our joint involvement in an international circumpolar study of youth resilience in five arctic indigenous communities (Rasmus et al. 2014a; Ulturgasheva et al. 2010, 2011; Ulturgasheva 2013, 2014; Ulturgasheva et al.2014). During the course of the study, our perspectives and experiences in conducting collaborative and community-based research came to align with our shared background as social anthropologists of indigenous[2] descent. We found ourselves critically examining the notion of community collaboration, de-colonization and participation in arctic research (Ulturgasheva et al. 2015). Our examination of the concept of collaboration in research prompted us to develop strategies to more thoroughly analyze and describe community-based and participatory approaches in arctic indigenous communities.

Too often the story of the Arctic is told from the single perspective of a typically European or American male in the form of a hero/explorer monograph that, as recent evidence shows, often leaves much of the story missing (see Pálsson 2001 and Nagy 2008 for critical discussions of the missing stories). Recent methodological and epistemic trends are moving this discourse more towards dialogic and multi-level perspectives (e.g., Akpik and Bodenhorn 2000; Williamson 2011; Ulturgasheva 2012). This shift towards the production of multi-faceted and multi-voiced research in the Arctic is an important step away from the trope of the polar explorer/hero or "lone wolf" that has characterized many contributions to circumpolar ethnology (e.g., Krupnik 2012; Muller-Wille and Barr 1998; Stefansson 1971).

---

[1] Yup'ik (pl. Yupiit) are the indigenous inhabitants of southwest and southcentral Alaska and parts of eastern Siberia. Yup'ik is part of the Inuit language family and the Yup'ik speaking peoples are traditional hunters and gatherers who would move on a seasonal round with a primary focus on salmon, sea mammals and tundra flora and fauna.

[2] The term 'indigenous' has been widely used in reference to all sorts of people, including dominant European nations interchangeably and has no straightforward and clear-cut definition. Here, we use the term 'indigenous' in reference to the endogenous populations of a particular geographic region, such as the Arctic, who have experienced colonial and de-colonial occupation and oppression by a dominant population.

The power dynamics and singularity of the "lone wolf" mentality in anthropology have been thoroughly deconstructed and critiqued by many researchers working in the field (Galman 2007; LeCompte and Schensul 2010; Nagy 2008). Collaborative and participatory ethnography has emerged over the past two decades, and many current anthropological approaches take into account community member observations and perspectives as a critical part of the co-production of knowledge in social research (Fluehr-Lobban 2003; Lassiter 2005; Bodenhorn 2012; Fienup-Riordan 2007; Sumsion 2014; Ulturgasheva et al. 2015). Yet, in arctic social science, the lone wolf myth and mentality seems to persist; perhaps aided by the collective efforts to preserve the North as our global "last frontier." The harsh climates and remote conditions of the North make the research endeavor, in many ways, more challenging than it might be in other areas of the world; researchers that brave these conditions are often heralded for their strength and durability, given the nature of the environment (e.g. Cole 1986). We learn about the indigenous inhabitants of these regions through the voice of the hero researcher who lived to tell the tale of the Arctic and its people (Brown 2015).

There remains a certain mystique given to arctic researchers who "survive" their fieldwork in the North, particularly for those spending long durations in the remote settlements. Aspects of this exoticism persist (e.g. Brody 2001) even with increasing numbers of indigenous scholars emerging from arctic communities (Fast 2002; Ulturgasheva 2012; Williamson 2011).

Historically, for a non-indigenous scholar, acclimatizing to the Arctic was one of the most primary and often highlighted process steps in the ethnographic inquiry (Briggs 1970; Stefansson 1971). Although the Arctic has changed significantly in the past 40 years and has become more hospitable and accommodating for those unfamiliar with arctic climates, there remains an emphasis on non-indigenous adaptations to arctic indigenous settings in contemporary ethnographic works (e.g. Collings 2014). In comparison, scholars native to the Arctic may be more able to skip a step in the research process or may experience a different kind of 'acclimatizing' experience.

All of the above raise questions about the ethics and accountability of the anthropological endeavor in the Arctic, and prompt us to examine the periphery in the construction of an anthropological accounting of arctic social life and culture. Although we do not deny the general physical challenges presented by arctic environmental conditions, we argue that the anthropological endeavor presents no less of a journey or challenge for researchers from the region who themselves must attend to both the professional and personal demands of indigenous arctic community life. Demonstrating and validating anthropological methods and practices has long been a weakness to our field (LeCompte and Schensul 2010). When it is only the anthropologist out alone in remote regions conducting research, indigenous or not, the production of knowledge is limited to a single perspective. Even when we attempt to address this limitation through approaches such as community-based participatory research (CBPR), participatory action research (PAR), and indigenous research methodologies (IRM), collaboration, writing, and research is still most often conducted through the perspective of a single PhD-level researcher engaging in research activities with indigenous community collaborators or co-researchers as opposed to the older and outdated "key informants" or "subjects."

Yet even in the more collaborative enterprises in anthropology it is still mainly the academic co-researcher in CBPR/PAR who writes the story representing the process and findings from the research (Rasmus 2014). This is one of the reasons that anthropologists, and social and cultural anthropologists in particular, are often criticized for being "anti-scientific" in their methods and have the validity of their research findings questioned (Kuznar 1996). Recent steps have been taken within the field to address this limitation, with an increased emphasis on inter-subjectivity as the basis for objectivity in ethnographic research (e.g. Duranti 2010). There is a need for methods that can capture the multi-layered and rapidly changing complexity of contemporary indigenous lives and knowledge.

In this chapter, we propose a new type of research engagement that involves multiple ethnographers engaged in research within settings, both new and accustomed in the Arctic, with the goals of observing and gathering knowledge while attending to the subtle details of engagement in fieldwork. By employing a new method in research practice, we attempt to respond to cutting-edge epistemological shifts in the growth of global contexts for research and co-learning that calls for revising old colonial narratives and developing new and decolonizing research methods (Konrad 2012; Bodenhorn 2012). The latter ultimately requires us to attend to the ways our subjectivities and positionalities (being at times both Native and non-Native, insider and outsider, observer and observed, the subject and object of research) are unfolding in a joint, collaborative effort of knowledge production. Here, we also refer to the latest scientific trend that invests in the ethos of collaboration. The practice of collaborative ethnography (Humphrey and Onon 1996; Lowe 2012; Marcus and Mascarenhas 2005; Strathern 2012) that challenges the 'lone wolf' narrative can potentially generate new understandings, especially given its emphasis on and recognition of shared intellectual 'returns' between equal knowledge producers. However, the methods and practices of collaborative ethnography remain unclear, and new methodologies that seek to decolonize and indigenize community-based research processes are beginning to emerge (e.g. Denzin et al. 2008; Kovach 2009; Medicine and Jacobs 2001; Smith 1999; Wilson 2008).

In a joint effort to achieve greater transparency in our collaborative ethnographic approach we combined two elements, namely the classical anthropological method of participant observation and borrowing from our Eveny[3] co-researchers. The latter framework of collaboration and mutual accountability between humans and reindeer, called *nyamnin*[4], can be translated into English as 'coordination of movement', 'coordination of rhythm' and 'joint or collective effort' to move with a purpose. Our approach involves a particular kind of *nyamnin*, focused on coordination and col-

---

[3] The Eveny are an Asiatic numerically-small people that number around 17,000 and speak a Tungus-Manchu language. Most Eveny live in northern Siberia and engage in a double economy that combines reindeer herding and hunting.

[4] We thank our knowledgable Eveny elder, Vasilii Spiridonovich Keimetinov-Bargachan for informing us on the concept of *nyamnin*.

laboration that we have back-translated into *peer observation of research*. We use the term *peer observation of research* to refer to our team of two anthropologists engaged simultaneously in social research activities with community members as well as with each other. In turning the anthropological gaze back on ourselves as the collaborative interlocutors, we illuminate the participant observation method as it is in practice, rather than as it should or could be in theory. Using an illustrative case example, we show how our peer observation of research methods were applied during our fieldwork and how the practice led to productive points of orientation and disorientation in the ethnographic encounter. Following the case example we discuss how our peer observation of research approach is useful in opening-up and demystifying the research process not only for other social scientists but also for community members and participants who may come to understand and appreciate research and see a future for themselves in the field.

17.2   Methods

The peer observation of research component is based on established "peer observation of teaching (PoT)" methods traditionally used in classroom and educational settings (Fullerton 1999; Hammersley-Fletcher and Orsmond 2005). In education, peer observation is a novel expansion of the more classical participant observation method developed within the discipline of anthropology. By having everyone involved in the experience take up both subject and object roles, peer observation of research levels the field. For indigenous researchers working in their own or other indigenous community contexts, peer observation of research has the potential to play a particularly important role in establishing the validity and ethics of the research. In our experience, indigenous researchers must manage their own constantly shifting personal and professional roles in the community while also balancing social expectations with research standards and ethics.

As stated earlier (see footnote 2) the term "indigenous" is itself problematic, complex and unclearly defined. Here though, we use 'indigenous' to refer to a shared status or descent from a group of people who, as traditional inhabitants of a geographic region, have been subject to colonization and subjugation by outside groups or forces. While the very broad term 'indigenous' has some clear utility connecting vastly different peoples and regions together in a shared cause for equity and social justice on a global level, it brings the potential for essentializing and politicizing peoples and cultures.

The complexity of contemporary indigeneity is demonstrated within our own research team, with one of us being native Eveny while the other is of mixed-blood American Indian background. Our shared indigenous descent may accelerate our acceptance in other indigenous community contexts and provide initial trust that would take much longer for a non-indigenous researcher to gain. It may also move us into the potentially problematic conditions in an indigenous community context that have been associated negatively in anthropology with "going Native" (Tresch

2001). The more closely a researcher is perceived as related or indigenous "like us" the more important it is to create or establish a method to allow for a productive distance as part of the research study.

While conducting research in Alaska and Siberia we have observed that our appearance and our ethnic affiliation and background would instantly inspire people from both sites to look for a kinship connection. Both Eveny and Yup'ik were trying to figure out how we are related through time and distance to one another. And we have realized that by making an attempt to figure out the kinship relationship we are acknowledging each other as extended family in a way that invokes mutual and reciprocal obligation and accountability. Hence, it is important in these cases of close relational research collaborations to have a way for placing checks in the process that reduce potential for bias and increase peripheral perspective, illuminating possible "blind-spots" that can develop within long-term relationships, research or otherwise. One of the main goals of this study is to illuminate those details that remain unchecked and unnoticed by a single researcher. The fieldwork in Alaska and Siberia which involved our participation in the social lives of both communities was aimed at illustrating how the practice of peer observation of research helped us in shedding new light on seemingly obvious things and reflecting relations from a different angle.

As a starting point, we take the main parameters of indigenous research methodologies (Denzin et al. 2008; Mutua and Swaderner 2004; Seale et al. 2004; Smith 1999, 2005; Wilson 2008), and redefine them to revolve around the relational accountability of *nyamnin*. In Eveny terms, indigenous research methodologies would mirror the coordination and negotiation of roles and relationships within reindeer herding to reflect the research engagement. The process would involve a coordination of movement and awareness of each other at all times in an effort to prevent procedural crisis from occurring that would disallow both humans and reindeer, or in this case indigenous researcher and interlocutor, to get to their desired outcome.

For our study, we traveled first to Alaska, where Rasmus has been working with Yup'ik communities to develop community-level, cultural interventions to reduce suicide and alcohol use disorder for over a decade (Mohatt et al. 2004; Rasmus et al. 2014b). We went next to northeast Siberia, where Ulturgasheva has been studying Eveny adolescence and social resilience for the past 15 years (Ulturgasheva 2012, 2013, 2014, 2016; Ulturgasheva et al. 2010). We spent a month at each site engaging in ethnographic fieldwork together as members of a team. As part of an exploratory methods study we sought to understand how both our own positions and perspectives and those of the community could be deployed in a process of shared knowledge production; specifically, in identification of 'blind-spots' from different and multiple angles.

We present results in the form of fieldwork episodes that represent how peer observation of research can increase our peripheral vision and allow us to see what may not have been observable to us based on our positionalities and subjectivities. Anthropologists have noted the potential for events and encounters to be observed and represented in many different ways, even by the same researcher returning to observe the same event or practice (e.g. Bateson 1958). Peer observation of research has the potential to expand the anthropological gaze in a way that shrinks the reflection to a single fieldwork experience but involves multiple simultaneous perspectives of the same event or encounter to occur. We present two encounters from our fieldwork in Alaska and Siberia in a dialogic fashion, beginning with Ulturgasheva's perspectives drawn from fieldnotes of her first day in an Alaskan Yup'ik community, followed by Rasmus' observations and initial interactions in a Siberian Eveny community. The two episodes demonstrate productive points of orientation and disorientation in our fieldwork and provide a glimpse into our collaborative ethnographic process. We conclude with a critical discussion of the role of peer observation in demystifying and describing more empirically the anthropological endeavor.

17.3   Results

17.3.1   Fieldwork Episode 1: Washing-Up

17.3.1.1   Ulturgasheva's Perspective

My very first day in a Yup'ik community ended up with me washing up after all guests/relatives have left the house of our local collaborator. Somehow it was assumed that I would take on this responsibility unquestioningly. To my Siberian understanding of hospitality this gesture would have been read as an expression of disrespect and even hostility towards myself as a guest. But I felt strongly that, just like Jean Briggs (1970), I should take my dismay as the first step in learning about contemporary Yup'ik sociality with its ebbs and flows.

In that particular situation, I thought it was perhaps expected that a newly arrived female guest must take this duty on her shoulders because, first, she is a university researcher (who will disappear as soon as the study is over) and, second, she is an indigenous, non-white woman who is "just like us"— expected to wash up after the dinner with other outsiders/guests is over. In contrast, no one assumed or offered this task to my indigenous Siberian collaborator, Sayan, who arrives as a male guest. As per my observations from the very start, there was a marked contrast of intergender behavior within the household structure, with males seeming to have more dominance and a preferred position. From our informal exchanges prior to our fieldwork, I learned that Stacy, as a woman, was also subject to gendered expectations of behavior, especially, when she was in the community with other male, non-indigenous university researchers.

In my initial observation, our assigned roles in the household as women from similar ethnic backgrounds and university researchers reflected the perspective of the community members affording males, and those perceived as academic males of non-indigenous background in particular, higher status with full guest privileges. In this regard, upon initial entry in the community, non-indigenous male researchers would experience less hassled and troubled relations with community members, particularly as this asymmetrical model has become more ubiquitous and entrenched since the colonial era. Asymmetrical relations and gendered norms have the potential to profoundly impact us and our capacities as indigenous female researchers working in the Arctic. For both of us, this study is also an attempt to contest a well-established, asymmetrical, 'top down' model of relations that we have experienced, not only within academia but also within the indigenous communities themselves.

17.3.1.2 Rasmus' Perspective

At the same time that Olga is documenting her experience in a Yup'ik community, I am observing a Yup'ik community as they encounter and share experiences with Olga. What I observe with the initial washing-up interaction was the immediate acceptance of Olga as a member of the household and kinship unit. Sayan, too, is immediately obliged to take a steambath with the men. Olga, since she enters the community with me and is experiencing the community initially through my established connections, is conferred a status adjacent to mine (which is something like a daughter-in-law) and is treated as returning family rather than as new-coming guest or visitor. While I am only conferred this type of family association in this particular Yup'ik household in the community, Olga is frequently assigned kinship status in other interactions with community members and often as part of the formal research process as well.

I witness with Olga something I have never yet witnessed in conducting interviews with my non-indigenous research colleagues in the same community: community members listen much more attentively to Olga's questions and respond much more thoroughly and thoughtfully. Community members clearly want to share more of themselves and their lives with Olga. They are very eager to speak to Olga about their way of life and culture and often say "well you must have this too" or "you must have heard about that in your community." Interestingly, this association occurs much less frequently with Sayan, also an indigenous person from Siberia, but from a different region. In observational contexts, Yup'ik community members are much more likely to say that Olga looks more Yup'ik and that Sayan looks perhaps more "like an (American) Indian." The community's perception of Sayan's appearance as "not Yup'ik" appears to give him more of an outsider status akin to my own in a Yup'ik community. And, in this way, Sayan receives the "benefit" of being perceived as guest or visitor at first entry into the village.

17.3.2 Fieldwork Episode 2: Finding Geronimo

17.3.2.1 Rasmus' Perspective

We arrive in an Eveny village after two days of long travel by caravan over vast fields, mountains, rivers. It is very late and we are immediately given rest in the homes of our host families. The next day is full of many meetings and greetings with the people of the community, but one interaction stands out. Among the first questions that I am asked after introductions are made and my background as "Native American" is disclosed, is whether or not I have yet met with Geronimo. At first I am confused by the question, and ask if they mean whether I know who Geronimo was. I confirm through Olga as an interpreter, that yes I do know of Geronimo as an important American Indian figure. Olga clarifies for me that no, that is not what they mean. Rather, the community members are wondering if I had met the Eveny man 'Geromino' who lives in the village and carries the great Apache leader's name. I have not yet met this Geronimo but am asked several more times if I have and am encouraged to meet him and hear the story of how he got his name.

When I do come to meet with Geronimo it is at the reindeer herders' celebration and we are all warming up in one of the tents set up for the participants. Olga introduces me to Geronimo and tells him that I am Native American too but from a different tribe. The man called Geronimo is quite tall and stately and looks at me with an intensity to his gaze as he searches my face for what I assume to be signs of likeness to his namesake. He turns to Olga and states that he does not think that I look very much Apache, and rather thinks I look instead more "Mexican." Initially, I feel a bit upset. My immediate response is to point out the connection between indigenous populations of southwest United States and Mexico, to defend my position as indeed related, but yet not really, to the American Indian of Geronimo's imagination. But I do not have the chance as he turns to walk away. In the days following, Olga explains to me why I am so suddenly disowned by Geronimo and perhaps other Eveny, who do not see in me their own struggles against power and colonization, and ideals and longing for a glorious past. I realize then too that while the questions and comparisons to American Indian idealizations and national stereotypes might be a disorienting factor for me, just listening to a story of a familiar and much respected hero in a remote community in northeast Siberia makes me also feel a little less far from home, and I am glad to have found Geronimo.

17.3.2.2   Ulturgasheva's Perspective

During our first days in the village I was utterly surprised by the local Eveny's response to my introduction of Stacy. The first thing I mentioned during the introduction is that she comes from the United States, she is an anthropologist working for the University of Alaska Fairbanks and is also an American Indian. They would completely skip over my mention of her professional profile and would instead focus with great excitement and fascination on her ethnic affiliation. Suddenly, after hearing about her being one of the American Indians everyone would start speaking to her about Apache and Huron people for whom they have had special respect and admiration. One of the local women, Nadezhda, mentioned a research project that took place in early 1980s in the village when Soviet geneticists arrived, took blood samples and few years later discovered a genetic connection between Athabascan groups (like Apache) and Siberian Eveny. Nadezhda suggested that Stacy was not randomly visiting the village as any other foreign visitor or researcher would, but she was there because of her special (genetic) connection. That was Stacy's destiny! According to Nadezhda, Stacy was there as a distant relative of all Eveny people. Indeed, everyone viewed her as some distant relative who travelled through time and space from a distant land of the Apache because there was a unique kinship connection that powerfully pulled her there. They wanted to acknowledge that connection in various ways: some hosted a generous meal as a tribute for their special and rare guest, others literally showered her with gifts made out of reindeer skin and fur, some insistently offered to share their home and vodka, in case she needed to be entertained.

Only one, 'Geronimo,' received Stacy with surprising skepticism, as apparently she failed to meet his own 'standards' and 'parameters' of real and honorable Apache people, who he had never met in his life. It was at the Reindeer Herders' Day, celebrated on a flat surface of the frozen river where we met 'Geronimo'. He looked at her and to my own shock pronounced: "No, she doesn't look properly Apache! She looks more like Mexican Pueblo." He turned away and left. My surprise came from an understanding that local Geronimo's imagery of American Indians could have only come from the socialist/Soviet 1970s movies in a genre of 'western' that he saw in his childhood, which would have depicted Apache and Huron in a particularly romantic, 'noble savage' style. Those movies would portray noble warriors, brilliant hunters and gracious friends who would always defeat the pale-skinned, canny and insidious colonizers. There was no chance for Stacy to meet such expectations, as she did not look like that cinematic, noble Apache, a daughter of the historic Geronimo. That was an extremely funny moment. Three of us were caught in our own understandings of the situation, completely different from each other. Stacy was upset, 'Geronimo' was disappointed and I was completely amused. The three of us shared a moment of sheer disorientation.

17.4 Discussion

The addition of peer observation into our ethnographic fieldwork process brought certain aspects of the research into higher resolution. Despite the tensions inherent in negotiating our relational accountabilities, or *nyamnin*, in the communities as indigenous researchers, the fact that we are both indigenous turned out to be quite an important point for creating a productive context for social bonding at both sites. That we share not only cultural connections but also greater potential for familial connections with our hosts made us welcome in many homes and social circles. At the same time we understood that, depending on the situation, people, including ourselves, would manipulate a line between relatedness and unrelatedness. This manipulation, in some cases, would lead to what we term 'productive points of disorientation' in our ethnographic encounters. Productive points of disorientation can occur when indigenous people from different parts of the Arctic or different parts of the world, with different socio-economic, political and cultural contexts have discrepant or divergent views of a similar or shared social practice or norm. We use this term as a parallel to the experience of "culture shock" often described by non-Native, non-indigenous anthropologists coming to know a new and exotic community and cultural setting for the first time. We have both been questioned in our careers about our decision to conduct our anthropological research "at home" or with "our own" people. Results from our peer observation of research demonstrate how we too, as indigenous researchers, can go through similar processes of discovery and negotiation as non-indigenous scholars.

The points of disorientation emerged as our disagreements, puzzlement, dismay and vulnerability became productive and interlocking in unpredictable ways. The fact that we have had different perspectives on what happened to us in Alaska and Siberia brought up the need to attend to differences in our perceptions of the same social situations. The differences in researchers' perspectives can be deployed as an analytical and social resource for understanding how access to the local knowledge can be gained methodologically. In this sense, while speaking differently about our experience in Alaska and Siberia neither of us is incorrect or disingenuous. Peer observation can serve productively to capture the complexity and multi-faceted nature of indigenous social and cultural contexts while also verifying the process through which indigenous knowledge is accessed and produced. But our starting point has always been our acceptance of the premise that we shall endeavor to discover multiple viewpoints that enrich our attempts to develop a complex and coherent picture, something impossible, or ill-advised, to do alone.

The examples above show how a multi-faceted perspective emerged as part of the peer observation process, where you see one side, but then have to consider all of the other sides of the event/episode. One person alone has more limited range for viewing all the angles, but two people together can achieve greater distance and range in view and illuminate blind-spots.

Our peer observation of research method brings into full relief our collaborative ethnographic methods and practices, and reveals a research process that historically has been hidden in lone wolf and arctic hero narratives of the North. Peer observation also has potential to de-essentialize the indigenous person or community, by moving us beyond representing the dual-perspective (or insider-outsider debate) to register, with more complexity and acuity, indigenous peoples' realities. We are only able here to share two short moments from a much longer sequence of events and interactions. More work is needed to understand the utility and practice of peer observation of research and its relationship to the production of knowledge within indigenous communities. What we can propose here based on our preliminary findings is that the lone wolf approach, no longer sustainable in a new Arctic, will give way to the model of the relational reindeer, working together in conscious collaboration to move towards new places, new practices and new knowledge in the North.

References

Akpik, F., & Bodenhorn, B. (2000). *Pilgallasiniq Ivalupianik—Learning to braid 'real' thread.* Barrow: Illisagvik College.

Bateson, G. (1958). *Naven: A survey of the problems suggested by a composite picture of the culture of a New Guinea tribe drawn from three points of view.* San Francisco: Stanford University Press.

Bodenhorn, B. (2012). Meeting minds; encountering worlds: Sciences and other expertises on the North Slope of Alaska. In M. Konrad (Ed.), *Collaborators collaborating: Counterparts in anthropological knowledge and international research relations* (pp. 225–244). Oxford: Berghahn Books.

Briggs, J. (1970). *Never in anger: Portrait of an Eskimo family.* Cambridge: Harvard University Press.

Brody, H. (2001). *The other side of Eden: Hunters, farmers and the shaping of the world.* New York: North Point Press.

Brown, S. (2015). *White Eskimo: Knud Rasmussen's fearless journey into the heart of the Arctic.* Boston: De Capo Press.

Cole, D. (1986). Franz Boas in Baffin-Land: The great anthropologist began his career with arctic explorations and hardships. *The Beaver: Exploring Canada's History, 66,* 4–15.

Collings, P. (2014). *Becoming Inummarik: Men's lives in an Inuit community.* Montreal: McGill-Queen's University Press.

Denzin, N. K., Lincoln, Y. S., & Smith, L. T. (Eds.). (2008). *Handbook of critical and indigenous methodologies.* Thousand Oaks: Sage.

Duranti, A. (2010). Husserl, intersubjectivity and anthropology. *Anthropological Theory, 10*(1), 1–20.

Fast, P. (2002). *Northern Athabaskan survival: Women, community and the future.* Lincoln: University of Nebraska Press.

Fienup-Riordan, A. (2007). *Yuungnaqpiallerput. The way we genuinely live: masterworks of Yup'ik science and survival.* Seattle/London: University of Washington Press.

Fluehr-Lobban, C. (Ed.). (2003). *Ethics and profession of Anthropology: Dialogue for ethically conscious practice.* New York: Alta Mira Press.

Fullerton, H. (1999). Observation of teaching. In H. Fry, S. Ketteridge, & S. Marshall (Eds.), *A handbook for teaching and learning in higher education.* London: Kogan Page.

Galman, S. C. (2007). *Shane, The lone ethnographer: A beginner's guide to Ethnography.* New York: AltaMira Press.

Hammersley-Fletcher, L., & Orsmond, P. (2005). Reflecting on reflective practices within peer observation. *Studies in Higher Education, 30*(2), 213–24.

Humphrey, C., with Onon, U. (1996). *Shamans and elders: Experience, knowledge and power among the Daur Mongols.* Oxford: Clarendon Press.

Konrad, M. (Ed.). (2012). *Collaborators collaborating: counterparts in anthropological knowledge and international research relations.* Oxford: Berghan Books.

Kovach, M. (2009). *Indigenous methodologies: Characteristics, conversations, and contexts.* Toronto: University of Toronto Press.

Krupnik, I. (2012). The 50-year arctic career of Ernest S. Burch, Jr.: A personal ethnohistory, 1960–2010. *Arctic Anthropology, 49*(2), 10–28.

Kuznar, L. (1996). *Reclaiming a scientific anthropology* (Cambridge cultural social studies). New York: AltaMira Press.

Lassiter, L. E. (2005). *The Chicago guide to collaborative ethnography.* Chicago/London: University of Chicago Press.

LeCompte, M., & Schensul, J. (2010). *Designing and conducting ethnographic research* (Ethnographer's toolkit 2nd ed.). New York: AltaMira Press.

Lowe, C. (2012). Recognizing scholarly subjects in the politics of nature: Problematizing collaboration in Southeast Asian area studies. In M. Konrad (Ed.), *Collaborators collaborating:*

Counterparts in anthropological knowledge and international research relations (pp. 245–268). Oxford: Berghahn Books.

Marcus, G., & Mascarenhas, F. (2005). *Ocasião: The marquis and the anthropologist: a collaboration*. Walnut Creek: Altamira Press.

Medicine, B., & Jacobs, S. E. (2001). *Learning to be an anthropologist and remaining "Native": Selected writings*. Ubana: University of Illinois Press.

Mohatt, G. V., Rasmus, S. M., Thomas, L. R., Hazel, K., Allen, J., & Hensel, C. (2004). "Tied together like a woven hat:" Protective pathways to sobriety for Alaska Natives. *Harm Reduction Journal, 1*(10). http://harmreductionjournal.com/content/1/1/10.

Muller-Wille, L., & Barr, W. (1998). *Franz Boas among the Inuit of Baffin Island, 1883–1884: Journals and letters*. Toronto: University of Toronto Press, Scholarly Publishing Division.

Mutua, K., & Swaderner, B. B. (Eds.). (2004). *Decolonizing research in cross-cultural contexts: Critical personal narratives*. Albany: SUNY Press.

Nagy, M. (2008). Sex, lies and northern explorations: Recent books on Peary, MacMillan, Stefansson, Wilkins and Flaherty. *Etudes Inuit Studies, 32*(2), 169–185.

Nelson, E. W. (1900). *The Eskimos about Bering Strait*. Washington, DC: Smithsonian Institution Press.

Pálsson, G. (2001). *Writing on ice: The ethnographic notebooks of Vilhjalmur Stefansson*. Hanover: Dartmouth College Press.

Rasmus, S. M. (2014). Indigenizing CBPR: Evaluation of a community-based participatory research process implementation of the Elluam Tungiinun (Towards Wellness) program in Alaska. *American Journal of Community Psychology, 54*(1–2), 170–179.

Rasmus, S. M., Allen, J., & Ford, T. (2014a). "Where I have to learn the ways how to live": Youth resilience in a Yup'ik village in Alaska. *Transcultural Psychiatry, 51*(5), 713–734.

Rasmus, S. M., Charles, B., & Mohatt, G. V. (2014b). Creating Qungasvik (a Yup'ik toolbox): Case examples from a community-developed and culturally-driven intervention. *American Journal of Community Psychology, 54*(1–2), 140–152.

Seale, C., Gobo, G., Gubrium, J. F., & Silverman, D. (Eds.). (2004). *Qualitative research practice*. London: Sage.

Smith, L. T. (1999). *Decolonizing methodologies: Research and indigenous peoples*. New York: Zed Books Ltd.

Smith, L. T. (2005). On tricky ground: Researching the Native in the age of uncertainty. In N. K. Denzin & Y. S. Lincoln (Eds.), *The SAGE handbook of qualitative research* (3rd ed., pp. 85–108). Thousand Oaks: Sage Publications, Inc.

Stefansson, V. (1971). *My life with the Eskimo*. New York: Collier Books.

Strathern, M. (2012). Currencies of collaboration. In M. Konrad (Ed.), *Collaborators collaborating: Counterparts in anthropological knowledge and international research relations* (pp. 109–119). Oxford: Berghahn Books.

Sumsion, J. (2014). Opening up possibilities through team research: An investigation of infants' lives in early childhood education settings. *Qualitative Research, 14*, 149–165.

Tresch, J. (2001). On going Native: Thomas Kuhn and anthropological method. *Philosophy of the Social Sciences, 31*, 302–322.

Ulturgasheva, O. (2012). *Narrating the future in Siberia: Childhood, adolescence and autobiography among Eveny*. Oxford: Berghahn Books.

Ulturgasheva, O. (2013). Schastlivoye budusheye antiutopiyi postsotsializma [Happy futures in Post-Socialist anti-utopia]. In N. Ssorin-Chaikov (Ed.), *Topografiya schastya: Ethnograficheskye karty moderna* (pp. 219–33). Moscow: Novoe Literaturnoe Obozrenie.

Ulturgasheva, O. (2014). Attaining Khinem: Challenges, coping strategies and resilience among Eveny adolescents. *Transcultural Psychiatry, 51*, 632–650.

Ulturgasheva, O. (2016). Spirit of the Future: Movement, kinetic distribution and personhood among Siberian Eveny. *Social Analysis, 60*(1), 56–73.

Ulturgasheva, O., Grotti, V. & Brightman, M. (2010). Personhood and 'frontier' in contemporary Amazonia and Siberia. In *Laboratorium. Russian review of social research. Thematic issue:*

"Russia/CIS/Latin America: Comparative Studies in Post-Authoritarian Transformation", 2(3), 348–365.

Ulturgasheva, O., Wexler, L., Kral, M., Allen, J., Mohatt, G. V., & Nystad, K. (2011). Navigating international, interdisciplinary, collaborative inquiry: Phase 1 process in the Circumpolar Indigenous Pathways to Adulthood Project. *Journal of Community Engagement and Scholarship, 4*(1), 50–59.

Ulturgasheva, O., Rasmus, S., Wexler, L., Nystad, K., & Kral, M. (2014). Arctic Indigenous youth resilience and vulnerability: Comparative analysis of adolescent experiences and resilience strategies across five arctic indigenous communities. *Transcultural Psychiatry, 51*, 735–756.

Ulturgasheva, O., Rasmus, S., & Morrow, P. (2015). Collapsing the distance: Indigenous youth engagement in a circumpolar study of study of youth resilience. *Arctic Anthropology, 52*(1), 50–60.

Williamson, K. J. (2011). *Inherit my heaven: Kalaallit gender relations. Inussuk Arctic Journal 1.* Nuuk: Department of Culture, Education, Research and Church, Government of Greenland.

Wilson, S. (2008). *Research is ceremony*. Winnipeg: Fernwood Publishing.

Chapter 18
Building Relationships in the Arctic: Indigenous Communities and Scientists

Heather Sauyaq Jean Gordon

Abstract Climate change, pollution, and resource extraction have heightened interest in arctic research, and communities now encounter scientists more often. This chapter addresses how to build sustainable research relationships through the perspectives of Indigenous Greenlanders and arctic researchers. Collaborative research, community-based participatory research, and Indigenous methodologies provide examples of research partnerships but do not explain the initial steps of building relationships. I examined results from fifteen interviews with arctic researchers, nineteen interviews with Inuit Greenlanders, and four focus groups—working with fourteen of the original Greenlandic interviewees. Through a grounded theory approach, I found the central theme to a sustainable relationship was trust, surrounded by eight prominent actions necessary to create, build, and sustain trust. The actions for trust include: knowing extensive community history; developing strong local contacts; communicating openly about the project; treating the community members as equals; displaying manners and etiquette through honesty and reciprocity; acting ethically in Indigenous cultures outside of the academic world; exchanging knowledge to build social capital; and giving project results to the community so they can be put to practical use.

Keywords Community based participatory research (CBPR) • Greenland Inuit • Trust and relationship building • Arctic • Indigenous methodologies

H.S.J. Gordon (✉)
Indigenous Studies, University of Alaska Fairbanks, Fairbanks, AK, USA
e-mail: gordon.heather.j@gmail.com

© Springer International Publishing Switzerland 2017
G. Fondahl, G.N. Wilson (eds.), *Northern Sustainabilities: Understanding and Addressing Change in the Circumpolar World*, Springer Polar Sciences, DOI 10.1007/978-3-319-46150-2_18

18.1   Introduction

Climate change dramatically affects built and social environments, heightening interest in arctic research. Cemeteries are eroding into the ocean, subsistence species' migration routes are changing, and entire arctic communities are relocating (Bronen 2008). In Greenland, natural resources become accessible as the ice melts, bringing people seeking profits in oil, gas, and minerals (Rosenthal 2012). Drawing on data from North American arctic researchers and Inuit Greenlanders, I focus this chapter on how social, natural, and physical scientists can create collaborative relationships in research that are in the interest of both communities and scientists.

Arctic communities are becoming a part of the global economy. While conducting research in the Arctic is not new, scientists need to recognize that people live in the communities, their land is private (owned by individuals or tribes), and they do not accept continued "helicopter" research (flying in, taking data, and leaving) (Berkes 2002). Ethical research practices point toward community collaboration, regardless of whether the project involves interviewing people or measuring ice depth. The *Arctic Climate Impact Assessment* states: "Further development of the collaborative model, from small projects to large research programs and extending from identifying research needs to designing response strategies, is an urgent need" (Huntington and Fox 2005: 94). My project addresses this directly by first analyzing existing literature on different methods of researching with Indigenous people through collaboration, community based participatory research, and Indigenous methodology. Next, I explain the historic interactions between outsiders and Inuit Greenlanders. I then explain how using a grounded theory methodology led to researcher interviews further deepening my research question. Finally, I turn to my interviews and focus groups to explore the dynamics of relationship building and develop an eight-part thematic model. I began the project asking, how can researchers and arctic Indigenous communities partner in social, natural, and physical science research for mutual benefit? This chapter will illuminate the process through which arctic Indigenous communities and researchers can engage in sustainable research relationships.

18.2   Researching *with* Indigenous Peoples

Researchers working in and around arctic communities can promote contention or collaboration. Contention often leaves the community inaccessible to future research due to researchers or outsiders committing ethical abuses or cultural mishaps and ignoring historical traumas. This has included colonizing research, forced boarding schools with reprehensible treatment (Hirshberg and Sharp 2005); radioactivity studies conducted on people without consent (Hodge 2012); and helicopter research (Berkes 2002). Collaboration, on the other hand, can develop into sustainable working relationships. Community-based participatory research (CBPR) is a key type of research conducted in partnership with the community and researcher or university/ campus. This literature review first explains how well arctic collaborative research and CBPR address relationship building in research projects. Second, it addresses how Indigenous methodologies provide additional necessary information on building relationships with Indigenous people.

18.2.1 Collaborative Research

Collaborative projects range from minimal community involvement in a project, to CBPR, to the community initiating the project (Huntington and Fox 2005). Huntington and Fox (2005) declare that the collaborative model is the most promising model for involving arctic Indigenous people in research. First, the model offers the opportunity to engage community knowledge,[1] bring communities and researchers together, and build a project based on the needs of each community. Second, without collaboration and Indigenous participation, traditional knowledge is not accessible to scientists and scientific knowledge is not accessible to the community in other types of research. Third, doing collaborative research empowers communities and enables community development and capacity building (Berkes 2002).

However, articles on arctic collaborative work focus on the challenges of collaboration rather than on a model of collaboration and/or successful relationship building (Weatherhead et al. 2010; Huntington and Fox 2005). Collaboration in arctic research varies and lacks a larger framework for researchers to begin building research relationships. If articles do talk of a relationship building application and methodology, the process is widely varied and is seen as a unique experience in each project (Gearheard et al. 2010). Huntington et al. (2011: 437) emphasize that building "strong personal relationships, can increase the likelihood that collaborative fieldwork will be productive, enjoyable, and rewarding." Although collaboration in arctic research varies by each project, it lacks a larger framework for the researchers to begin building trusting relationships for research.

18.2.2 Community-Based Participatory Research (CBPR)

Community-based participatory research (CBPR) seeks to address and move beyond colonizing research by recognizing the relevance of knowledge outside of traditional academic research. CBPR is collaborative but also establishes a framework for the research process that includes community and researcher input in the

---

[1] Community knowledge, traditional knowledge, Indigenous knowledge, traditional ecological knowledge, and local knowledge are used interchangeably within this paper, without judgment, malintent, Westernization, or any disrespect, even though some of these terms were contended by Ray (2012).

project. CBPR is built on eight principles: (1) recognize the community is the unit of identity; (2) build on community strengths; (3) facilitate equitable partnerships; (4) engage in co-learning and capacity building; (5) achieve mutual benefit; (6) use locally relevant issues for research; (7) conduct cyclical and iterative research; and (8) disseminate the findings back to the community (Israel et al. 1998).

CBPR is partnership based, so the first step in a project is building relationships. When CBPR articles specifically talk about relationship building, they identify what ideal partnerships include. These studies come from the researcher perspective (Fisher and Ball 2003), community perspective (Leiderman et al. 2003), or perspectives of both communities and researchers (Fondahl et al. 2009). All three types of studies accentuate the elements of relationships, most paying particular attention to trust and relationship building. They all look back at what went well in the project, or what needed to happen, rather than analyzing the relationship process.

The CBPR Conceptual Logic Model walks the reader through a CBPR project visually from 'Contexts to Consider', 'Group Dynamics & Equitable Partnerships', 'Intervention & Research', and lastly 'Outcomes' (Wallerstein et al. 2008). The context section points to culture and historic trust as the outside influences that may affect the partnership. The Relational Dynamics section under Group Dynamics lists values, personal beliefs, power dynamics, trust, and self-reflection as important. However, an important aspect of CBPR is the relationship building between researchers and community members. This model assumes a relationship already exits.

The CBPR literature that does examine the relationship process focuses on organized communities ready to co-manage a research project (Examining ... 2006). These perspectives do not capture how to build relationships with people who are not part of a highly organized community with high levels of capacity and social capital. People not prepared for research and meeting researchers for the first time do not build relationships with researchers in the same way; the relationships are individual and not always group-oriented when a project does not follow a full CBPR methodology. When explaining how to build a research relationship, explaining how to build it between individuals and across cultures is vital.

18.2.3   Indigenous Methodology

Indigenous methodologies add to the importance of collaborative relationships and the framework outlined by CBPR. Indigenous people often experience the research relationship as unequal collaborators. Current ethical research relationships are defined by Institutional Review Boards and research funders, emphasizing professional codes of conduct and minimizing the needs of "marginalized and vulnerable communities" (Smith 2005: 96). The United Nations Declaration of the Rights of Indigenous Peoples (UN General Assembly 2007) made official the rights of Indigenous peoples to practice self-determination and to protect and own their knowledge. Many Indigenous groups worldwide protect themselves in research by developing their own ethical views of research and requirements (e.g. Alaska Federation of Natives 1993). There are no such guidelines in Greenland (Holm et al. 2010).

The most prominent research principles in working with Indigenous peoples are: respecting Indigenous peoples' right to self-determination; benefiting the researched community by building capacity and determining the research agenda with the community; valuing Indigenous culture and people as researchers by acknowledging, respecting, and advocating non-Western worldviews; and protecting Indigenous knowledge and sharing all knowledge (Kennedy and Wehipeihana 2006). In addition, defining respect through a "community-up" format, instead of a "top-down" researcher definition, brings community definition of ethical behaviors to the forefront (Smith 2005). Burnette and Sanders (2014) explored cultural sensitivity from a researcher perspective. They found many important themes in trust building between researchers and Indigenous people in the United States, similar to those identified by Smith (2005), Kennedy and Wehipeihana (2006), and this study. The authors emphasize the importance of asking the Indigenous members of the community their opinions. That is where my study adds to the understanding of trust in relationship building. I ask community members how to build lasting research relationships.

18.3 Historical Context

I turned to Greenland to explore sustainable relationship building because its history of contact with outsiders differs from the research abuses experienced by North American Indigenous people (Hodge 2012; Hirshberg and Sharp 2005). The current Indigenous people in Greenland, the Inuit, arrived to Greenland around 1000CE. Except for minimal trade with the Norse, contact between Inuit Greenlanders and outsiders began in 1720 with missionary ships coming from Denmark. Moravian and Lutheran missionaries established colonies, learned Greenlandic, and converted Greenlanders who then spread the gospel (Gad 1973). Greenland became a good resource for Denmark once the fisheries were opened; this led to Greenland becoming a permanent interest of the Danish crown (Gad 1973). Greenland officially became a Danish colony in 1814. From 1817 to 1917, the Inuit interacted with English and Scottish whalers and arctic explorers, many of whom did not treat Greenlanders as equals (Vaughan 1991). Explorers established trading posts, which made outsider contact more regular. Yet, even with increased formal contact, Greenlanders expressed feelings of a lack of equal treatment from whites (Rasmussen 1908).

The Danish "civilized" the Greenlandic people by colonizing them both physically and mentally (Lynge 2011). After years of being a Danish colony, colonialism ended in 1953 and, by 1979, Greenland had achieved Home Rule. Home Rule came about as Greenlanders cultivated increased pride in their ethnicity (Dahl 1988).

However, Home Rule did not change the economic relationships between Greenland and Denmark, and on June 21, 2009, Greenland and Denmark passed the Self Rule Act. Under Self Rule, Greenlandic became the official language of Greenland, the Greenlandic government gained additional autonomy, the Danish froze their yearly block grant, and Greenland gained some rights to their mineral, oil, and natural gas resources (Embassy of the United States 2012).

With the ice cap receding, many outsiders are interested in Greenland's exposed resources. In response to foreign interest in oil, gas, and minerals, Jens Fredericksen, the Vice Premier of Greenland, said: "We are treated so differently than just a few years ago…We are aware that is because we now have something to offer [natural resources], not because they've [outside actors] suddenly discovered that Inuit are nice people" (Rosenthal 2012: 1–2).

As such, the feeling of unequal treatment by outsiders is a part of Greenlandic history that has been perpetuated over the last few 100 years. Additionally, with Greenland beginning to take charge of their government and mining permits, research is becoming a shared endeavor between Greenland and outside researchers.

18.4   Methods

This study asks how researchers and arctic Indigenous communities partner in social, natural, and physical science research for mutual benefit. I used theoretical sampling to choose participants through a grounded theory research design (Glaser 1978; Coyne 1997; Strauss and Corbin 1990). I also used an interview guide with specific questions asking participants about how to build a relationship between researchers and Indigenous people. I conducted 15 individual interviews with North American arctic researchers who worked with Indigenous people. Out of the 15 arctic researchers, 14 interviewees had PhDs and one did not. One of the fifteen researchers was an Alaska Native and the other fourteen were whites from the U.S. and Canada; five were men and ten were women; twelve were social scientists and three were biophysical scientists. I asked questions from an interview guide to get comparative answers on some questions, but I also let the interviewee tell long stories and elaborate.

Using grounded theory, I found that researchers were most concerned with building trusting relationships so that the research relationship would be mutually beneficial. I used this finding to further develop my original question: *How can researchers and arctic Indigenous communities partner in social, natural, and physical science research for mutual benefit?* This question became: *how can researchers and arctic Indigenous communities build trusting relationships in order to partner in social, natural, and physical science research for mutual benefit?* After the researcher interviews, I conducted 19 individual interviews with Greenlanders through a translator. I met my translator through a series of contacts in the community. In addition to specifically addressing trust, I adjusted the cultural elements of the survey by talking with community members during an exploratory trip to the community, working with my translator, talking to Greenlandic officials in Nuuk, and addressing revisions by academic professionals. The interview had 12 main open-ended questions and some of the questions had set follow-ups.

My translator in Greenland handed out the interview guides to potential interviewees ahead of time to familiarize them with materials so they could determine if they were comfortable participating. During the interviews, my translator led the questioning with some back translation to me so I could answer questions from the interviewees. We recruited self-identified Inuit Greenlanders with experience in relationships with outsiders (nine women, ten men; many were retired). The women were from 32 to 73 years old and the men from 44 to 83 years old. Most had never left Greenland; some had traveled to Denmark and a few to other countries in Europe. The retired women were involved in sewing, playing bingo, working on sealskins, and entertaining friends, while retired men hunted and fished for their families, but not commercially. The younger women I talked to were involved in raising families and had jobs in small shops in the town while the young men were hunters and fishermen by trade. All participants were longtime residents of the community.

On a later trip to Greenland, I met again with my translator and we conducted four, one-hour focus groups with fourteen of the original nineteen Greenlanders interviewed. We used focus group guides to ensure that the issues in need of clarification from the individual interviews were addressed, and so we had guiding topics for expansion. My translator led the focus groups while I observed facial expressions and the order of conversation. I conducted focus groups to give the interviewees the opportunity to construct knowledge together and lead the discussion (Madriz 2000). Additionally, the focus groups clarified the English translations of the Greenlandic individual interviews, ensuring respect for the Greenlandic culture and allowing me to "member check"[2] for information accuracy and validity.

18.5 Results

Asking questions about how to build research relationships revealed that trust is paramount and eight important actions are needed to build a trusting relationship: (1) Knowing community history; (2) Having local contacts; (3) Communicating openly; (4) Respecting each other; (5) Having good manners; (6) Following institutional ethics; (7) Exchanging knowledge; and (8) Giving back. Researchers addressed actions they felt contributed to relationship and trust building while Greenlanders confirmed some of those ideas and added others (such as good manners). The interviews revealed that trust is paramount if a relationship is to be

---

[2] Member checks are a way to get feedback from the interviewee about the validity and accuracy of researcher interpretations of the interviewee's responses (Morse et al. 2002).

sustainable, and that trust is accessible through the eight actions.[3] One researcher explained how to approach a community:

> People are not interested in my credentials. They are interested in why I am there, about my family, if I am married, where I come from. People will see through you if you are
>
> fake...Listen, look, observe much longer than you think, before you start asking for or proposing things, but be honest with who you are and what you want to do from the start but do not pursue until there is a relationship.

18.5.1 Knowing Community History

Three Greenlanders elaborated on the importance of researchers knowing the history of the community before arriving, so they know how to treat people and know about cultural differences and ways-of-life. Five researchers mentioned how knowing community history not only helps in terms of interacting with the community members, but it also supports the research process. By being in the community prior to writing a research proposal, the researcher can work toward gaining community trust, learn about community needs, and get community input for the proposal and research methodology. Issues arise when researchers do not want to put in the time to work with communities. It hurts the research and the process of relationship building with communities. As one researcher commented:

> Social, natural, and physical science projects are able to involve communities. Some researchers who have been around a long time still do not take the community into consideration. They do not put in time to find out about the community makeup and the social and political structure, which will ultimately help the research.

18.5.2 Having Local Contacts

All of the researchers and the Greenlander focus groups highlighted the importance of having a local contact prior to entering the community. This action facilitates communication and introduces the researcher to community members, thus giving the researchers some credibility from an introduction by a member of the community. Researchers talked about the local contact in a variety of ways. Some called this person a local contact, advisor, a gatekeeper, or a key informant. As one Greenlander explained: "*It is better to have prior knowledge before they [the researchers] come here, and that they know who to contact before coming to town,*

---

[3] Interviews and focus groups with Greenlanders were conducted in Greenlandic and were translated into English. The quotations are English translations of Greenlandic, seeking to translate the meaning instead of just words. The interviews with researchers were typed into a laptop as the interviewee spoke. As I did not use a tape recorder with the researchers, the quotes should be considered a close approximation.

to give a heads up to the people they want to talk to, so the people know in advance and are not surprised."

18.5.3 Communicating Openly

There are various reasons why it is important to foster open communication. Every Greenlander and researcher emphasized that, to conduct a research project, there must be communication flowing between the researcher and community. Both researchers and community members found communication important and talked about face-to-face interaction as the best form of communication. The researchers tended to worry about communicating along the proper channels, starting with the tribal or community government, the elders, the informal leaders, etc. Although this is both important and necessary for research approval, community members want researchers to interact with regular people, build relationships with, and get information from those who have firsthand experience with the research topic. A Greenlander explained the importance of communicating with locals: *"It is important to ask the right group of people. All have importance. My opinion is, if one wants to know something, getting close to the local people is where one gets better results."*

18.5.4 Respecting Each Other

In all four of the Greenlandic focus groups, respondents underlined how Greenlanders immediately give respect to outsiders. However, the Greenlanders said that researchers should be careful not to take advantage of this immediate respect. The researchers also need to reciprocate the respect. They can demonstrate this by acting non-offensively in their behavior or speech. Researchers can learn these behaviors through learning about community history, working with their contact person, and communicating openly. Researchers and Greenlanders can demonstrate respect by treating each other as equals. One Greenlander explained respect towards foreigners: *"Greenlanders are polite but are modest and have huge respect for foreigners who come here. It is because most Greenlanders think that the foreigner knows best or has better qualifications, then therefore is to be respected."*

Five researchers recognized the need to respect the communities they worked with by not using people and just leaving with data through the traditional research model of "helicopter research." One researcher summed this up by saying:

> *The key is ageing with the communities in a way that you are not doing research on the community but WITH them. This is a subtle difference but one in which people can quickly make a distinction when hearing about a research project...There is a history of people coming in and gathering data and running off with it and not coming back to the community or misrepresenting it in their reports.*

18.5.5 Having Good Manners

Ultimately, actions shape trust. Acting with good manners by community standards is very important for people to feel comfortable around outsiders. Greenlanders spoke of themselves as if they inherently had good manners, learning it in their upbringing. Good manners are tied to food and hosting someone, where the goal is to build a strong reciprocal relationship. One Greenlander explains that good manners are tied to communication: *"Good manners are intertwined with communication. It is the how of communication. It involves smiling and greeting, language, behavior, and the body. One also can have good manners through language, behavior and body."*

Three researchers emphasized that even if someone in the community is not explicitly involved in the project, the research team can say hello and engage in a brief conversation to maintain a positive presence in the community. A researcher explained how to have good manners:

> *You have to be constantly vigilant about nurturing relationships. Always go say hi, even if you are not working directly with administration people, etc. It gets people to be more responsive when they see you as a person. It is vital to build person relationships... Being an objective observer is not as important when you need to demonstrate to the community you are involved with them.*

18.5.6 Following Institutional Ethics

Institutional ethics,[4] also called contractual ethics, are the legal provisions universities and funders require in research involving human subjects (Brydon-Miller 2009). Two researchers mentioned how institutions, such as universities and funders, are changing; they expect more than traditional research and require dialogue with the communities. One researcher explained:

> *A lot of scientists get outraged since they used to be able to do what they wanted and now have to get permission for their work or talk to local people about their projects. The scientists just want to walk in and do whatever they want regardless of the effects on the community.*

The four Greenlandic focus groups and all fifteen researchers all spoke of the importance of involving the community from the beginning of the project conception. One Greenlander explained:

> *At the beginning, local people who are interested in the project should be invited to learn about the research; what is the goal? By the midway, there should be an evaluation and an explanation of the upcoming work process. In the end, the results need to be explained to people who were involved.*

---

[4] Scientists and researchers often use the term ethics to talk about how to treat people in research. This term does not easily translate into Greenlandic, and the translator had to use Danish for "ethics" and the words respect and good manners in Greenlandic.

18.5.7 Exchanging Knowledge

Exchanging local knowledge with the knowledge brought in by outside researchers results in both co-created knowledge and new knowledge for each side. The community provides a long history of knowledge about the area and year-round observation. Over half of the 19 Greenlanders interviewed advocated for exchanging knowledge and outlined the benefits available for both researchers and the community. A Greenlander explained:

> We should give our knowledge to these people. It is when this happens that we will develop and our knowledge will grow. They won't learn anything if we don't talk to them – like now – and we won't learn anything if we don't talk to them. That is an exchange of knowledge.

A researcher emphasized the value of traditional knowledge from a community: *"By utilizing traditional knowledge, talking to, and partnering with communities existing for centuries, much depth and context can be added to the data where scientists usually only collect field data for a few months."*

One Greenlander had trouble with climate researchers coming in with their own data and not listening to local observations. This person emphasized the importance of getting to know someone and their experiences, beyond just gathering climate observations from them:

> I have met many researchers. For example, I have spoken to German, Russian, or Danish climate researchers. They already have presumptions about how Greenland has never been this warm. When they have these opinions, it is hard to explain things to them – of course; they want evidence, as they are researchers. When they have presumptions, it creates barriers. I have problems with that. And as we have lived here for many years, our knowledge should be greater than theirs. This interview is the first time that someone has asked how we are experiencing things; how we experience things can be hard to explain. And this search for the background information, which should be included in most research, is often missing.

18.5.8 Giving Back

At the end of the project, nothing is more important than bringing the data, project findings, and results back to the community. All 19 Greenlanders spoke about this issue. One Greenlander summed up their experience with Danish researchers:

> A lot of other Danish researchers come and leave, without further notice; they find what they are interested in, buy it, take it, and never come back. They took so many rocks with them, without returning them. These belong to Greenland…That is the behavior of Danish and Europeans. They come here, take things and leave, without saying anything. So many Danish professors get their degrees writing about Greenland. They never thank us; they never give any information to the Greenlanders, after making research in Greenland.

Researchers talked about different ways to give back data: presentations, face-to-face interactions, and conferences. However, while the researcher stays in the community, Greenlanders would like them to be involved in the schools, teaching an activity, or helping out with something possibly completely unrelated to their research project. Over half of the 15 researchers interviewed talked about this as well. They make the point that not giving back may close the community to further research relationships. One researcher explained that: *"Research cannot be done for a random research argument. It needs to matter and be understood and important to the community…What the researcher is doing actually needs to be helpful for the community."*

18.6   Coming Full Circle to Trust

Building trust as a researcher is like working towards becoming a member of the community (Fig. 18.1). Actions taken by both researchers and Indigenous community members lead to a trusting relationship, but a trusting relationship takes time to develop. Lucero (2013) links the CBPR framework with ethics and trust building. She defines trust through her project results as, "trust is having an emotionally safe and respectful environment based on shared values to promote a sense of responsibility to the partnership while working toward shared goals" (Lucero 2013: 120). While Lucero *defines* trust, I sought *how* to arrive at trust. I started out by wondering how to collaborate equitably; researcher interviews demonstrated that trust is necessary. When turning to previous collaborative research, I found that, yes, collaboration is best (Huntington and Fox 2005), but how do researchers do it? I then turned to CBPR literature and found a model showing what relationships need to consider, and Lucero's (2013) definition of trust (Wallerstein et al. 2008). Finally, I turned to Indigenous methodologies for guidance in working with Indigenous people, which revealed unique types of collaboration (Smith 2005). After I examined the literature, I still wondered about how to arrive at the trust needed for mutually beneficial relationships.

Fig. 18.1   Trust

From the interview data, I found that maintaining open communication with the community, being transparent about the project, and being accountable to community issues goes a long way to building trust. To help manage this issue of trust, researchers cannot just show up with their project money and start doing their work. They need to first learn about the community and develop local contacts. Fondahl et al. (2009) suggest scheduling time for socializing so that the community and researchers can get to know one another. Additionally, it is important for researchers to consider their exit from the community before even arriving; continuity in research projects builds long-term relationships and trust with communities through personal friendships. A principal investigator who maintains long-time ties with the community and even lives there year-round develops strong bonds with people in the community. This year-round contact helps maintain interest in the project (Gearheard and Illauq 2009).

Researchers who do not follow community cultural norms can cause confusion, discomfort, and distrust among community members. When working in Indigenous communities, researchers need to remember the historical trauma people endured and how this is passed down between generations (Minkler 2004; Fondahl et al. 2009). In all types of relationships and projects the researcher needs to remember to be honest about who they are and reflexive on their position in the community. This does not mean the researcher assumes they are a part of the community. They do not share the same history, trauma of colonization, or experiences as community members. Making the commitment to act according to how community members treat one another is a respectful and ethical way to conduct research, observing definitions of behavior established at the community level. This includes researchers meeting people face-to-face; paying attention and listening; sharing and hosting; being reflexive and non-judgmental; being respectful of dignity; demonstrating reciprocity and responsibility; and avoiding flaunting knowledge (Smith 2005; Hart 2010). Researchers must be humble and responsible.

Many ethical guidelines that are designed specifically for research with Indigenous people articulate these responsibilities. However, Greenland is a unique case in that it currently lacks ethical guidelines for research specific to Greenland and its Indigenous peoples (Holm et al. 2010). There are guidelines for research in Denmark based on Danish laws. These guidelines, however, are not followed or enforced in Greenland, as it has different laws (Olsen et al. 2004). There is a need to develop ethical guidelines for Greenland. Holm et al. (2010) recommend that the Greenlandic government begin to develop guidelines through consultation with arctic agencies that already have guidelines and with Greenlanders that would be involved in the research. It is important to take into account the history of Greenland as a colony, a country with Home Rule, and as an Inuit community connected to other Inuit communities by the Inuit Circumpolar Council.

Exchanging knowledge by collaborating to gather information creates trust and a feeling of shared community between researchers and community members (Huntington et al. 2002). Greenlanders and researchers interviewed in this study emphasized this point. Additionally, the researchers cannot just take the data they gather. The researcher "has a debt to the communities in which he or she works and must spend a certain amount of time doing practical work at the behest of the community in addition to carrying out fieldwork to meet his or her personal goals" (Penfield et al. 2008: 189). Perhaps researchers can think about what type of project the community may want and from there take on the role of grant writing and data analysis to build community capacity and ultimately complete the work.

CBPR and the literature on Indigenous methodology often identifies trust-building aspects during or after the project. This paper demonstrates that trust is an active process built prior to and during the research process. It also shows that researchers have a lot expected of them when they work with Indigenous peoples in the Arctic. However natural it is, relationships take work. Oetzel (2005: 366) explains, "Many people believe that good communication skills are "common sense"…the problem with common sense is that it is not all that common…" As this study demonstrates, relationship building is not common sense, and the eight actions identified are vital to creating, building, and sustaining a trusting research relationship.

Acknowledgements I wish to thank the Washington Internships for Native Students, ASRC Federal, National Science Foundation Office of Polar Programs, NSF ARC #0908151, International Arctic Social Sciences Association, Association of Polar Early Career Scientists, International Arctic Science Committee, International Arctic Research Center, Arctic-FROST: NSF PLR #1338850, University of Wisconsin-Madison Sociology Department, Dr. Randy Stoecker, Dr. Elizabeth Rink, the researchers willing to describe their experiences, and all of the Greenlanders who welcomed me into their community and shared their stories with me.

References

Alaska Federation of Natives. (1993). *Alaska Federation of Natives guidelines for research.* Retrieved June 3, 2016, from http://ankn.uaf.edu/IKS/afnguide.html

Berkes, F. (2002). Epilogue: Making sense of arctic environmental change? In I. Krupnik & D. Jolly (Eds.), *The earth is faster now: Indigenous observations of arctic environmental change* (pp. 334–349). Fairbanks: Arctic Research Consortium of the United States.

Bronen, R. (2008). Alaskan communities' rights and resilience. *Forced Migration Review, 31,* 30–32.

Brydon-Miller, M. (2009). Covenantal ethics and action research: Exploring a common foundation for social research. In D. M. Mertens & P. E. Ginsbergm (Eds.), *Handbook of social research ethics* (pp. 243–258). Newbury Park: SAGE Publications.

Burnette, C. E., & Sanders, S. (2014). Trust development in research with Indigenous communities in the United States. *The Qualitative Report, 19*(44), 1–19.

Coyne, I. T. (1997). Sampling in qualitative research. Purposeful and theoretical sampling; Merging or clear boundaries? *Journal of Advanced Nursing, 26*(3), 623–630.

Dahl, J. (1988). From ethnic to political identity. *Nordic Journal on International Law, 57*(3), 312–315.

Embassy of the United States. (2012). *Greenland: Embassy of the United States Copenhagen Denmark.* Retrieved from http://denmark.usembassy.gov/gl/about.html

[The] Examining Community-Institutional Partnerships for Prevention Research Group. (2006). *Developing and sustaining community-based participatory research partnerships: A skill-building curriculum.* Retrieved from www.cbprcurriculum.info

Fisher, P. A., & Ball, T. J. (2003). Tribal participatory research: Mechanisms of a collaborative model. *American Journal of Community Psychology, 32*(3/4), 207–216.

Fondahl, G., Wright, P., Yim, D., Sherry, E., Leon, B., Bulmer, W., Grainger, S., & Young, J. (2009). Co-managing research: Building and sustaining a First Nation – University partnership, CDI Publications Series, University of Northern British Columbia Retrieved from http://www.unbc.ca/cdi/publications.html

Gad, F. (1973). *The history of Greenland II: 1700–1782*. London: C. Hurst & Company.

Gearheard, S., & Illauq, N. (2009). The Ittaq Heritage and Research Center: Inuit led research in Nunavut. *Meridian*,12–15.

Gearheard, S., Pocernich, M., Stewart, R., Sanguya, J., & Huntington, H. P. (2010). Linking Inuit knowledge and meteorological station observations to understand changing wind patterns at Clyde River, Nunavut. *Climatic Change, 100*(2), 267–294.

Glaser, B. G. (1978). *Theoretical sensitivity*. Mill Valley: Sociology Press.

Hart, M. A. (2010). Indigenous worldviews, knowledge, and research: The development of an Indigenous research paradigm. *Journal of Indigenous Voices in Social Work, 1*(1), 1–16.

Hirshberg, D., & Sharp, S. (2005). *Thirty years later: The long-term effect of boarding schools on Alaska Natives and their communities*. Anchorage: Institute of Social & Economic Research.

Hodge, F. S. (2012). No meaningful apology for American Indian unethical research abuses. *Ethics & Behavior, 22*(6), 431–444.

Holm, L. K., Grenoble, L. A., & Virginia, R. A. (2010, October). *Toward a new research ethic for Greenland*. Paper presented at the 17th Inuit Studies Conference, Val-d'Or, Québec.

Huntington, H., & Fox, S. (2005). The changing Arctic: Indigenous perspectives. In C. Symon, L. Arris, & B. Heal (Eds.), *Arctic climate impact assessment* (pp. 61–98). Cambridge: Cambridge University Press.

Huntington, H.P., Brown-Schwalenberg, P.K., Frost, K.J., Fernandez-Gimenez, M.E., Norton, D.W., & Rosenberg, D.H. (2002). Observations on the workshop as a means of improving communication between holders of traditional and scientific knowledge. *Environmental Management, 30*, 778–792.

Huntington, H., Gearheard, S., Mahoney, A. R., & Salomon, A. K. (2011). Integrating traditional and scientific knowledge through collaborative natural science field research: Identifying elements for success. *Arctic, 64*(4), 437–445.

Israel, B., Shulz, A., Parker, E., & Becker, A. (1998). Review of community-based research: Assessing partnership approaches to improve public health. *Annual Review of Public Health, 19*, 173–202.

Kennedy, V., & Wehipeihana, N. (2006). A stock take of national and international ethical guidelines on health and disability research in relation to Indigenous People (Unpublished report), The National Ethics Advisory Committee Te Kahui Matatika o te Motu.

Leiderman, S., Furco, A., Zapf, J., & Goss, M. (2003). *Building partnerships with college campuses: Community perspectives* (Monograph: A publication of the Consortium for the Advancement of Private Higher Education's Engaging Communities and Campuses Program). The Council of Independent Colleges.

Lucero, J. (2013). *Trust as an ethical construct in community based participatory research partnerships*. Unpublished doctoral dissertation. University of New Mexico, Albuquerque.

Lynge, A. E. (2011). Mental decolonization in Greenland. *Intern-Nord, 21*, 273–276.

Madriz, E. (2000). Focus groups in feminist research. In N. K. Denzin & Y. S. Lincoln (Eds.), *The Sage handbook of grounded theory* (pp. 835–850). Thousand Oaks: Sage.

Minkler, M. (2004). Ethical challenges for the "outside" researcher in community-based participatory research. *Health Education & Behavior, 31*(6), 684–697.

Morse, J. M., Barrett, M., Mayan, M., Olson, K., & Spiers, J. (2002). Verification strategies for establishing reliability and validity in qualitative research. *International Journal of Qualitative Methods, 1*(2), 13–22.

Oetzel, J. G. (2005). Effective intercultural workgroup communication theory. In W. B. Gudykunst (Ed.), *Theorizing about intercultural communication* (pp. 351–371). Thousand Oaks: Sage.

Olsen, J., Mulvad, G., Pedersen, M. S., Christiansen, T., & Sørensen, P. H. (2004). An ethics committee for medical research in Greenland: History and challenges. *International Journal of Circumpolar Health, 63*(2), 144–146.

Penfield, S. D., Serratos, A., Tucker, B. V., Flores, A., Harper, G., Hill, J., Jr., & Vasquez, N. (2008). Community collaborations: Best practices for North American Indigenous language documentation. *International Journal of the Sociology of Language, 191*, 187–202.

Rasmussen, K. (1908). *The people of the polar North: A record* (G. Herring, Trans.). London: Kegan Paul, Trench, Trübner & Co.

Ray, L. (2012). Deciphering the "Indigenous" in Indigenous methodologies. *Alternative: An International Journal of Indigenous Peoples, 8*(1), 85–98.

Rosenthal, E. (2012). Race is on as ice melt reveals arctic treasures. *New York Times*, September 19. Retrieved September 21, 2012, from http://www.nytimes.com/2012/09/19/science/earth/arctic-resources-exposed-by-warming-set-off-competition.html

Smith, L. T. (2005). On tricky ground: Researching the Native in the age of uncertainty. In N. K. Denzin & Y. S. Lincoln (Eds.), *The Sage handbook of grounded theory* (pp. 85–107). Thousand Oaks: Sage.

Strauss, A., & Corbin, J. (1990). *Basics of qualitative research: Grounded theory procedures and techniques*. Newbury Park: Sage.

UN General Assembly, United Nations Declaration on the Rights of Indigenous Peoples: resolution/adopted by the General Assembly, 2 October 2007, A/RES/61/295, Retrieved 16 October 2014, from http://www.refworld.org/docid/471355a82.html

Vaughan, R. (1991). *Northwest Greenland: A history*. Orono: The University of Maine Press.

Wallerstein, N., Oetzel, J., Duran, B., Tafoya, G., Belone, L., & Rae, R. (2008). What predicts outcomes in CBPR? In M. Minkler & N. Wallerstein (Eds.), *Community based participatory research for health* (2nd ed., pp. 371–394). San Francisco: Jossey Bass.

Weatherhead, E., Gearheard, S., & Barry, R. G. (2010). Changes in weather persistence: Insight from Inuit knowledge. *Global Environmental Change, 20*(3), 523–528.

Chapter 19
Beginnings of a Rural Sustainability Paradigm: The Arctic as Case in Point

Susan A. Crate

Abstract Developing sociocultural, economic, and environmental strategies for sustainability is a major challenge facing the world's rural and urban populations alike. Rural areas lag severely behind their urban counterparts in addressing sustainability milestones yet, as nexuses of biological, cultural and ethnic diversity, they play a crucial role in planetary sustainability. Therefore, there is great need for rigorous research with rural communities to define issues, exchange necessary knowledge and synthesize nascent initiatives exploring rural sustainability. This paper lays a framework for one possible research approach by reflecting on insights from a comparative case, involving long-term research in two arctic contexts: Viliui Sakha settlements of northeastern Siberia, Russia and Nunatsiavut settlements in Labrador Canada. Despite their location on opposite sides of the Arctic, communities in both regions struggle with contemporary issues of a changing climate, an unpredictable economic basis, outmigration of their young people to the urban areas and issues of environmental contamination from past and projected resource extraction.

Keywords Rural sustainability • Complexity of change • Sakha Republic, Russia • Nunatsiavut, Labrador, Canada • Climate change

19.1 Introduction

Developing sociocultural, economic, and environmental strategies for sustainability is a major challenge facing the world's rural and urban populations alike. Rural areas severely lag behind their urban counterparts in addressing sustainability milestones yet, as nexuses of biological, cultural, and ethnic diversity, they play a crucial

S.A. Crate (✉)
Department of Environmental Science and Policy, George Mason University,
Fairfax, VA, USA
e-mail: scrate1@gmu.edu

© Springer International Publishing Switzerland 2017   253
G. Fondahl, G.N. Wilson (eds.), *Northern Sustainabilities: Understanding and Addressing Change in the Circumpolar World*, Springer Polar Sciences,
DOI 10.1007/978-3-319-46150-2_19

role in planetary sustainability. Therefore, there is great need for rigorous research with rural communities to define issues, exchange necessary knowledge and synthesize nascent initiatives exploring rural sustainability. This article lays a framework for one possible research approach by reflecting on insights from a comparative case, involving long-term research in two arctic contexts: Viliui Sakha settlements of northeastern Siberia, Russia and Nunatsiavut settlements in Labrador Canada. Despite their location on opposite sides of the Arctic, communities in both areas struggle with contemporary issues of a changing climate, an unpredictable economic basis, outmigration of their young people to the urban areas, and issues of environmental contamination from past and projected resource extraction.

This article begins by laying a foundation of the larger rural sustainability literature and a focus on sustainability efforts in the Arctic. It then introduces a model of the complexity of change, based on two arctic cases. Although varying in specific human-environment relationships, socio-cultural particularities, and historical and economic trajectories, the two cases are challenged by at least three similar drivers of change that work to undermine sustainability at the local level. These three drivers, in turn, make up the foundation of a hypothetical model of the complexity of change. The complexity of change model can then be used as an adaptable model for broader use in other global rural contexts. The article emphasizes the need to bolster rural sustainability via community-collaborative efforts, highlighting two approaches, to not only inform research but also local understandings of the complexity of change in order to facilitate self-empowerment for those communities into the future.

19.2 The Larger Case for Rural Sustainability

Since the mid-twentieth century, the diversity of earth's ecological and social systems has been declining due to climate alterations and multiple socio-economic forces of change (c.f. Andersson 2010). This decline, in turn, threatens planetary sustainability (Díaz et al. 2006; Dietz et al. 2003; Hooper et al. 2012; Turner et al. 2003). Most amelioration efforts focus on technological approaches, urbanization, and ecologically sound industrialization (Raven 2002; Stokstad 2005; Steffen et al. 2007). However, "for millions of people around the world, there are issues more immediate, more urgent, than addressing political economic equity or designing some new social or technological utopia" (Rajan and Duncan 2013: 70). Rural areas (regions that are both low in population and have most land in agriculture and/or subsistence production) are nexuses of biocultural diversity and are thereby critical to planetary health. They are also undergoing socio-cultural, economic, and environmental crises (Tilman 1999; Tilman et al. 2006; Zavaleta et al. 2010; Kay et al. 2012; Bradshaw et al. 2007; Stedman et al. 2012; IPCC 2012). The drivers of change in the global diversity of rural contexts are socio-cultural, environmental, and economic, and referred to from here on as the 'complexity of change' (Crate 2014). Research approaches that are local, and designed to identify how the complexity of change interacts with local communities, are bringing solutions in local contexts (Rasmussen 2009; Burke and Shear 2014; Rajan and Duncan 2013; Muldonado 2014; West 2013) and they do so by understanding those drivers within the 'bottom-up complexity,' or the specific ecosystem and sociocultural interactions in the locale (Hastrup 2009). Because rural communities' perceptions, understandings and responses to the complexity of change are critical to facilitating sustainable futures, rural areas and the communities inhabiting them are a needed focus of comprehensive efforts to understand and address global sustainability.

Rural communities represent *complex adaptive systems*, a dynamic interplay of 'biophysical systems' (e.g. mechanisms of change and communication) and compatible key features of social systems: "the holistic nature of culture, knowledge sharing through the senses, and the formative power of traditions, structures/materials, strategies and habits of mind' (Crumley 2012: 305). Diversity is a generative feature of complex adaptive systems (Gregory 2006). Through the continued use of local plant, animal, and natural resources, rural community inhabitants manage and steward biological diversity. Additionally, by maintaining a diversity of languages and dialects, basing use and management of local resources on situated local knowledge, utilizing a variety of effective and robust institutional arrangements for community-based resource management (Ostrom 1990, 2009), and practicing historically-founded and place-based cultural and spiritual practices, rural community inhabitants support biocultural diversity (Maffi and Woodley 2010). In this way, many researchers argue that rural areas are 'biocultural refugia,' that safeguard not only biodiversity towards global food security, but also the culture-specific practices that maintain that biodiversity (Barthel et al. 2013).

The Arctic represents one region of the world where these rural community characteristics are readily apparent and where the complexity of change is clear: "For the Arctic, social, political and economic drivers of change may be of equal or greater importance, bringing about thresholds whether or not there are changes in the climate" (Arctic Council 2013: 56). Rural arctic communities continue, to a greater or lesser extent, to rely on community/extended kin networks of interdependence, labor, and resource reciprocation as well as on informal economies and barter. While all societies are vulnerable to significant perturbations in the equilibrium of their socio-ecological systems (Smith and Wishnie 2000), rural arctic inhabitants have long utilized biocultural diversity to ensure their adaptability and resilience to a changing world (Berkes et al. 2007; Sveiby 2011).

However, the larger mix of changes present in contemporary rural arctic communities is increasingly threatening to diversity and necessitates approaches with the analytical power to discern between them. The local and regional effects of climate change result in a 'cascade' of other ecological changes within the ecosystem. Once identified, the cascading ecosystem effects need to be considered within the context of other changes, not climate related. For example, O'Brien and Leichenko (2000) demonstrated how the synergisms between two global processes, climate change and globalization, which the authors coined as "double exposure", need to be reconsidered together because their joint impact in analysis creates different results than each taken alone. Others term the host of impacts 'multiple-stressors' in determining the vulnerability of arctic peoples and ecosystems to potentially harmful impacts (McCarthy and Martello 2005); this interdisciplinary climate science often engages ecologists, climatologists, modelers, and social scientists to discern multi-stressors (Stenseth et al. 2002). Using the frameworks of adaptation and resilience, interdisciplinary groups working in the Arctic have emphasized the need to investigate this larger mix: from the Arctic Climate Impact Assessment, which argued that understanding the impact of climate change "requires a holistic understanding of multiple drivers of change and their interactions" (McCarthy and Martello 2005: 947) to the Arctic Resilience Interim Report which states "[t]he need for 'integrative concepts and models' that can aid systemic understanding of the Arctic, including the cumulative impacts of a diverse suite of interconnected changes, is critical in the current period of rapid ecological, social, and economic change" (Arctic Council 2013: 4).

Attempts to account for the complexity of change using modeling (Pielke et al. 2012) often fall short in terms of integrating human aspects, despite the fact that the "development and testing of interactive models that can simulate the evolving nature of interactions among social and environmental states is a major research priority" (Dearing et al. 2010). Research efforts need to contribute to this priority by focusing on social science approaches that are holistic and that investigate epistemology in its most basic form: the ways humans know. They can take an inter-epistemological approach and contemplate the "upstream" questions related to understanding the foundations of the knowledge system(s) being considered, and focus on *how* things are known rather than *what* is known (Murphy 2011: 492). In addition to other considerations, these are key components of the holistic approach needed to bring climate science and other socio-ecological analyses to full realization.

19.2.1 The Need for a Holistic Approach

Efforts to integrate a 'human dimension' in climate science, although having made some progress in recent years, remain lacking (IPCC 2001, 2007, 2012; Hovelsrud et al. 2011; Krupnik et al. 2005; ACIA 2005; AHDR 2004). There is a need for a systematic understanding of the social science of climate change because, "studies of the functioning of social systems generally take more diverse perspectives, recognize the contingency of social change, and seek to respect and reflect the specificity of local concerns" (Arctic Council 2013: 56). What is needed is a holistic approach, an appreciation and accommodation for how indigenous/ local knowledges are synthetic and holistic (Agrawal 1995, 2002; Berkes 2008; Cajete 1999; Sillitoe 1998; Aporta 2010).

Arctic research is one arena that increasingly uses a holistic and human-inclusive approach by down-scaling to the community level and engaging communities in the documentation and monitoring process to understand change (Berkes et al. 2007; Pearce et al. 2009). The last decade has seen a significant increase in community engagement in and ownership of communicating local change (Hovelsrud et al.

2011). Such efforts in arctic climate change research began with community-based research discerning how the global phenomenon of climate change was affecting long-standing human adaptive strategies (Krupnik and Jolly 2002). These types of investigations led towards efforts to substantiate local knowledge by corroborating it with scientific data, to integrate the two knowledge systems towards new understandings (Danielsen et al. 2009; Crate and Fedorov 2013a) and to build community capacity (Krupnik et al. 2010; Gearheard et al. 2011).

One major hurdle to adopting holistic approaches that comprehensively integrate qualitative and quantitative data that are scaled down to the local level where effects on humans and ecosystems are occurring (Arctic Council 2013:4) is that such approaches require significant time and resources. Qualitative approaches require in-depth and often multiple local studies, fulfillment of proper ethical considerations, additional time to gain entry, time to learn local protocols, the accommodation of community expectations, and sensitivity to political alliances (Heikkilä and Fondahl 2012: 62). Research can address these issues by working in communities with well-established research relationships, and by planning for the time and resources required to accomplish tasks. Multiple levels of society benefit when researchers use such a holistic approach. Most obviously, there are immediate benefits to the affected communities and the researchers collaborating with them. Less apparent are the more latent benefits, since the effects of such an approach gradually transform ways of thinking among larger groups of stakeholders associated with the project, and eventually, "the dominant society that suffers a shortage of wisdom" (Herman 2008: 77). Critical to the effectiveness of holistic approaches is engaging the next generation of stakeholders, the youth.

19.2.2   Working Across the Generations by Engaging Youth

One of the key challenges of retaining the next generation and reversing the trends of young people's out-migration is countering external forces of change. These include economic globalization and the impacts of consumer culture that work to disrupt local economies and increase unemployment thus setting youth sights away from home prospects. Adolescents and youth are poised to play a pivotal role in rural contexts, despite finding themselves at the center of major transformations observed indistinctively throughout the world that include depopulation, de-agrarianization, and the loss of the previous generations' knowledge (Virtanen 2012; Steward 2007; Bryceson 1996). There are also internal forces at work such as the older generations' limited acceptance of youth's abilities and interests which lead to high unemployment rates and overall lack of future perspectives within rural communities (Rasmussen 2009; Lundholm and Malmberg 2006; Hamilton and Seyfrit 1993; Garcia and Gonzalez 2004). The severity of this situation is further exacerbated by youth's lack of skills and qualifications and the lack of adequate educational resources to respond to new demands in rural areas (Rasmussen et al. 2010; Carson et al. 2011; Virtanen 2012). Most prominent is the breakdown in

knowledge transmission from elders to youth because knowledge is not being transmitted in pace with social change in rural areas. This is further complicated by the
fact that knowledge has moved from being local and activities-based to being global
and virtually-based.

Promoting opportunities for the upcoming generations to actively participate in
their culture and livelihood while at the same time guaranteeing access to high-
quality educational and training opportunities are key elements to the development
of human resources in rural areas (NORDEN 2006). Although youth out-migration
from rural areas is highly variable, with some areas experiencing one-way movement (rural to urban) and others demonstrating patterns of rural-urban circulation
(Hansen et al. 2012), the global trend is an overall decrease of youth populations in
rural areas (McGranahan et al. 2010; Argent and Walmsley 2008; Stockdale 2006).
There are also wide divergences in local, regional and state efforts to retain youth
with a pattern of developed countries having retention programs, while less developed countries do not have programs (Wyn and Harris 2004; Glendinning et al.
2003). Because youth play a pivotal role in shaping the future of rural communities,
understanding and enacting rural youth's values, aspirations, and future perspectives in addressing short- and long-term effects of change is key to the global transformation towards sustainability. Research efforts need to place youth at the
forefront with questions such as: How can knowledge systems be best integrated?
What and where are the disconnections between youth and elders' different ways of
knowing, understanding, and communicating? And where are the disconnections
within the varied understandings of the usefulness of knowledge? The article now
considers these aspects of rural sustainability and the challenges to it in the context
of two cases in the Arctic that informed a hypothetical model of the complexity of
change.

19.3 Understanding Rural Sustainability via the Cases

The two cases are Viliui Sakha communities (horse and cattle-breeding agropasto-
ralist communities) of northeastern Siberia, Russia and in Inuit, Settler, and Métis
communities of Labrador, Canada. The Viliui Sakha inhabit the Viliui River regions
of western Sakha Republic (Yakutia), an area twice the size of Alaska, which is
landlocked except for its northern border on the Arctic Ocean. The region has an
extreme continental climate with annual temperatures ranging from −60 °C in the
winter to +40 °C in summer. The Sakha's Turkic ancestors transmigrated to the
Viliui and Lena regions of the area in the eleventh and twelfth centuries from southern Siberia. They adapted their horse and cattle breeding to the extreme environment mostly by protecting their cattle in barns in the cold seasons and harvesting
copious amounts of hay in the brief summer for their fodder. The area is one of
continuous permafrost, with over 25 types of permafrost within its borders.

Similarly, the inhabitants of coastal Labrador communities have adapted their
lifestyle to an extreme environment; albeit in a coastal context, relying upon the

annual establishment of sea-ice for sealing and fishing. Although Inuit had already inhabited and thrived in these areas for centuries, Moravian missionaries began establishing communities in the mid-1700s. The cod industry became the nexus of economic growth and development followed by salmon fisheries. Beginning in the 1970s, Inuit founded the Labrador Inuit Association and worked the next 35 years to settle land claims and establish the Nunatsiavut Government. Today communities practice mixed economies, continuing to depend on subsistence hunting and fishing and also working jobs in wage economy.

Both cases illustrate how sustainability is affected at the local level. The project's methodological foundation was grounded theory, an inductive approach to collecting and analyzing data in order to induce theory from the data (Mayan 2009: 47). This approach was the best fit for a project that intended to make the efforts community-owned because, "[t]o generate theory … we suggest as the best approach an initial systematic discovery of the theory from the data of social research. Then one can be relatively sure that the theory will fit the work. And since categories are discovered by examination of the data, laymen involved in the area to which the theory applies will usually be able to understand it …" (Berg and Lune 2012: 358).

The research used qualitative methods of field study with components such as interviews, focus groups, participatory action research, community modeling, grounded theory, and secondary data analysis. The main objective of these methods was to discern the communities' perceptions, and understandings of and responses to the changes and their opinions about the causes of change. The translation of those change components was used to test and refine a hypothetical model and a knowledge exchange process, utilizing a variety of 'tools' to integrate understandings and refine the model. The project worked with several communities in each of the two research sites.

Although situated in very different geopolitical contexts, the communities face similar challenges of being 'ice-dependent'; with the local effects of global climate change being shorter ice seasons in Nunatsiavut and degrading permafrost in Sakha (Crate 2012). The longer ethnographic picture in Sakha shows how inhabitants have been impacted and have responded to these changes in land-ice conditions and, at the same time, how other drivers of change are at work that are more urgent to life on a daily basis (Crate 2014). In the process of tracking the local effects of climate/ seasonal change, local change beyond climate and seasonality, including socio-cultural, economic, demographic, historical and geopolitical change, were perceived as more immediate and urgent to the affected communities (Crate 2015, 2014). Indeed although climate and phenological change are clearly affecting local contexts – most notably by the absence of important resources for major parts of the year and the 'wrong timings' of the seasons – communities perceive the other drivers of change to have a more immediate and urgent impact on daily life. The main concerns are the effects of economic globalization that undermine the local subsistence economy by increasing the demand for imported products and encouraging the one-way out-migration of youth to regional and urban centers for education and employment (see Fig. 19.1).

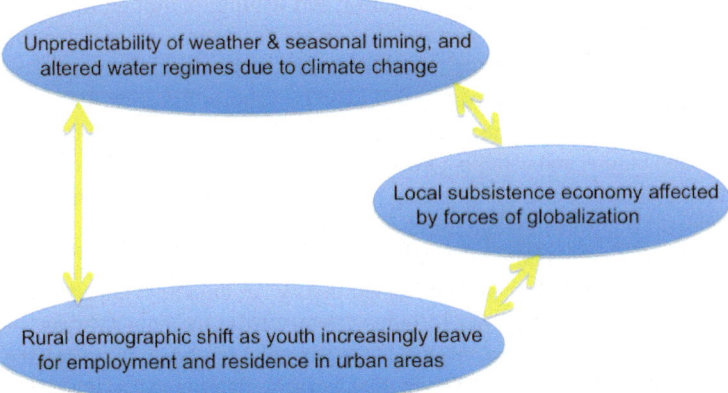

Fig. 19.1 Key drivers of change

Furthermore, in a similar way that ice dependence was a consistent issue across arctic communities (Crate 2012), the complexity of change affecting rural communities has broad similarities between regions. However, the sociocultural responses and the ways in which the communities perceive, interpret, and respond to complex social changes are highly variable. For example, in the Viliui Sakha context, middle-aged and older inhabitants view success for their youth as being able to leave the home village to get a higher education and make it elsewhere. Conversely, the majority of the same age group in the Nunatsiavut communities considered their young people's success as their ability to get a higher education and return to their home village to work. Much of that variation is due to geopolitical, ethno-historical, socio-cultural, and economic differences. This understanding is represented in a hypothetical model seen in Fig. 19.2. This model demonstrates how climate change and its accompanying phenological alterations act as one of several main drivers of change affecting and interacting with local communities.

These contrasting contexts not only clarify the various drivers of local change and their effects, but also the biocultural particularities that shape how communities perceive, understand, and respond to changes. The juxtaposed contexts also illuminate the extent that knowledge exchange can inform all relevant stakeholders, the best 'tools' to promote communication and information exchange, and the ways in which cross-generational interaction can proceed.

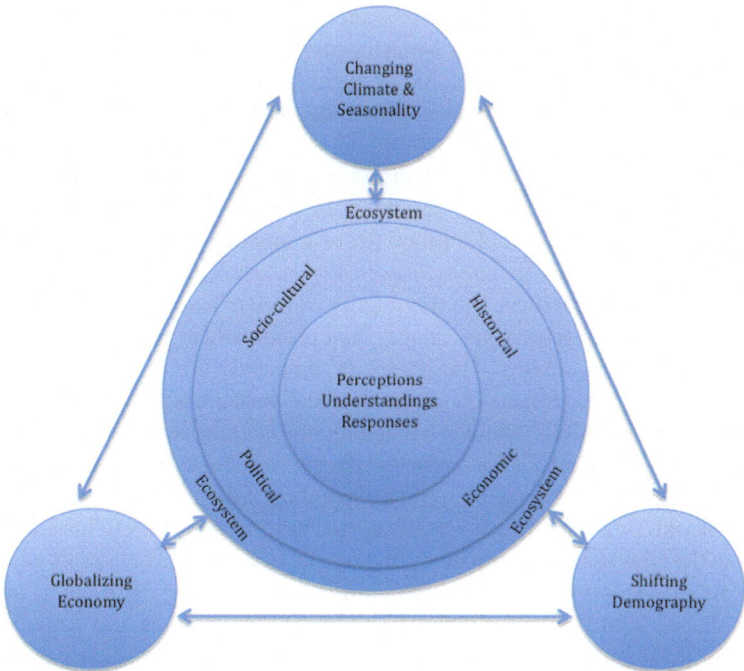

Fig. 19.2 Hypothetical model of the complexity of change

19.3.1 Effective Tools for Grounding Rural Sustainability Work in the Affected Communities

This research created two effective tools for community collaborative sustainability efforts: a knowledge exchange process and an online atlas project. Alexander Fedorov, permafrost scientist with the Melnikov Institute in Yakutsk, and I, developed a methodology for exchanging knowledge across stakeholder groups in northeastern Siberia (Crate and Fedorov 2013a). Although the project was aimed at informing two main stakeholder groups (regional scientists and local communities) about the local effects of climate change, the knowledge exchange activities worked to bridge greater understandings across the diversity of stakeholders involved, including regional administrators, policymakers, and environmental protection representatives. Evaluations and direct feedback from the eight knowledge exchanges in the summer of 2010 showed that participants wanted the process to reach a wider audience, preferably all inhabitants of the Viliui watershed regions. The research team's first thought was to create an interactive website that replicated as much as possible the knowledge exchange process. However, considering that the internet is not the chosen vehicle of communication and information access for the majority of the inhabitants of this region, the team developed a booklet in a community-collaborative writing process. Although it took 3 years, this iterative process produced 3000 copies which were distributed throughout the regions with the participation of the Sakha Republic's Ministry of Ecology (Crate and Fedorov 2013b). Similarly, other research efforts could implement the knowledge exchange process during both workshops in order to inform understandings amongst and between all relevant stakeholder groups.

The research also initiated two prototype online atlases. Work began in Makkovik, one of the Nunatsiavut, Labrador communities with whom we collaborated, where the community quickly settled on the idea of a map-based display of information about fishing stages and locations historically associated with local families and subsistence fishing activities. The community wanted to associate each fishing stage with audio and video interviews, images, and performances to capture information about their heritage/culture. These aspects included information about people, important events, and artifacts both from the local museum and found locally. Based on community decisions about what additional information should be captured about each type of object, a draft suite of schema was created and a few points were populated in order to provide community members with a product to critique. One of the outcomes of that critique was the desire by the community to have a timeline added to the atlas such that only items that overlapped the time range selected would be displayed. This would require the custom programming of the atlas and was accomplished with ELOKA (Exchange of Local Observations and Knowledge in the Arctic)[1] funding. The community nominated a local school teacher to facilitate entries into the atlas to ensure that they were suitable, and to demonstrate how to add information to the atlas and involve other members of the community, especially students. A month after the initiation of the project, I demonstrated the pilot atlas to inhabitants of Kutana, one of the Sakha communities where we worked. Here the community gravitated towards emulating the schema defined by the partner community of Makkovik. However, instead of fishing stages, the relevant unit of subsistence historically had been the *alaas*,[2] an ecosystem complex historically used by extended families in pre-Soviet times. A Sakha-specific atlas was developed based on the fields defined by the community. One major technical difference between Makkovik and Kutana was that Kutana preferred to host the atlas locally and manage a local copy of the atlas. In the meantime, the community has added several items to the atlas, most notably using the Cyrillic alphabet. Both demonstration atlases remain in community use today.

Such research efforts can set up ways for communities to share findings via an independent website, in both academic and popular publications and also at conferences. The project will reach at least four other populations: the immediate stakeholder groups that have direct relationships with the local communities, including policymakers; the wider social science community to more fully integrate qualitative information into global change science; the climate science community to

---

[1] See https://eloka-arctic.org/ for more information on ELOKA.
[2] Alaas are shallow thermokarst lakes, surrounded by hayfields that transition to taiga. They are a unique ecosystem type characterized by large areas of subsided ground surface resulting from thawing permafrostf, found in the permafrost regions of Sakha Republic (Yakutia).

understand how social science approaches such as this can complement climate/ seasonal change understandings; and that part of the global community eager to gain a better understanding of how climate and other change is understood, responded to, and how local inhabitants are finding ways to adapt and thrive.

19.4   Conclusion

This article drew upon insights from two arctic contexts, Viliui Sakha settlements in northeastern Siberia, Russia and Nunatsiavut settlements in Labrador Canada. Both of these contexts are challenged by a complexity of change involving not only issues of a rapidly changing climate and loss of ice and permafrost, but also of an unpredictable local economic basis and of out-migration of young people to urban areas. This paper lays a framework for a rural sustainabiity paradigm, using an understanding of these three key drivers of change in the context of the investigation of local cultural, geopolitical, historical, economic and geographical particularities. Considering the extent to which rural communities and their biocultural diversity are threatened in our rapidly changing world, this paradigm has the potential to contribute to the identification of pathways towards amelioration.

References

ACIA. (2005). *Arctic climate impact assessment*. Cambridge: Cambridge University Press.

Agrawal, A. (1995). Indigenous and scientific knowledge: Some critical comments. *Development and Change, 3*(3), 7–8.

Agrawal, A. (2002). Indigenous knowledge and the politics of classification. *International Social Science Journal, 173*, 287–297.

AHDR. (2004). *Arctic human development report*. Akureyri: Stefansson Arctic Institute.

Andersson, M. (2010). Provincial globalization: The local struggle of place- making. *Culture Unbound: Journal of Current Cultural Research, 2*, 193–214.

Aporta, C. (2010). The sea, the land, the coast and the winds: Understanding Inuit sea ice use in context. In I. Krupnik, C. Aporta, S. Gearheard, G. J. Laidler, & L. Kielsen-Holm (Eds.), *Siku: Knowing our ice, documenting Inuit sea ice knowledge and use* (pp. 165–182). Dordrecht: Springer.

Arctic Council. (2013). *Arctic resilience interim report 2013*. Stockholm: Stockholm Environment Institute and Stockholm Resilience Centre.

Argent, N., & Walmsley, J. (2008). Rural youth migration trends in Australia: An overview of recent trends and two inland case studies. *Geographical Research, 46*(2), 139–152.

Barthel, S., Crumley, C., & Svedin, U. (2013). Bio-cultural refugia—Safeguarding diversity of practices for food security and biodiversity. *Global Environmental Change, 23*(5), 1142–1152.

Berg, B., & Lune, H. (2012). *Qualitative research methods for the social sciences* (8th ed.). Boston: Pearson.

Berkes, F. (2008). *Sacred ecology* (2nd ed.). New York: Taylor and Francis.

Berkes, F., Berkes, M. K., & Fast, H. (2007). Collaborative integrated management in Canada's North: The role of local and traditional knowledge and community-based monitoring. *Coastal Management, 35*(1), 143–162.

Bradshaw, C. J. A., Sodhi, N. S., Peh, K. S. H., & Brook, B. W. (2007). Global evidence that deforestation amplifies flood risk and severity in the developing world. *Global Change Biology, 13*, 2379–2395.

Bryceson, D. F. (1996). Deagrarization and rural employment in sub-Saharan Africa: A sectoral perspective. *World Development, 24*(1), 97–111.

Burke, B. J., & Shear, B. W. (Eds.). (2014). Introduction: Engaged scholarship for non-capitalist political ecologies. *Journal of Political Ecology, 21*, 127–144.

Cajete, G. (1999). *Native science: Natural laws of interdependence*. Santa Fe: Clear Light Publishers.

Carson, D., Rasmussen, R., Ensign, P., Huskey, L., & Taylor, A. (2011). *Demography at the edge: Remote human populations in developed nations*. Farnham: Ashgate.

Crate, S. A. (2012). Climate change and ice dependent communities: perspectives from Siberia and Labrador. *Polar Journal, 2*, 61–75.

Crate, S. A. (2014). An ethnography of change in northeastern Siberia. *Sibirica, 13*(1), 30–74.

Crate, S. A. (2015). Chapter 11: Towards imagining the big picture and the finer details: Exploring global applications of a local and scientific knowledge exchange methodology. In J. Tischler & H. Greschke (Eds.), *The challenges of global climate change: Contributions from the social and cultural sciences* (pp. 155–171). Dordrecht: Springer.

Crate, S. A., & Fedorov, A. (2013a). A methodological model for exchanging local and scientific climate change knowledge in Northeastern Siberia. *Arctic, 66*(3), 338–350.

Crate, S. A., & Fedorov, A. (2013b). *Alamai tiin: Buluu ulustarigar klimat ularitigar uonna atin kihalghar* [Alamai tiin: Climate change and other change in the Viliui regions]. Yakutsk: Bichik.

Crumley, C. L. (2012). A heterarchy of knowledges: Tools for the study of landscape histories and futures. In T. Plieninger & C. Bieling (Eds.), *Resilience and the cultural landscape: Understanding and managing change in human-shaped environments* (pp. 303–314). Cambridge: Cambridge University Press.

Danielsen, F., Burgess, N. D., Balmford, A., Donald, P. F., Funder, M., Jones, J. P. G. ... & Yonten, D. (2009). Local participation in natural resource monitoring: a characterization of approaches. *Conservation Biology, 23*(1), 31–42.

Dearing, J. A., Braimoh, A. K., Reenberg, A., Turner, B. L., & van der Leeuw, S. (2010). Complex land systems: the need for long time perspectives to assess their future. *Ecology and Society, 15*(4), 21. [online] URL: http://www.ecologyandsociety.org/vol15/iss4/art21/

Díaz, S., Fargione, J., Chapin, F. S., III, & Tilman, D. (2006). Biodiversity loss threatens human well-being. *PLoS Biology, 4*, e277. doi:10.1371/journal.pbio.0040277.

Dietz, T., Ostrom, E., & Stern, P. C. (2003). The struggle to govern the commons. *Science, 302*, 1907–1912.

Garcia, J., &Gonzalez, M. (2004, 25–29 August). Rural development, population ageing and gender in Spain: The case of rural women in the autonomous community of Catilla Y Leon. In *44th European Congress of the European Regional Science Association*, Oporto.

Gearheard, S., Aporta, C., Aipellee, G., & O'Keefe, K. (2011). The Igliniit project: Inuit hunters document life on the trail to map and monitor Arctic change. *Canadian Geographer/Le Géographe Canadien, 55*(1), 42–55.

Glendinning, A., Nuttall, M., Hendry, L., Kloep, M., & Wood, S. (2003). Rural communities and well-being: A good place to grow up? *The Sociological Review, 51*(1), 129–156.

Gregory, T. A. (2006). An evolutionary theory of diversity: The contributions of grounded theory and grounded action reconceptualizing and reframing diversity as a complex phenomenon. *World Futures, 62*, 542–550.

Hamilton, L. C., & Seyfrit, C. L. (1993). Town-village contrasts in Arctic youth aspirations. *Arctic, 46*(3), 255–263.

Hansen, K., Rasmussen, R., & Roto, J. (Eds.). (2012). *Nordic perspectives on demography – A background report for the Project on Coastal Societies and Demography* (Nordregio Working Paper 2012, 12), Nordregio.

Hastrup, K. (Ed.). (2009). *The question of resilience: Social responses to climate change.* Copenhagen: Royal Danish Academy of Sciences and Letters.

Heikkilä, K., & Fondahl, G. (2012). Co-managed research: Non-indigenous thoughts on an Indigenous toponymy project in northern British Columbia. *Journal of Cultural Geography, 29*(1), 61–86.

Herman, R. D. K. (2008). Reflections on the importance of indigenous geography. *American Indian Culture and Research Journal, 32*(3), 73–88.

Hooper, D. U., Adair, E. C., Cardinale, B. J., Byrnes, J. E. K., Hungate, B. A., Matulich, K. L., et al. (2012). A global synthesis reveals biodiversity loss as a major driver of ecosystem change. *Nature, 486*, 105–109.

Hovelsrud, G., Krupnik, I., & White, J. (2011). Human-based observing systems. In I. Krupnik, I. Allison, R. Bell, P. Cutler, D. Hik, J. López-Martínez, V. Rachold, E. Sarukhanian, & C. Summerhayes (Eds.), *Understanding Earth's polar challenges: International Polar Year 2007–2008* (pp. 435–456). Edmonton: CCI Press.

IPCC. (2001). Climatic change 2001: The scientific basis, contribution of Working Group I to the *Third Assessment Report of the Intergovernmental Panel on Climatic Change* (J. T. Houghton, Y. Ding, D. J. Griggs, M. Noguer, P. J. van der Linden, X. Dai, K. Maskell, & C. A. Johnsen, Eds.). Cambridge: Cambridge University Press.

IPCC. (2007). *Climate change 2007: Impacts, adaptation and vulnerability.* Working Group II Summary for Policymakers. Geneva: IPCC Secretariat. Retrieved December 10, 2013, from http://www.ipcc.ch/pdf/assessment-report/ar4/wg2/ar4-wg2-spm.pdf

IPCC. (2012). *Managing the risks of extreme events and disasters to advance climate change adaptation.* A Special Report of Working Groups I and II of the Intergovernmental Panel on Climate Change. (C. B. Field, V. Barros, T. F. Stocker, D. Qin, D. J. Dokken, K. L. Ebi, M. D. Mastrandrea, K. J. Mach, G.-K. Plattner, S. K. Allen, M. Tignor, & P.M. Midgley, Eds.). Cambridge/New York: Cambridge University Press.

Kay, R., Shubin, S., & Thelen, T. (2012). Rural realities in the post-socialist space. *Journal of Rural Studies, 28*, 55–62.

Krupnik, I., & Jolly, D. (Eds.). (2002). *The earth is faster now: Indigenous observations of Arctic environmental change.* Fairbanks: Arctic Research Consortium of the United States.

Krupnik, I., Bravo, M., Csonka, Y., Hovelsrud-Broda, G., Muller-Wille, L., Poppel, B., Schweitzer, P., & Sorlin, S. (2005). Social sciences and humanities in the International Polar Year 2007–2008: An integrating mission. *Arctic, 58*(1), 91–101.

Krupnik, I., Aporta, C., Gearheard, S., Laidler, G., & Kielsen Holm, L. (Eds.). (2010). *SIKU: Arctic residents document sea ice and climate change.* Berlin: Springer.

Lundholm, E., & Malmberg, G. (2006). Gains and losses, outcomes of interregional migration in the five nordic countries. *Geografiske Annaler. Series B, Human Geography, 88*(1), 35–48.

Maffi, L., & Woodley, E. (2010). *Biocultural diversity conservation: A global sourcebook.* London: Earthscan.

Mayan, M. (2009). *Essentials of qualitative inquiry.* Walnut Creek: Left Coast Press.

McCarthy, J. J., & Martello, M. B. (Eds.). (2005). Ch 17: Climate change in the context of multiple stressors and esilience. In *Arctic climate impact assessment (ACIA)* (pp. 945–988). Cambridge: Cambridge University Press.

McGranahan, D., Cromartie, J., & Wojan, T. (2010). *Nonmetropolitan outmigration counties: Some are poor, many are prosperous.* Economic Research Report No. ERR-107.

Muldonado, J. (2014). A multiple knowledge approach for adaptation to environmental change: Lessons learned from coastal Louisiana's tribal communities. *Journal of Political Ecology, 21*, 61–92.

Murphy, B. (2011). From interdisciplinary to inter-epistemological approaches: Confronting the challenges of integrated climate change research. *The Canadian Geographer/Le G'eographe Canadien, 55*(4), 490–509.

NORDEN. (2006). *Strategy for children and young people.* Nordic Council of Ministers, Copenhagen. ANP 2006,723.

O'Brien, K. L., & Leichenko, R. M. (2000). Double exposure: Assessing the impacts of climate change within the context of economic globalization. *Global Environmental Change, 10*(3), 221–232.

Ostrom, E. (1990). *Governing the commons: The evolution of institutions for collective action.* Cambridge: University Press.

Ostrom, E. (2009). A general framework for analyzing sustainability of socio-ecological systems. *Science, 325*, 419–422.

Pearce, T., Ford, J., Laidler, G. J., Smit, B., Duerden, F., Allarut, M., ... & Wandel, J. (2009). Community collaboration and climate change research in the Canadian Arctic. *Polar Research, 28*, 10–27.

Pielke, R. A., Wilby, R., Niyogi, D., Hossain, F., Dairuku, K., Adegoke, J., Kallos, G., Seastedt, T. & Suding, K. (2012). Dealing with complexity and extreme events using a bottom-up, resource-based vulnerability perspective. In A. Surjalal Sharma, A. Bunde, V. P. Dimri & D. N. Baker (Eds.), *Extreme events and natural hazards: The complexity perspective.* Geophysical Monograph Series, 196, Published Online: 2 APR 2013. doi:10.1029/2011GM001086.

Rajan S. R., & Duncan C. A. M. (Eds.). (2013). Introduction: Ecologies of hope. *Journal of Political Ecology, 20*, 70–79.

Rasmussen, R. (2009). Gender and generation: Perspectives on ongoing social and environmental changes in the Arctic. *Signs, 34*(3), 524–532.

Rasmussen, R. O., Barnhardt, R., & Keskitalo, J. (2010). Education. In J. N. Lasen & G. Fondahl (Eds.), *Arctic social indicators* (pp. 67–90). Akureyri: Stefansson Arctic Institute.

Raven, P. H. (2002). Science, sustainability, and the human prospect. *Science, 297*, 954–958.

Sillitoe, P. (1998). The development of indigenous knowledge. A new applied anthropology. *Current Anthropology, 29*(2), 223–252.

Smith, E. A., & Wishnie, M. (2000). Conservation subsistence in small-scale societies. *Annual Review of Anthropology, 29*, 493–524.

Stedman, R. C., Patriquin, M. N., & Parkins, J. R. (2012). Dependence, diversity, and the well-being of rural community: Building on the Freudenburg legacy. *Journal of Environmental Studies and Sciences, 2*, 1–11.

Steffen, W., Crutzen, P. J., & McNeill, J. R. (2007). The Anthropocene: Are humans now over-whelming the great forces of nature? *Ambio, 36*, 614–621.

Stenseth, N. C. et al. (2002). Ecological effects of climate fluctuations. *Science* 297, 1292–1295. [online] URL:http://www.sciencemag.org/cgi/content/full/297/5585/1292

Steward, A. S. (2007). Nobody farms here anymore: Livelihood diversification in the Brazilian Amazon, a historical perspective. *Agriculture and Human Values, 24*, 75–92.

Stockdale, A. (2006). Migration: Pre-requisite for rural economic regeneration? *Journal of Rural Studies, 22*, 354–366.

Stokstad, E. (2005). Taking the pulse of the Earth's life support systems. *Science, 308*, 40–43.

Sveiby, K.-E. (2011). Collective leadership with power symmetry: Lessons from aboriginal prehis-tory. *Leadership, 7*, 385–414.

Tilman, D. (1999). Diversity by default. *Science, 283*, 495–496.

Tilman, D., Reich, P. B., & Knops, J. M. H. (2006). Biodiversity and ecosystem stability in a decade long grassland experiment. *Nature, 441*, 629–632.

Turner, B. L., Kasperson, R. E., Matson, P. A., McCarthy, J. J., Corell, R. W., Christensen, L., et al. (2003). A framework for vulnerability analysis in sustainability science. *Proceedings of the National Academy of Sciences of the United States of America, 100*, 8074–8079.

Virtanen, P. K. (2012). *Indigenous youth in Brazilian Amazonia: Changing lived worlds*. New York: Palgrave Macmillan.

West, C. T. (2013). Documenting livelihood trajectories in the context of development interventions in northern Burkina Faso. *Journal of Political Ecology, 20*, 342–360.

Wyn, J., & Harris, A. (2004). Youth research in Australia and New Zealand. *Young, 12*(3), 271–289.

Zavaleta, E. S., Pasari, J. R., Hulvey, K. B., & Tilman, G. D. (2010). Sustaining multiple ecosystem functions in grassland communities requires higher biodiversity. *Proceedings of the National Academy of Sciences of the United States of America, 107,* 1443–1446.

Chapter 20
Urbanisation and Land Use Management in the Arctic: An Investigative Overview

Ryan Weber, Rasmus Ole Rasmussen, Lyudmila Zalkind, Anna Karlsdottir, Sámal T.F. Johansen, Jukka Terräs, and Kjell Nilsson

Abstract This chapter investigates the role of land use planning in the context of arctic urban development through six city-profiles. Urbanisation in the Arctic is driven by a range of socio-economic and political factors. Not least, these include political processes to concentrate public services and a withdrawal from state-led socio-economic planning, economic processes that have led to the development of labour markets and new social institutions that are needed for reproducing the labour force. While these processes create development opportunities, they also have a wide range of spatial impacts and these require increased attention toward urban land-use planning. Through the city-profiles, we suggest that effective urban land use planning in the Arctic is highly context dependent. Significant issues appear to include: the importance of preserving relationships between society and the natural environment; the necessity in some cases for planning measures in response to significant urban sprawl; the recognition of complex governance structures that influence development strategies; and even the necessity for planning responses to suburban sprawl.

Keywords Urban • Land use • Arctic • Planning • Development

R. Weber (✉) • R.O. Rasmussen • J. Terräs • K. Nilsson
NORDREGIO – Nordic Centre for Spatial Development, Stockholm, Sweden
e-mail: ryan.weber@nordregio.se; rasmus.ole.rasmussen@nordregio.se;
jukka.teras@nordregio.se; kjell.nilsson@nordregio.se

L. Zalkind
Department of Urban Socio-Economic Development, Kola Science Centre,
Apatity, Murmansk Oblast, Russian Federation

A. Karlsdottir
Department of Geography and Tourism Studies, University of Iceland, Reykjavik, Iceland
e-mail: annakar@hi.is

S.T.F. Johansen
Søvn Landsins (The Faroese National Archive), Hoyvík, Faroe Islands
e-mail: samaltrondurj@savn.fo

© Springer International Publishing Switzerland 2017
G. Fondahl, G.N. Wilson (eds.), *Northern Sustainabilities: Understanding and Addressing Change in the Circumpolar World*, Springer Polar Sciences,
DOI 10.1007/978-3-319-46150-2_20

20.1 Introduction

Land use is a heavily researched planning topic in the context of sustainable urban development, where diverse challenges and responses associated with urban sprawl dominate many discussions. Problems include the loss of productive land for agriculture, the disruption of ecosystems services, pollution associated with car dependency, and reduced quality of life due to increased commuting distances (Nilsson et al. 2014). Solutions often include densification strategies such as the compact city, brownfield development, infilling, or transport oriented development. However, most of the discussion continues to take place within continental European and North American perspectives, bound to the territorial contexts of mid-latitude, developed countries. In reality however, the interplay between growth and land use impacts requires an understanding of the local context to create resilient, attractive, and sustainable urban settlements. Local factors include issues of scale, environment, society and culture, politics, and economy – each of which engrain both explicit and hidden agendas into the development discourse.

This chapter focuses on the urbanisation of arctic settlements. It uses historical planning documents and the situated perspectives of the authors to investigate how socio-economic development has impacted urban land use processes in selected arctic urban areas. The main focus is on urban morphology within settlements, and, vis-à-vis the hinterland, governance factors associated with urban land management. The chapter begins by briefly examining the main socio-economic factors influencing urbanisation in the Arctic, and follows up with an analysis of land use development in six different urban settings. Each city profile was developed in a semi-structured way, where individual authors used general guidelines to present key trends in the historical development and current land use management of each city.

20.2 Urbanisation in the Arctic

By 2010, over 50 % of the world's population was living in urban areas (EC 2011) and Fig. 20.1 shows that the Arctic is no exception to this trend. With the exception of Russia (depopulation) and Alaska (high birth rate), larger urban centres in the Arctic are growing compared to the stagnation or depopulation being experienced in smaller settlements. This pattern is particularly clear in Anchorage, Whitehorse, Nuuk, Reykjavik, Akureyri, Tromsø, Bodø, Luleå, and Rovaniemi.

While urbanisation is a complex phenomenon, some common drivers in the Arctic include:

- *Urbanisation as a political process:* Changes in the payments of transfers, concentrated public service provision, and the withdrawal of the state from social and economic planning have led to the continued concentration of populations in larger settlements (Southcott 2010).

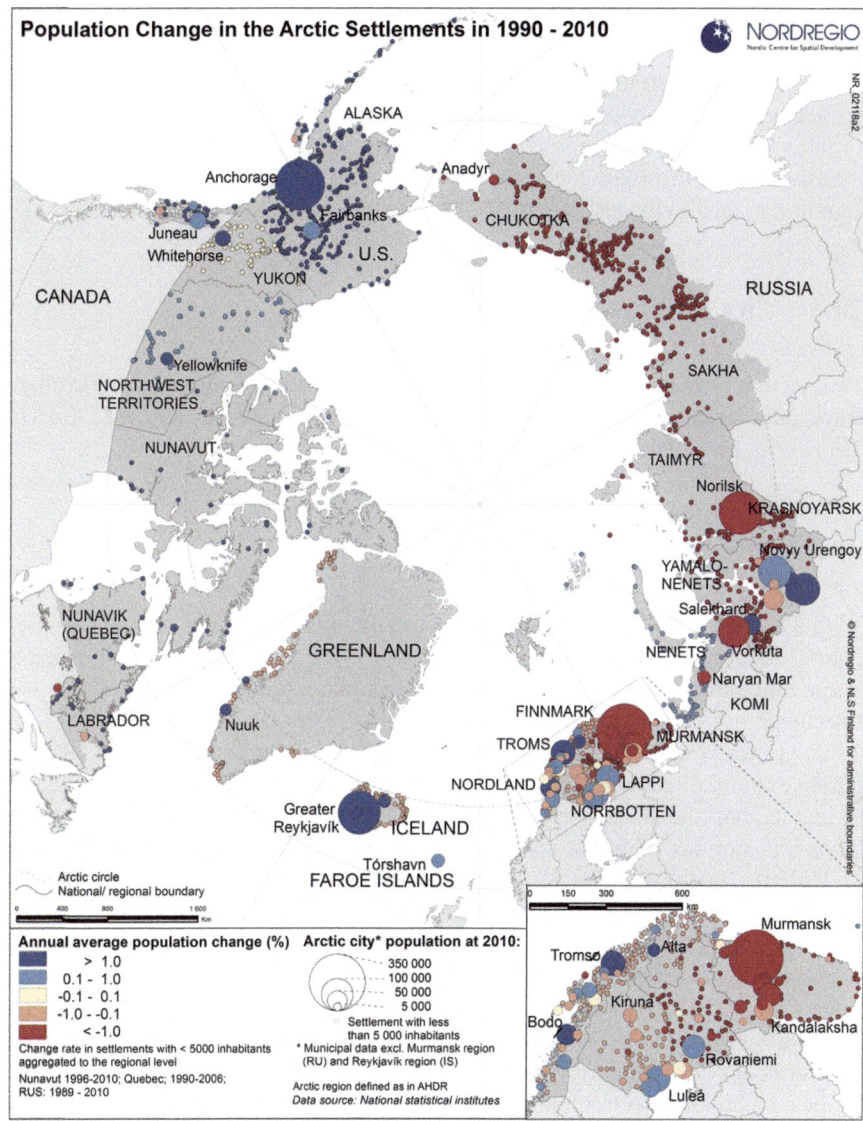

Fig. 20.1 Change in settlement structure in the Arctic

- *Urbanisation as an economic shift:* Traditional arctic resource dependencies required a dispersed and flexible settlement structure. However, as a result of modernisation and industrialisation, it became necessary to create labour markets. This especially includes economic development via natural resource exploitation in the post-war period, which has coincided with the increased influence of national and international institutions in terms of political and economic management (Hansen and Rasmussen 2013). This growth has continued to evolve to the present day. Arctic urban economic expenditure tends to be dominated by administrative and educational activities.

- *Urbanisation as opportunity:* Alongside new labour structures, new social institutions are needed for reproducing the labour force. For instance, young people are attracted to urban centres for educational purposes and are less likely to return to small villages after they complete their education (Rasmussen 2011).

20.3 City Profiles

This section provides a short synthesis of land use development in six arctic cities: Reykjavik, Iceland; Rovaniemi, Finland; Nuuk, Greenland; Kirovsk, Russia; Thorshavn, Faroe Islands; and Kiruna, Sweden.[1]

20.3.1 Reykjavik, Iceland

With a current population of 201,000, greater Reykjavik consists of seven municipalities and covers 1062 km^2 – a territory that is ten times larger than the size of Paris. Since the Icelandic Republic gained independence from Denmark in 1944, the capital has incrementally grown in several different phases. Between 1945 and 1965, growth was driven mainly by the region's role as the commercial and administrative centre of the country. Spatially, greater Reykjavik's population grew by 70 % and the built landmass increased by 700 %, and thus experienced typical North American notions of urban sprawl and automobile dependence (Bjarnason and Gylfadóttir 2004).

In the 1970s and 1980s, modernisation and the expansion of Reykjavik towards neighbouring municipalities further contributed to urban sprawl. By the 1980s, the neighbouring municipalities surrounding Reykjavik were hardly discernible as separate towns. Development over the past two decades has reinforced urban sprawl even further. Average population development by municipality in Fig. 20.2 shows little or no growth in Reykjavik proper, whereas populations are increasing in many of the municipalities surrounding the city. Public housing has had very limited or no impact on development. After the 1970s, favourable mortgage conditions made home buying realistic and, by the early 1980s, upwards of 91 % of people in suburb municipalities "owned" their own home (Sveinsson 2007).

Alongside the acceptance of long commutes and the private car as a way of life in Reykjavik, urban sprawl seems to be predominantly driven by inter-municipal competition for growth through land bids and residential growth. This challenge is

---

[1] More detailed presentations of each city profile are presented in an elaborated version of this paper: www.nordregio.se/publications

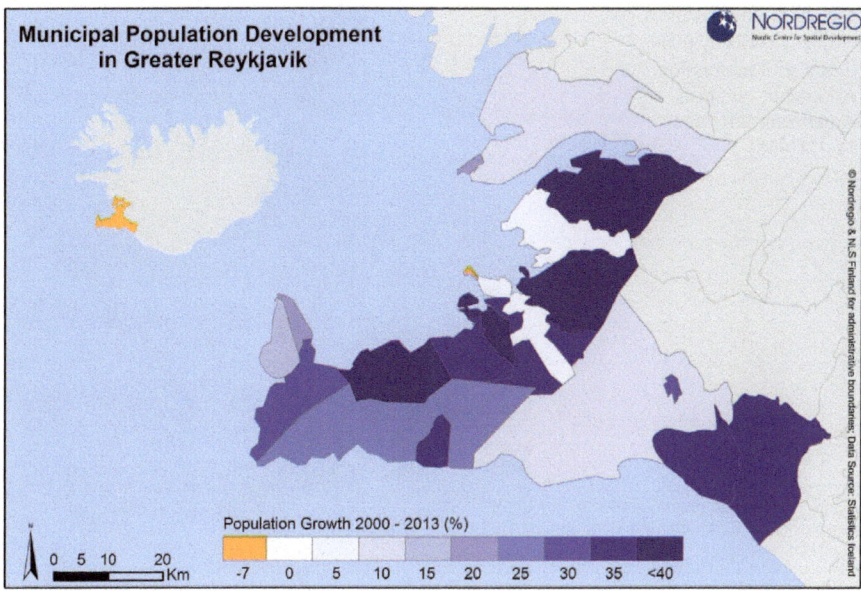

Fig. 20.2 Population development (%) in municipalities around the capital 2000–2010 (Source: Ryan Weber)

well understood by local authorities and it raises questions about the need for a regional and/or national policy on urban growth boundaries and sustainable land use, including the question of regional allocation of new housing units between the different municipalities (Theodórsdóttir et al. 2012).

20.3.2   Rovaniemi, Finland

The municipality of Rovaniemi has 61,000 residents, making it the 13th most populous municipality in Finland. However, it also covers an area of 8016 km², making it the largest municipality in Europe by area. Rovaniemi has a relatively balanced economy based on its role as the political and administrative capital of Lapland, alongside contributions from the education, tourism, and energy (hydro) sectors. Slow and stable population growth between 1985 and the late 1990s led to a short downturn, but steady growth has returned since 2003 and trends suggest a continuation of slow and steady growth in the future (Rovaniemi Municipality 2012).

The town suffered near complete destruction during the Second World War, and redevelopment was based on a town plan conceptualised by the Finnish architect Alvar Aalto (Fig. 20.3). The implementation of the plan resulted in extensive, low density development and current planning documents regard this development as limiting the ability for improving Rovaniemi's resource efficiency through compact

Fig. 20.3 Alvar Aalto's
"Antler Like" city plan
(Used with permission,
Arkkitehti – Finnish
Architectural Review
11-12/1945)

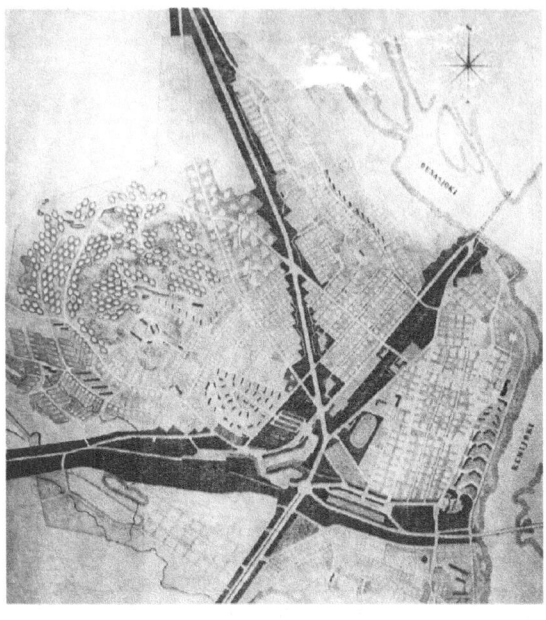

city planning principles of density, mixed land use and accessibility (Rovaniemi
Municipality 2012).

In Finland, land use planning is integrated through multiple levels of government. For instance, Rovaniemi's Local Land Use Strategy must comply with
Lapland's Regional Land Use Plan, which presents the general principles of land
use and the community structure that guide development for the region. As such,
regional councils in Rovaniemi have greater planning authority than regional governments in other Nordic countries (Smas et al. 2012).

Rovaniemi's Climate Adaptation Strategy (Järviluoma and Suopajärvi 2009)
concentrates on three themes: (1) tourism; (2) energy and waste management; and
(3) urban planning and traffic. The third theme highlights how Aalto's plan resulted
in a sprawled urban form, and how smaller villages around Rovaniemi have grown
since they were consolidated into one municipality. This has put further strain on
services and infrastructure, increased private car use and raised questions regarding
urban sustainability (Järviluoma and Suopajärvi 2009). As a result, the strategy
describes the importance of integrating the community structure through the strategic densification of already built-up areas and the promotion of more eco-friendly
means of travel, including public transport, walking and cycling.

Yet, public opinion is divided regarding the priority for city-centre development.
Some people favour densification around the existing urban centre, while others are
concerned over the loss of urban green space and the preservation of the current
building qualities near the city centre (Järviluoma and Suopajärvi 2009). Residents
of outer villages also fear that strategic densification will mean a prioritisation of the
town centre over their non-central neighbourhoods, forcing residents to seek basic

services in distant locations. As such, existing villages face the dual challenge of a lack of services and effective public transport to Rovaniemi city centre. While these villages may not be large enough to sustain a comprehensive public transportation network, there does seem to be the need for improved local bus service as well as some concentrated, mixed-use development within these suburban communities.

20.3.3   Nuuk, Greenland

Founded in 1728 as a Danish-Norwegian colony, Nuuk's current population is approximately 16,000 residents (29 % of Greenland's population). Its role as the political and administrative centre was supported by the G50 and G60 socio-economic development plans[2] (Petersen and Rasmussen 2007; Rasmussen 1998; Kjær-Sørensen 1983) which included the modernisation and growth of the fishing industry. The establishment of Home Rule in 1979 brought another wave of urban growth in Nuuk, as many decision-making and management duties were transferred from Denmark to Greenland.

Figure 20.4 shows the evolution of Nuuk since 1945. The settlement was first concentrated around the original western harbour, but the harbour's limited growth capacity led to the construction of a new harbour to the southeast. By 1965, development evolved to include power facilities, a hospital, a fish processing plant and, in particular, new housing. However, private investment sought by the G50 and G60 plans did not materialise and the Danish Government assumed responsibility for large-scale housing investments. This resulted in rapidly built, industrial-style concrete housing blocks built close to today's city centre.

The population growth through 1975 coupled with the low quality of existing housing led to the construction of higher quality apartment blocks and colonial-style detached houses with bright colours. Three new schools, two churches and extended central administration buildings were also built. By 1985 a new airport was built, along with 45 km of roadways and three bus routes, and by 1995 rock blasting helped development move into previously unsuitable areas. Since 2000, the concrete apartment blocks built in the 1970s have been incrementally demolished and replaced by smaller-scale and higher quality apartment buildings. Furthermore,

---

[2] The G50 and G60 (the report of the Greenland Commission of 1950 and Greenland Commission of 1960 respectively) were long-term development plans for Greenland and its people under the direction of the Danish Government. G50 had the ambition to establish a similar economic system, the same civil rights and the same standard of living as in Denmark. The G60 aimed at continuing the plans of G50 and furthermore improve the health conditions and living conditions to standards comparable to those in Denmark by improving the health care, the education system, the electrification of the country and increasing the concentration of housing and job creation in the major towns along the west coast of Greenland close to the major fishing opportunities. Furthermore, the G60 had the ambition to raise the political, social and cultural status of the Greenlandic people and improve their standard of living.

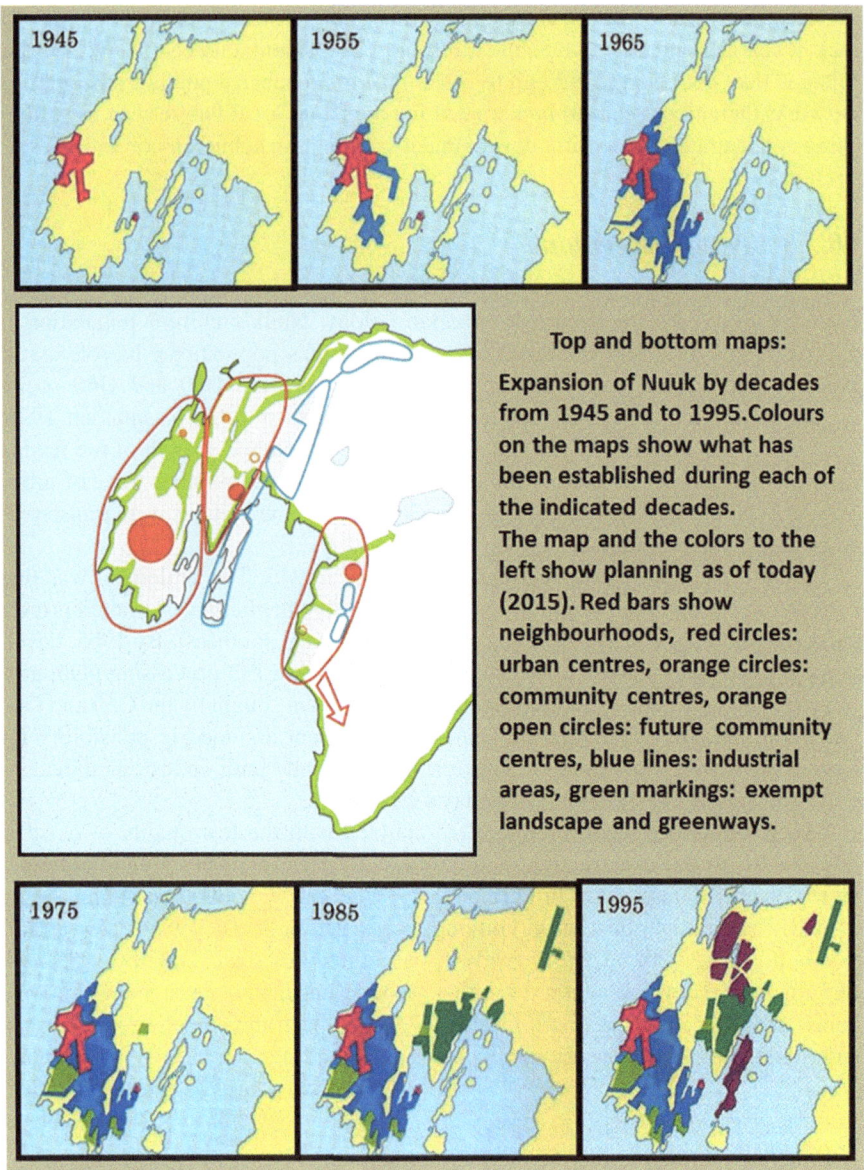

Fig. 20.4 Territorial development of Nuuk since 1945. Maps for 1945 to 1995 based on Berthelsen et al. (1993). 2015 plans based on kommuneqarfik sermersooq (2015) (Design: Rasmus Ole Rasmussen)

a main focus of growth is currently to the southeast in Qinngorput, a completely new urban district east of Malene Bay.

In 2009, Greenland's 18 municipalities were amalgamated into four municipalities, a move which has strengthened land use planning by concentrating and formalising urban planning competencies. In Sermersooq (which includes Nuuk), this includes development of the new "Digital Municipal Plan" promoting "robust and sustainable" development. "Robust" largely refers to growth driven development by a planned airport expansion to enable access by trans-Atlantic flights, development possibilities in the raw materials sector and the development of the harbour. "Sustainable" reflects Nuuk's goal to strengthen the special qualities of the community in an environmentally and economically sound manner.

In the new municipal plan "open countryside" is a collective term for all areas not zoned for settlement. In the past, the development of open space was undertaken on a case-by-case basis by the national government (i.e. no comprehensive strategy) because of the impression that there would always be plenty of room available. Reflecting important social and economic connections to nature, however, the open countryside can no longer be considered an area where there are no potential land use conflicts as a wide range of actors (and their associated interests) are increasingly vying for space. For example, a number of abandoned settlements surrounding Nuuk have been re-born with important new land use functions and these settlements need to be planned and managed accordingly. When these are considered in relation to national decisions on further large-scale natural resource exploitation there appears to be a need for further revisions of the open land planning strategy, toward more consistent and integrated approaches to spatial planning, reflecting the socio-economic interests of various different stakeholders.

20.3.4 Kirovsk, Russia

Kirovsk, in Murmansk *Oblast'* (region) in northwestern Russia, was founded as a natural resource town based on apatite-nepheline deposits in the Khibiny mountains. Initial mining operations started in 1929 and by 1931 Kirovsk had a population of 32,000. While this grew to nearly 44,000 by 1990, the city has declined in population in a similar manner to other towns in arctic Russia and currently has a population of roughly 28,000 (Fig. 20.5).

Strategic land use planning was initiated in 1938, but was challenged from the outset. A plan for a new main street was developed that did not consider local topography, making it impossible to use the street during the winter. In the meantime, new buildings had been constructed along the previous route so it was impossible to restore it. A new master plan was developed, but this too was only partially implemented due to poor coordination and a lack of funding. Nevertheless, successful mining operations resulted in a strong population profile between the late 1950s and 1990. As a result, 93 % of the current residential building stock was constructed in the 1970s and 1980s.

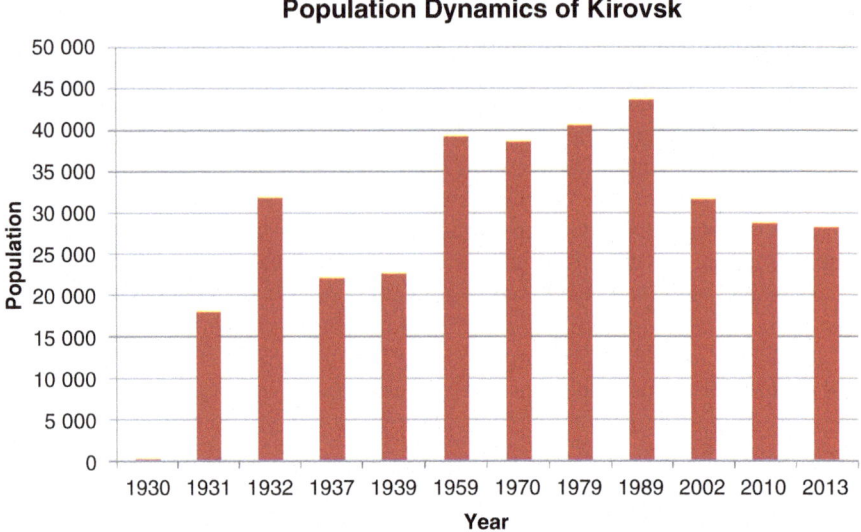

Fig. 20.5 Population dynamics Of Kirovsk, Russia (Data source: Kirov History Museum 2014)

Until the 1990s, the state was the sole land owner, and it granted rights to individuals as land users. In 1991, rights of individuals increased and by 2001, a new Land Code had legalized private ownership and set general zoning. This included the development of new municipal master plans, local land use acts and multipurpose investment plans, iterations of which are still in place today. Thus, the municipality is the main authority for land use matters and it currently promotes two directions of development: industrial growth via mineral extraction and tourism growth, particularly skiing.

Reflecting the attachment to nature, the town's population actively uses the forests and foothills surrounding Kirovsk for recreation and production of agricultural goods in private plots. During the overall socio-economic crisis and food shortage in the early 1980s, the local government started a gradual allocation of land to employees for dachas and gardens. Nowadays, people continue to actively use these plots for holiday stays and for growing produce for home consumption or informal market sale.

Yet, the main land use conflict is in regards to mining development versus the formation of a national park. Proposed mining development around Kirovsk conflicts with the national government's plan to create the Khibiny National Park. Establishing the park would mean that a considerable portion of land currently owned by the Kirovsk municipality would be transferred to national control. While the municipality understands that both local residents and domestic tourists use the surrounding landscape for recreation, they are against plans for the park because of the impact it would have on mining expansion. The municipality prefers a balanced approach where nature preservation and the development of tourist infrastructure coincide with additional mining activities in the area, which they view as critical to sustaining the livelihoods of the current population. We see, therefore, a clear power struggle between local and national planning authorities, each of which is interpreting a different best course of action due to the different stakeholder and development perspectives they favour.

20.3.5 Tórshavn, Faroe Islands

The abolishment of the Danish Trade Monopoly by the mid-nineteenth century presented new growth opportunities in the Faroe Islands, at the time a colony of Denmark, especially in fisheries. As a result, villages where fish trade houses were located bloomed, including Tórshavn. Shifts towards a modern economy took place in the immediate post-war period, mainly resulting in a more productive fishing fleet and fish processing industry. In parallel, infrastructure was developed throughout the country as roads were paved and the first tunnel that began connecting the 18 islands opened in 1963. By 1975, more than 11,000 people lived in Tórshavn (28 % of the Faroese population) and development over the next 15 years expanded out toward Hoyvík and Argir, which were originally small settlements in the vicinity of Tórshavn. Today, Tórshavn has a population of more than 12,600.

During the 1960s and 1970s 'village-development' was the main focus of public policy that steered spatial development. This policy supported the fishing industry in smaller settlements in an effort to prevent migration away from these settlements. It was kept alive through different forms of financial transfers until it collapsed in the early 1990s, leading to outmigration toward the larger towns. However, as part of the village development policy, investments were made in roads, tunnels, bridges and other infrastructure components to reduce travel time from small settlements to Tórshavn (Fig. 20.6). While the intention was to improve village accessibility to services in larger cities (thus maintaining the attractiveness of small settlements) the opposite happened – people moved within commuting distance to Tórshavn.

Today, 72 % of the total land area and 86 % of the population in the Faroe Islands are linked to a fixed road. The new traffic plan for the Faroe Islands (2008–2020) is the spatial backbone of the Faroe Islands; enabled by the main transportation network linking various regions of the country to Torshavn.

20.3.6 Kiruna, Sweden

Sweden's most northern municipality was established following the Swedish Parliament's decision to extend the railway to exploit iron ore in the Kirunavaara and Luosavaara mountains. Greater Kiruna has more than 23,000 inhabitants, 18,000 of whom live in the town itself. The population peaked during the 1970s, but after a long period of decline, it has begun to increase again with the development of additional mining opportunities.

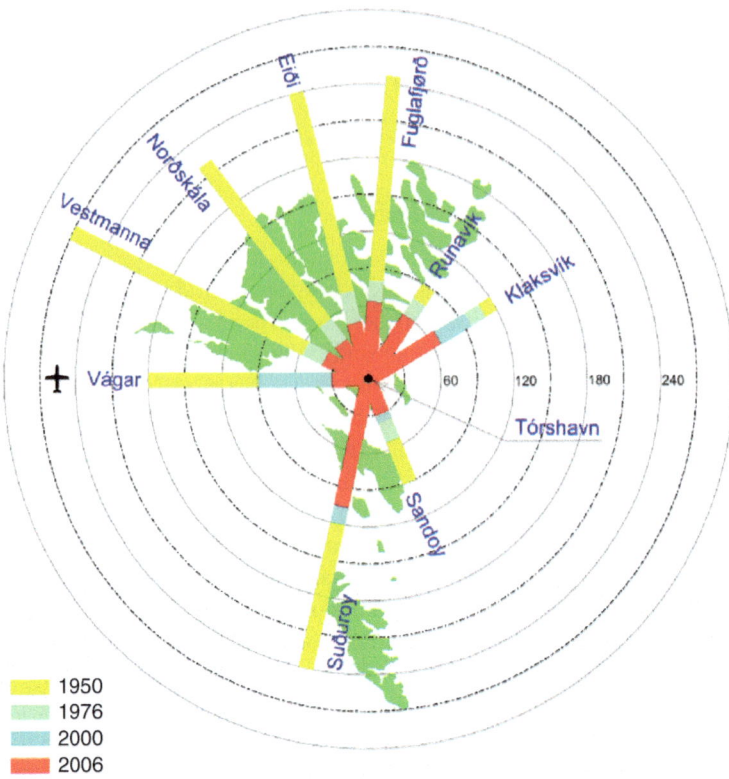

Fig. 20.6 Development of transport time from 1950 to 2006. The *colours* show how long time it would take to move from Tórshavn to other towns/villages and islands (Source: Steinhólm 2007, used with permission)

Kiruna represents an extraordinary case of land use planning. Plans to extend and expand existing mining activity until at least 2030 have forced the relocation of the city centre due to ground level deformations (Fig. 20.7).

The first phase of the construction of the new city centre started in 2014, and by 2033, approximately 3000 apartments, 750 beds in hotels and hostels, and nearly 200,000 m^2 of retail and service offerings will be relocated (Zackrisson and Cars 2013). Kiruna's residents also realise the necessity of moving the city, and have contributed to the planning and design of the new city centre. Zackrisson and Cars (2013) summarise the residents' four main demands:

- Accessibility to the centre by car, but a centre that promotes walking, cycling and public transportation.
- Mixed land use with retail shopping on the ground floor of residential houses and good accessibility to work-places and public services.
- High quality public space and meeting places, including parks, squares, streets, libraries, art galleries, cafés and restaurants.

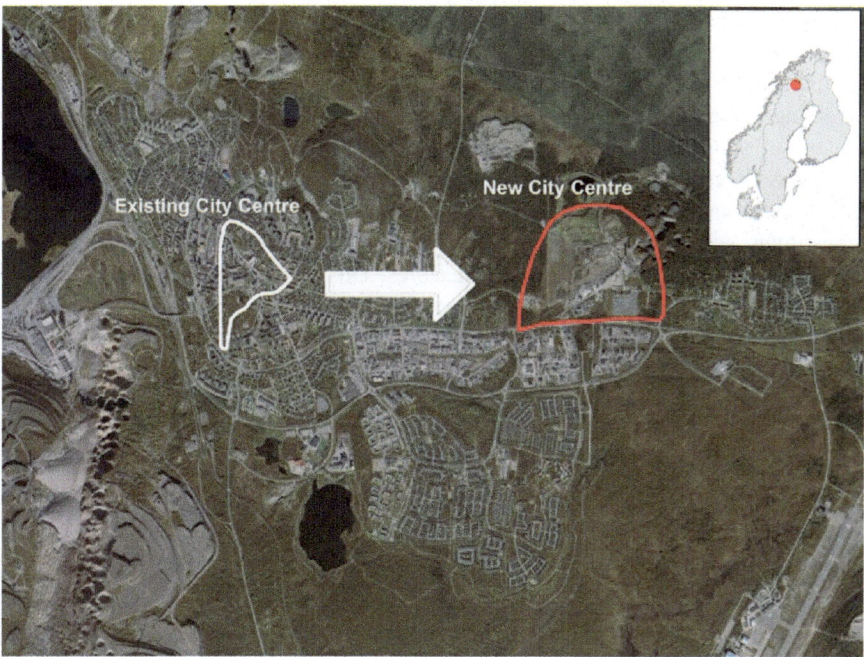

Fig. 20.7   Moving of the city centre Of Kiruna (Design: Ryan Weber)

- High architectural quality, including the preservation of cultural buildings. This includes the Church of Kiruna, which was recently voted as the most beautiful building in Sweden. As a result, it will be taken down and reassembled in the new city centre.

Kiruna has thus managed to turn a threating situation into an opportunity through open and transparent dialogue with its residents. Although many issues are still unresolved – primarily those of an economic nature since the costs for new home construction is much higher than the market values of existing single-family houses or apartments – Kiruna is facing the future with optimism. Therefore, the process of building a completely new, mixed-use city centre with high architectural quality might serve as a model for a new kind of sustainable urban development in the arctic region.

20.4   Conclusions

Based on this presentation of land use planning and development strategies in diverse arctic contexts, some key findings are identified. First, in many cases, arctic urbanisation is a new phenomenon and in communities such as Nuuk and Kirovsk it was shown how connections to nature are a critical component of the arctic culture in many regions, both through informal economic activities and for leisure. For instance, hunting, fishing, berry and mushroom picking, hiking and camping, sailing, skiing and even farming are all activities that are included as an integral part of living in the Arctic; not only as leisure time activities, but also in some cases as a means of subsistence, related to dietary needs and informal distribution mechanisms. As a result, arctic cities have unique and strong societal relationships to their natural surroundings and providing opportunities for regular interaction with nature constitutes an important dimension of human development, which can be met through proactive land use planning measures. This includes improved accessibility through green-space infrastructure investment and the preservation of natural assets within the built environment.

Second, some arctic cities face challenges associated with urban sprawl and strategic densification is a necessary planning response. In Reykjavik and Rovaniemi, the impacts of urban sprawl are clearly evident. In Reykjavik's case, the "Americanisation" of the city's buildings was perhaps an inherited outcome of the U.S. military presence and the main period of development took place while the car was seen as an ideal form of personal mobility. In both cases, the challenges associated with urban sprawl and the pursuit of urban sustainability is well understood by local planners.

Third, each case highlights how the division of roles and responsibilities between different levels of government has impacted land use planning in unique ways. In Kirovsk, the municipality's vision of what entails sustainable development is fundamentally different from that of the national government. In Reykjavik, the importance of governance scale was evident as a factor contributing to urban sprawl, where intense competition for investment between municipalities in the city-region has extended development away from the city centre. Planners acknowledge the need for a city-regional approach to comprehensive land use planning, but negotiating within the entrenched municipal barriers is a major hurdle. However, the Rovaniemi case demonstrated how urban sprawl increased following municipal consolidation. Seemingly counter intuitive to the necessary policy response for Reykjavik, it shows the unique, situational context of urban land use planning, regardless of where in the Arctic it is taking place.

Fourth, in Nuuk, Tórshavn, Kirovsk and Kiruna, urban development was founded upon the presence of a single natural resource based industry. In these cases, initial settlement structure hinged exclusively on maximising labour force accessibility to work locales. As a result, it seems that each city has maintained a relatively dense settlement structure considering that none of the existing planning documents describe challenges associated with sprawl. However, while primary sector interests continue their dominant economic role in Kirovsk and Kiruna, public sector roles such as administration and education are now key drivers of the economy in Nuuk and Tórshavn. In contrast, the lack of a single industrial presence in Reykjavik and Rovaniemi, coupled with larger overall population growth has coincided with the urban sustainability challenges associated with sprawl. This supports further investigation on the sustainability potentials associated with urban development surrounding single sector activities, along with more insight on how to resolve existing land use challenges in dispersed arctic settlements.

References

Berthelsen, C., Mortensen, I. H., & Mortensen, E. (1993). *Kalaallit Nunaat Greenland Atlas*. Nuuk: Greenland Home Rule Government.

Bjarnason, P. V., & Gylfadóttir, H. M. (2004). *Húsakönnun – Hamarsgerði, Langagerði, Sogavegur, Tungugerði* [Housing survey – Hamarsgerði, Langagerði, Sogavegur, Tungugerði]. Reyjavik: Minjasafn Reykjavíkur.

EC. (2011). *Cities of tomorrow: Challenges, visions, ways forward*. Luxembourg: DG Regional Policy; European Commission.

Hansen, K. G., & Rasmussen, R. O. (2013). New economic activities and urbanisation: Individual reasons for moving and for staying – case Greenland. In K. G. Hansen, R. O. Rasmussen, & R. Weber (Eds.), *Proceedings from the first international conference on urbanization in the Arctic* (pp. 157–182). Stockholm: Nordregio.

Järviluoma, J., & Suopajärvi, L. (2009). *Adapting to the anticipated impacts of climate change in the city of Rovaniemi:Summary. Clim-ATIC Project Report*. Rovaniemi: University of Lapland.

Kirov History Museum. (2014). Kirov History Museum. Retrieved May 14, 2014, from http://www.museum25km.ru/?p=post&m2=kirov

Kjær-Sørensen, A. (1983). *Danmark-Grønland i det 20. århundrede - en historisk oversigt* [Denmark-Greenland in the 20th century – A historical overview]. Copenhagen: Nyt Nordisk Forlag.

Kommuneqarfik Sermersooq. (2015). Kommuneplanen [Municipal plan for Sermersooq Municipality]. Retrieved November 27, 2014, from http://sermersooq.odeum.com/download/pdf/final/kp_u_bestemmelser_da.pdf

Nilsson, K., Nielsen, T. S., Aalbers, C., Bell, S., Boitier, B., Chery, J-P., Fertner, C., Groschowski, M., Haase, D., Loibl, W., Pauleit, S., Pintar, M. Piorr, A., Ravetz, J., Ristimäki, M., Rounsevell, M., Tosics, I., Westerink, J. & Zasada, I. (2014, March). Strategies for sustainable urban development and Urban-rural linkages, research briefings. European Journal of Spatial Development. Retrieved May 1, 2014, from http://www.nordregio.se/Global/EJSD/Research%20briefings/article4.pdf

Petersen, G. & Rasmussen, R. O. (2007). Grønlands erhvervsstruktur – den regionale dimension [Greenland's business structure – The regional dimension]. In T. Greiffenberg & L. G. Rasmussen (Eds.), *Nordiske perspektiver på Grønlands regionalisering, regional udvikling og regionalpolitik* [Nordic perspectives on the regionalization of Greenland, regional development, and regional policies] (pp. 87–137). Roskilde: Roskilde University.

Rasmussen, R. O. (1998). Settlement development and the formal, informal and subsistence sector in the Arctic. Geografisk Tidsskrift. *Geografisk Tidsskrift, Danish Journal of Geography (Special Issue), 1*, 171–180.

Rasmussen, R. O. (2011). *Megatrends*. Copenhagen: Nordic Council of Ministers.

Rovaniemi Municipality. (2012). *The Rovaniemi land use strategy – Strategy update 23 January 2012*. Rovaniemi: Rovaniemi Municipality.

Smas, L., Damsgaard, O., Fredricsson, C. & Perjo, L. (2012). *Integrering av översiktsplanering och regionalt tillväxtarbete: Nordiska och europeiska utblickar* [Integration of comprehensive planning and regional growth: The Nordic and European outlooks] (Nordregio Working Paper, 2012:5). Stockholm: Nordregio.

Southcott, C. (2010). Migration in the Canadian North: An introduction. In L. S. Huskey & C. Southcott (Eds.), *Migration in the circumpolar North: Issues and contexts* (pp. 35–55). Edmonton: CCI Press and the University of the Arctic.

Steinhólm, A. (2007, August). Færøerne – en by i verdenssamfundet – Trafikplan for Færøerne 2008–2020 [Faroe Islands – A city in the world community – traffic plan for the Faroe Islands 2008–2020] *Dansk Vejtidsskrift*. Retrieved May 1, 2014, from http://asp.vejtid.dk/Artikler/2007/08%5C4998.pdf

Sveinsson, J. R. (2007). *Meginþættir húsnæðisstefnu Íslendinga á 20. öld* [The main housing policy in Iceland in the 20th century.] Land Registry Annual Report 2005 (pp. 19–40). Reykjavík: Fasteignamat Ríkisins.

Theodórsdóttir, Á. H., Jónsdóttir, S., Guðmundsson, D., & Hreggviðsson, G. M. (2012). *Veðjað á vöxt – Byggðaþróun á stór-höfuðborgarsvæðinu* [Bet on growth – Urban development of the large-capital]. Reykjavík: Háskólinn í Reykjavík.

Zackrisson, K., & Cars, G. (2013). *Kiruna – A city in transformation*. Paper presented at The Royal Colloquium, 20 May, 2013, initiated and chaired by His Majesty King Carl XVI Gustaf of Sweden.

Chapter 21
"You Need to Be a Well-Rounded Cultural Person": Youth Mentorship Programs for Cultural Preservation, Promotion, and Sustainability in the Nunatsiavut Region of Labrador

Ashlee Cunsolo, Inez Shiwak, Michele Wood, and *The IlikKuset-Ilingannet Team*

Abstract Sustainability issues are an increasing concern across the Circumpolar North. The often-intense social, health, and cultural stressors from multiple pathways—including climate change, resource extraction, socio-economic shifts, and the enduring legacies of colonization—affect social cohesion, community wellness, sense of place and heritage, livelihoods, and many cultural structures. Indigenous peoples are at the frontline of these changes and, as a result, a priority of many communities is to develop strategies to support community wellness, foster livelihoods, maintain cultural values, enhance resilience, and preserve and promote cultural continuity. Responding to these stressors and needs, and building from previous research conducted in the region that indicated a desire to ensure cultural continuity, the Inuit Community Governments of Rigolet, Makkovik, and Postville, in the Nunatsiavut region of Labrador, designed and piloted the *IlikKuset-Ilingannet* (Culture-Connect!) Program. This program was premised on the Inuit relational

The *IlikKuset-Ilingannet Team*
Charlotte Wolfrey, Rigolet Inuit Community Government; Gemma Andersen and Herb Jacque, Makkovik Inuit Community Government; and Rebecca Brennen and Diane Gear, Postville Inuit Community Government

A. Cunsolo (✉)
Labrador Institute of Memorial University, Happy Valley-Goose Bay, NL, Canada
e-mail: ashlee.cunsolo@mun.ca

I. Shiwak
'My Word': Storytelling & Digital Media Lab, Rigolet Inuit Community Government, Rigolet, NL, Canada
e-mail: inezs@rigolet.ca

M. Wood
Nunatsiavut Department of Health and Social Development, Happy Valley-Goose Bay, NL, Canada
e-mail: michele.wood@nunatsiavut.com

© Springer International Publishing Switzerland 2017   285
G. Fondahl, G.N. Wilson (eds.), *Northern Sustainabilities: Understanding and Addressing Change in the Circumpolar World*, Springer Polar Sciences,
DOI 10.1007/978-3-319-46150-2_21

epistemology of *piliriqatigiinniq* ('working in a collaborative way for the common good'), and united five youth with five adult mentors per community (n = 30) to learn cultural skills, including trapping, snowshoe-making, carving, art, and sewing. This research found that participating in the program supported hands-on knowledge transmission, created new or enhanced relationships between and among the youth and mentors; revitalized cultural pride and wellbeing; promoted cultural preservation and promotion; and showed promise as a strategy for supporting cultural sustainability and resilience to change. This resonates with growing emphasis on Indigenous-led programs supporting cultural preservation, promotion, reclamation, and resurgence, and contributes to a wholistic understanding of, and strategies for, Northern sustainabilities.

Keywords Inuit • Cultural mentorship • Cultural sustainability • Resilience • Youth

21.1 Introduction

Recent and dramatic changes in culture, governance structures, economies, and environments across the Circumpolar North have left many Indigenous communities living with often-intense social, health, and cultural stresses as a result of the effects of climate change (Cunsolo Willox et al. 2012, 2013a, b; Ford et al. 2014; Ford 2012; IPCC 2013, 2014); resource extraction and industrial development (Ford et al. 2010a, b; McDowell and Ford 2014); economic development (Prowse et al. 2009); rapid urbanization and associated issues such as homelessness (Christensen 2013; Dybbroe et al. 2010); and the ongoing intergenerational impacts of colonialism on community well-being, with roots in the establishment of settlements resulting in changes in subsistence practices, the enduring legacies from residential schools, and continuing neo-colonial relations (Kirmayer et al. 2009; Lehti et al. 2009; Allen et al. 2014). These rapid and often-imposed social, political, economic and environmental transitions affect social cohesion, community wellness, sense of place and heritage, livelihoods, and many cultural structures (Allen et al. 2014; Cunsolo Willox et al. 2012; Kral et al. 2011; Kral and Idlout 2012; Wexler 2009). Indigenous peoples are at the frontline of these changes, and a major priority of many communities is to find ways to foster livelihoods, maintain cultural values, and promote resilience in people and culture (Kirmayer et al. 2011; Petrasek MacDonald et al. 2013a, b; Wexler 2014; Wexler et al. 2014).

These transitions have also created new and unfamiliar challenges not faced by previous generations that often affect young people most acutely. Indeed, these disruptions have compromised traditional knowledge exchange, traditional rites of passage, and intergenerational knowledge continuity for many Arctic Indigenous youth (Collings 2014; Ford et al. 2010a, b; Pearce et al. 2011), thus making it more difficult for some young people to develop positive personal and cultural identities.

These identity attributes have been identified as protective factors for youth development (Petrasek MacDonald et al. 2013a, b, 2015), and are essential to cultural continuity and sustainability. This underscores the importance of finding ways to engage youth in positive cultural experiences by building on the strengths of shared heritage and culture, and connecting youth with positive adult role models to understand present-day struggles, discover coping strategies, and foster hope for a positive and productive future that is connected to cultural continuity and new cultural expressions (Wexler 2014; Wexler et al. 2014). Furthermore, there is increasing evidence emerging that indicates the need for the creation of Indigenous-led programs and strategies to promote the reclamation, revitalization, and renewal of Indigenous cultural skills and knowledge in order to foster a resurgence in traditions, skills, and wisdom among the generations (Alfred 2014; Freeland Ballantyne 2014; Radu et al. 2014). These understandings can support Indigenous youth well-being, resilience, and cultural strength in the context of rapid, unpredictable changes affecting the Arctic and, increasingly, other Northern or remote communities to build sustainability in culture, which, we contend, can support and enrich other forms of sustainability in the North.

Responding to this need for fostering cultural-based strengths and resiliencies, in 2013–2014, the Inuit Community Governments of Rigolet, Makkovik, and Postville in the Nunatsiavut region of Labrador created and evaluated the *IlikKuset-Ilingannet* (Culture-Connect!) program. This was a pilot, culturally-based mentorship strategy that linked youth with positive adult role models and mentors in each community to learn cultural skills identified by the community as important to Inuit culture, heritage, and present-day subsistence living. The strategy was also designed to positively support mental health for resilience and to increase adaptive capacities to socio-cultural and environmental changes by drawing on the strengths of Inuit culture. This program was created to discover if culturally-based youth mentorship strategies and intercultural and intergenerational knowledge exchanges can foster, enhance, and expand resilience and support cultural continuity and cultural sustainability.

21.2 Research and Programming Location: The Nunatsiavut Region, Labrador, Canada

Nunatsiavut ('Our Beautiful Land' in Labrador Inuttitut) is the homeland of the Labrador Inuit, and is one of four Inuit regions in Canada that comprise the Inuit Nunangat (Inuit Homelands) (Fig. 21.1). Home to 4 % of Canada's Inuit population, the Nunatsiavut land claims settlement region was formed and achieved self-government in 2005, and encompasses five communities (North to South): Nain, Hopedale, Postville, Makkovik, and Rigolet.

Labrador Inuit and their ancestors have been living in the Nunatsiavut region for thousands of years, surviving through an intimate knowledge of and relationship to the land, and through cultural skills and wisdom developed over hundreds of years

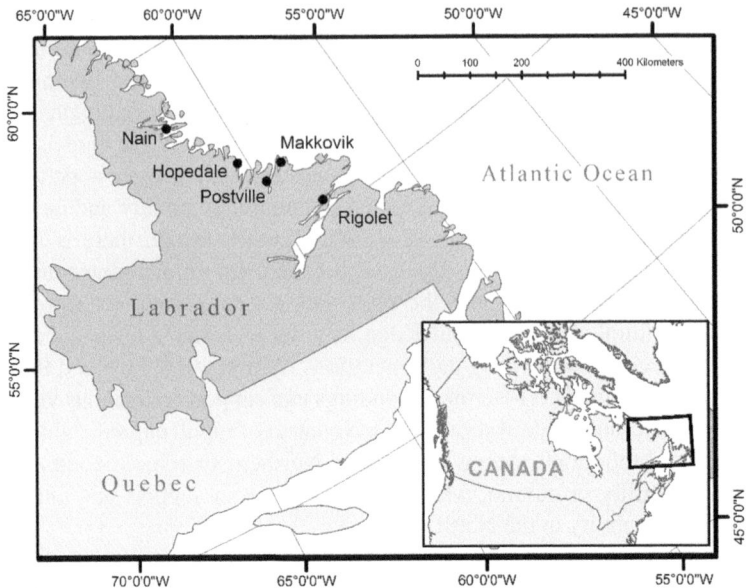

Fig. 21.1 A map of Nunatsiavut, Labrador, indicating all five communities

and passed down through intergenerational knowledge exchange (Natcher et al. 2012). Skills such as hunting, trapping, wild meat butchering and preserving, skin and fur preparation, sewing fur clothing, and Inuit arts and music have sustained the Labrador Inuit for generations. These skills are essential not only to culture and heritage, but for overall community wellbeing, sustainability, and life in the region. Labrador Inuit continue to lead semi-subsistence livelihoods, and harvest animals and plants from the abundant land and sea, including caribou (when available), seal, salmon, trout, ducks, geese, red berries and bake apples.

Rigolet, Makkovik, and Postville are all remote and coastal, and there is no road access to any of the communities. Travel options to the communities are weather-dependent, with year-round plane service, seasonal ferry service, and via personal transportation such as skidoos when ice forms and boats during the summer months.

As with other Indigenous communities in Canada, the Labrador Inuit have experienced a long history of colonization. Their history has included forced relocation from traditional homelands into settled communities, forced attendance of children in residential schools, ongoing acculturation and loss of language, and disruptions to knowledge transmission and cultural continuity (Lehti et al. 2009; Richmond and Ross 2009; Natcher et al. 2012). These processes have ongoing impacts on individual and community wellbeing. Impacts range from food insecurity, housing overcrowding, increased rates of chronic and infectious diseases, high rates of addiction and mental health issues, decreased life expectancy, and high suicide rates when compared to non-Inuit Canadian populations (Cunsolo Willox et al. 2014; Lehti et al. 2009; Natcher et al. 2012).

Over the last decade, these socio-economic, health, and cultural stressors have been compounded by rapid climate changes. The Inuit have faced loss of sea and freshwater ice, changes in freezing and melting seasons, increased seasonal temperatures, and changes to flora and fauna in the region. In recent years, some measurements indicate that the region has experienced increases in surface air temperature, with a projected temperature change of up to 8 °C compared to the twentieth century average if carbon emissions continue unabated (Way and Viau 2014). These changes are putting increased pressure on Inuit livelihoods, food systems, and travel safety; all of which are impacting Inuit culture in the region and adding increased challenges to the sharing of cultural practices (Cunsolo Willox et al. 2012, 2013a, b; Petrasek MacDonald 2013b, 2015).

Given these overlapping and intersecting challenges, Inuit communities in Nunatsiavut have been proactively conducting research and creating evidence-based programming to support resilience, sustainability, and wellbeing that is connected to Inuit culture by strengthening cultural opportunities to share knowledge between generations.

21.2.1 Community Leads and Partners

The Inuit Community Governments of Rigolet, Makkovik, and Postville, working in close collaboration with the Nunatsiavut Government's Department of Health and Social Development, led this research and program implementation. Rigolet (population 306, 95 % identify as Indigenous) is the Southern-most Inuit community in Nunatsiavut. In recent years, Rigolet has been proactive with research projects, and established the Inuit-run 'My Word:' Storytelling and Digital Media Lab, which has become a leader in Northern participatory research, with a particular focus on health and wellness. Makkovik (population 361, 90 % identify as Indigenous) is located between Rigolet in the South and Postville in the North. Makkovik is also home to the only available marine center on the North Coast of Labrador operated by the Torngat Fish Producers. Postville (population 206, 91 % identify as Indigenous) is the smallest community in Nunatsiavut, and has a long history as a Hudson Bay trading post, established in 1837 (Statistics Canada 2014). All three communities have been active in a wide range of research activities related to changing socio-cultural, climatic, and environmental conditions, and all three communities are working to support individual and community wellness through a variety of activities and supports, with an emphasis on continually strengthening cultural connectivity and continuity.[1]

---

[1] While there was interest from all Nunatsiavut communities in this program, during a regional meeting with all five community leaders and stakeholders from Nunatsiavut, it was collaboratively decided that Rigolet, Makkovik, and Postville were ideally suited to try a pilot program of this nature, as other youth-adult mentorship programming did not exist in the communities. At the time of publication, and based on the success of the program, funding opportunities are being explored to expand this program to all five communities as part of core funding through the Department of Health and Social Development.

21.3   The *IlikKuset-Ilingannet* Program

Conceptually, this program was developed based on Healey and Tagak's (2014) Inuit relational epistemology of *piliriqatigiinniq* ('working in a collaborative way for the common good'), which gives primacy to the relationships that people develop with each other, the environment, and their culture. Through *piliriqatigiinniq*, knowledge and understanding are located in *relationships*, not just within individuals, and knowledge construction accumulates over time through lived experiences and occurs through sharing, dialogue, trust-building, relationship-development, and connections with others. This program understands the epistemological power of 'making' or 'crafting', and supports the development of knowledge, skills, and ways of knowing that come from learning with others.

Operationally, this project was developed based on community ideas to support mental wellness and cultural continuity in Rigolet, Makkovik, and Postville in order to develop community resilience. From research conducted in Rigolet from 2009–2012[2] and in Rigolet, Makkovik, Postville, Hopedale, and Nain from 2012–2013,[3] residents reported that observed changes in precipitation, ice coverage and stability, storm patterns, temperature fluctuations, and changes in wildlife and vegetation related to climate change were negatively affecting mental and emotional health and wellbeing due to decreased access to the land and land-based activities (Cunsolo Willox et al. 2012, 2013a, b). This research also indicated that changes in climate and environment interacted with other mental health stressors, including loss of livelihoods, sense of self, connection to culture along with addictions, family stress, previous trauma, and already-present mental health challenges (Cunsolo Willox et al. 2012, 2013a, b). Finding ways to support mental wellness that reflect and celebrate Inuit culture has been a key priority identified by communities. As a result, finding ways to support more opportunities to learn cultural skills and participate in cultural activities was requested by the communities.

21.3.1   How It Worked

The *IlikKuset-Ilingannet* (Culture-Connect!) youth mentorship and cultural program united five youth (ages 15–25) in each community (n = 15) with five adult mentors/cultural role models (ages 35+) in each community (n = 15). The pairings aimed to teach the youth a wide range of cultural skills, including trapping, Inuttitut language skills, traditional music, sewing with fur and skins, snowshoe-making, Inuit art, carving, and traditional cooking and wild food preparation (Table 21.1). These skills were chosen by the communities based on their importance to Labrador Inuit culture, history, and heritage, their abilities to support subsistence living, their

---

[2] The *Changing Climate, Changing Health, Changing Stories* project.
[3] The *Inuit Mental Health Adaptation to Climate Change* project.

Table 21.1 A list of cultural skills by community from the *IlikKuset-Ilingannet* Program

Cultural Skill	Rigolet	Makkovik	Postville
Drawing & Art	O		
Inuttitut Language Learning		O	
Outdoor Living			O
Skin & Fur Sewing	O	O	O
Snowshoe Making		O	O
Stone Carving	O		
Traditional Music			O
Trapping & Fur Preparation	O	O	
Wild Food Preparation & Cooking		O	O
Wood Carving	O		

links to preserving, promoting, and sustaining Inuit culture, their continuing relevance to present-day life in the communities, and the expertise of available mentors in each community.

The program ran for 8 months from October, 2013 to May, 2014, with a 3-week break in December for holidays. During this time, each youth spent approximately 5–7 h a week for 4–5 weeks with each mentor in their home community, learning skills and participating in activities that support the transition of knowledge (each mentor was hired to provide mentorship in one skill to all five youth in the community). Once an activity was completed, youth would work with a different mentor for the same duration on a different skill or activity, until each youth had the opportunity to learn from each mentor in their community. Some sessions were conducted one-on-one, and some were done in small groups. The program was coordinated by an Inuit Local Research Coordinator (LRC) in each of the communities, with overall administrative oversight for all three communities situated in Rigolet. The LRCs worked to recruit youth and mentor participants, introduce and highlight the program to the community, create schedules, order supplies, and organize regular youth gatherings and mentor supports.

All necessary materials were provided, free of charge, to complete the activities and support the learning. Mentors were paid an honorarium in recognition of their valuable wisdom, knowledge, skills, and expertise they were sharing with the youth, and to honour the importance of their contributions to the program, their community, and to cultural preservation and promotion.

By the end of the program, each youth had spent approximately 30 h with each mentor, for a total of roughly 150 h of cultural programming for each youth. Across all three communities, the program provided an approximate total of 2250 youth-mentor contact hours and supported positive, healthy opportunities to fulfil the conceptual understanding of *piliriqatigiinniq* within youth and the community.

In addition to the youth-mentor activities, the youth in each community also had regular meetings to talk about their experiences, share their learning, and showcase what they made. Each community also had at least one 'on the land' trip, bringing youth together to learn land skills or 'field test' items made, such as snowshoes, mitts, and parkas. The mentors also got together as a group in each community, sharing tips and techniques for working with youth and sharing cultural skills.

During the activities, the youth and LRCs also took photographs and videos to document their experiences. With the editing support of Jordan Konek of Konek Productions in Iqaluit, Nunavut, these photos and videos were combined together to create a group-edited project video during a youth gathering in Rigolet to celebrate the end of the project and to share their experiences with others.[4]

Finally, each community also held an Open House and program celebration to share what the youth had made and learned, and to demonstrate some of the skills they developed with their communities. These Open Houses featured musical performances, wild food, sewing and snowshoe demonstrations, and screenings of the project film. They drew large crowds, with over 250 people attending across all three communities (Fig. 21.2).

21.4   Methods

In order to evaluate the effectiveness of this type of cultural mentorship programming for youth, mentors, communities, and cultural resilience and sustainability, semi-structured interviews with the youth, mentors, and key informants (n = 37: Postville = 12; Makkovik = 15; Rigolet = 10) were conducted in April and May 2014. The interview guide was broken into two sections: (1) an evaluation section that focused on strengths and weaknesses, benefits and concerns, areas for improvement, and locally-defined indicators of project success or impact; and (2) a research section aimed at understanding knowledge sharing, cultural continuity, and local ways of learning and teaching.[5]

Interviews were conducted in a conversational style (Kvale 1996), allowing the sharing of stories and ideas to flow freely and unencumbered by a rigid question guide. All interviews were conducted by the Local Research Coordinators (LRCs) in English, by participant choice, although Inuttitut translators were available if requested. Interviews ranged between 30 and 60 min, and were audio-recorded and transcribed by the LRCs.

Data were analyzed through an iterative and immersive constant-comparative process (Miles and Huberman 1994) that included reading through transcripts while listening to the audio-recording, and creating interview summaries that include key themes and ideas from each transcript. Preliminary themes were identified by the LRCs, who played an active role throughout the data analysis phase. These codes and themes were combined together within and among communities, until a final list of overarching themes was created. This list was continually checked with local researchers and key stakeholders to ensure accuracy of interpretations and ideas and to verify results with community members and participants.

---

[4] The video can be found at https://youtu.be/EAulcH3uXnc

[5] For more information about the knowledge transmission or formalization of knowledge sharing aspects of this program, please see Stephenson et al. (under review).

Fig. 21.2 A collage of photos from the *IlikKuset-Ilingannet* program. From *left* to *right*, *top* to *bottom*: (i) Kerry Pottle learning to carve soapstone (Rigolet); (ii) David Wolfrey teaching Anita Rich to tie a snare (Rigolet); (iii) Dillon Pottle constructing a marten box in trapping lessons (Rigolet); (iv) Cassie Jararuse learning Inuttitut from Katie Hayes (Makkovik); (v) Henry Jacque teaching Ocean Lane to make snowshoes (Makkovik); (vi) Megan Andersen displaying finished snowshoes (Makkovik); (vii) Kerry Pottle designing a pattern for Inuit art (Rigolet); (viii) Greg Jacque and Jordan Sheppard learning small engine repair from Bryce Gear (Postville); (xiv) Grant Gear showing off his new moccasins (Postville); (x) Pam Campbell with her finished art on moose-hide (Rigolet); (xi) finished sealskin and rabbit fur mitts made by Anita Rich from sewing lessons (Rigolet); (xii) Inuttitut lessons (Makkovik)

21.5   Results

From all accounts, this program was a strong success, with interviewees indicating very positive changes: new understandings from hands-on learning; new and enhanced relationships created between and among youth and mentors; revitalized cultural pride among youth and mentors; and increased cultural preservation, promotion and sustainability.

21.5.1   Epistemologies of Making and Doing: "Make Sure You Do More Hands on Work"

The majority of people interviewed emphasized the importance of participating in a program that not only promoted cultural knowledge and skills, but did so in a hands-on way that resonated with Inuit knowledge and historical forms of knowledge-sharing. Many of the mentors expressed concern over the ways in which the younger generations were being educated within formal educational settings and were losing connections to 'learning by doing'. As one mentor explained:

> Everything they done, they learned something from it. They had to do hands on, and that was real good, so they could really learn what to do, instead of just looking at it, and not really remember it, 'cause when you just look at something, it's easier to forget, but when you do hands on, then you'll have…you'll make it like a… good memory. Something that'll come easily to you.

The youth echoed this statement, with many of the youth sharing their excitement for being able to participate in hands-on learning, and to "be a well-rounded cultural person" by taking part in the activities. As one youth participant articulated:

> I learned a whole lot more and I did things that I didn't think I'd ever be able to do in my life, and it did make me feel better, because it was more than just an accomplishment, and I did learn about some like older ways and traditional things and it's good to be able to know and hopefully carry on.

21.5.2   Strengthening Relationships Through Culture: "It Strengthened Some Relationships in the Program. Maybe Even in the Community"

Building on the importance of creating a hands-on learning environment that brought generations together, many of the youth and mentor interviewees discussed the ways in which the *IlikKuset-Ilingannet* program supported the development or the deepening of relationships. As one youth participant described: "*it was a good opportunity to get to know people on a different level than you're used to seeing them within the community."* Another youth commented on the importance of connecting with the mentors to learn new skills and share experiences:

> *I was able to share the experience with someone else who was very knowledgeable in that area and it was a chance to share different perspectives and learn new things, share ideas, laugh, and it was a fun time, and I think one of the things that was underrated about the project, was getting the chance to know our mentors better than we knew them before, even though they are members of our community.*

Another interviewee commented on the level of trust that developed among the participants:

> *I noticed that there were youth coming to me and telling me all kinds of things that I don't think they would have ever come to me specifically about. It was kind of like a trust relationship there, where… it was almost like… I don't want to say like a priest, but it was almost like you had that confidentiality aspect there. Like what I say to you here is not gonna leave here. And there was trust built through this program that would never have been built any other way.*

Many people also discussed the positive benefits for the mentors as well, who were experiencing new and often-unexpected relationships with the youth, leading to feelings of pride and confidence, and to a new level of cultural sharing between generations. As one LRC described:

> *There was a relationship formed there [between the youth and mentors], where you kind of didn't expect it, and you really couldn't force it, it was just something that happened. They [the mentors] opened up and they shared things, they told stories, and they shared things about their pasts and about how they learned their skills and they were really proud, telling them how I had to do this all my life, and, you know, they became really open and I think it gave them some validity that what they had to share was important.*

21.5.3 Revitalized Cultural Pride: "Now That You Have a Skill that's Closely Tied with Your Culture, You Feel a Little More Pride and a Little More Confident in Who You Are"

The majority of youth participants and mentors commented that participating in this program revitalized or enhanced a sense of cultural pride. As one of the mentors commented: *"For me, the best part was the youth taking an interest in our local traditions or cultural connections, and seeing them get excited about accomplishing a craft or skill."* Echoing this, one of the LRCs explained that after sessions: *"the youth were coming out, they were saying things, like you know, they were proud. They were talking about how proud they were of their stuff and of each other."*

The youth all commented on the respect they felt for the mentors when they learned about their cultural skills and connections, and about the time, effort, attention, and skill it takes to create a craft. An interviewee explained that for mentors, being able to pass on skills and knowledge to the youth *"gave them a lot of pride. It was someone stopping and taking notice that 'hey you can do this', and we respect you for what you can do, we're grateful you can teach us, we think a lot of your skill, and we're glad you're taking the time out of your day to teach us."*

Finally, both youth and mentors identified that participating in the *IlikKuset-Ilingannet* gave a deep sense of connection to culture. As one of the LRCs explained:

> *It's a sense of self, it's important, it's a part of your cultural identity. It makes you proud to be an Inuit person and it makes you proud to be from Labrador. It makes you proud, and shows pride and hard work and it shows that you're dedicated to sustaining your life in the North.*

21.5.4 Cultural Preservation, Promotion, and Sustainability: "Making Them Understand How Important Making Crafts Is and Keeping Our Traditions Alive"

Connected to cultural pride was the strong theme of cultural preservation and promotion. For all participants in the *IlikKuset-Ilingannet* program, one of the motivating factors to initially get involved in the program was a desire to both preserve and promote Labrador Inuit culture through the passing on of crafts and skills. There was a strong concern among all interviewees about the potential loss of culture in Nunatsiavut. As one youth articulated: *"a lot of our Elders are passing away, and this is a good way for the youth and younger generations to learn other stuff, because not often you get to see families going off and doing things like hunting and trapping."*

Another youth expressed similar concerns, worried that without the older generations to teach the traditional skills, Labrador Inuit culture may begin to erode as fewer and fewer people maintain the cultural skills: *"So like once that person is gone, it's gone. …Once one person is gone, the next generation is not gonna know how to do it."* For most participants, these skills are essential for life in Nunatsiavut because, as one youth participant described: *"It's just survival. It just learns you to get ready for the world."*

Mentors were equally as concerned as youth about the loss of knowledge-sharing and skills-transmission in the region. As a mentor in the program described, *"That's why you share, so it's always passed down. Or it'll be lost, like people lose interest, 'cause they're not doing it…It's not important to them, 'cause they don't know about it."* Another mentor agreed, and expanded this idea:

> *I think that there's a loss of a connection to your culture when there's a gap as there has been, I feel, here between a generation who has knowledge and has wisdom and is able to teach, and a generation much younger that is perhaps intimidated, doesn't know who to ask, what to ask, what kinds of things people can do. There seems to be a gap, and this program that we had kind of bridged the gap between the culture and the connection that it needed, to be passed down between elders and youth.*

Despite this gap in knowledge-sharing and intergenerational knowledge transmission, many participants in the program explained that for youth, keeping these traditions alive and learning these skills is an important part of being Inuit, and is a natural and necessary skill for many people: *"I think people still need it. They just don't realize they need it."*

Many mentors also expressed hope and excitement about passing on the skills to the younger generation. As a result, the mentors were inspired to continue to hone their abilities and start to teach others, expanding the reach of not only the program, but also of the knowledge and experiences. As one mentor explained: *"when this program is over, those five students that I taught, I hope that they can go out and still do it and make money and produce and teach when they get older."*

Finally, beyond being necessary for survival and life in the region, one LRC explained the inherent connection that Inuit have to these skills and crafts:

> *I think it's kind of an innate thing [to do cultural activities]...I think it's kind of an inborn thing, something that's just in you to do...I saw in everybody a real connection to a certain thing, and it was just magic to watch it unfold and to watch the youth and mentor connect over this one thing. ...They realized what was absent and what almost they were destined to do.*

21.5.5 Suggestions for Improvement and Expansion

While participants spoke in overwhelmingly positive terms about this program, they also had valuable suggestions for improvement if it was to run again or replicated in other Northern Indigenous contexts.

1. *Increase the length of time to September to May for seasonal variations and to give more time for each skill.* Both the youth and the mentors all indicated that they would have liked more time to go deeper into the skills-sharing and the relationship-building. Expanding the time period of the program would also allow for greater access to seasonal variation in hunting, trapping, land-based activities, and outdoor learning.
2. *Increase the number of youth in each community from five to at least eight to ten, to provide more youth with the opportunity to participate.* Once the program began, there was increased demand in each community from additional youth who wanted to participate in the activities and learn the skills. It was widely suggested that, should this program run again, more spaces be opened to allow increased numbers of youth to be accommodated.
3. *Consider adding an additional mentor if the program is lengthened to encourage more skills learning.* As the program progressed, many youth suggested additional skills they would like to learn, many mentors shared that they had other skills they would like to also teach, and several new adults came forward in the communities to indicate they would like to participate as a mentor if the program were expanded.
4. *Have more small-group sessions (rather than one-on-one sessions).* When this program was originally designed, emphasis was placed on developing opportunities for the youth and mentors to develop relationships. Yet, as the program

went on, both the youth and the mentors naturally gravitated to small-group sessions, with one mentor and more than one youth. The group setting allowed for greater camaraderie and relationship-building to occur, and removed any awkwardness or pressure that some participants felt when trying one-on-one sessions. Small-group settings would also allow for the inclusion of more youth should the program be expanded.

5. *Incorporate opportunities to gain specific accreditation or certificates.* As the program progressed, several youth expressed interest in having specific certification opportunities through the program, such as taking a Bear Safety course or taking the trapping course, in order to encourage increased learning and enhance depth of skills and knowledge, while also providing accreditations that are transferable to other contexts.

6. *Add hunting to the skills in each community (especially if the program is lengthened).* There was a resounding request from youth and mentors to have hunting skills and safety added to the program, provided that the program was lengthened to allow for seasonal differences in wildlife. This was seen as essential for Inuit culture, food security, individual and family resilience, and for overall health and wellness.

7. *Have a dedicated space for some of the activities in each community, rather than mentors' homes.* Since this was a pilot project, most mentors ended up inviting youth into their homes or their sheds to learn the skills. Several participants suggested that, were this program to continue, securing dedicated spaces in the communities where people could gather to learn some of the activities would increase comfort levels of participants, while simultaneously creating a hub of activity for the program.

Despite these suggestions for improvement, all participants recommended that this program be continued, because it was an essential component of ensuring that Inuit cultural skills and knowledge could be passed on, learned, and flourish, while supporting individual and community betterment. As one Local Research Coordinator explained:

> when somebody learns how to do something, it only betters them. And they were people who had a place to go, who maybe never had one before, never had a place where they were respected and admired and people were proud of them and they were put to hard work. I'm sure there were times where they came to do something, you know, to build them up, instead of to go somewhere where someone might have torn them down.

21.6 Discussion and Conclusion: Supporting Cultural Sustainability

Sustainability issues are becoming an ever-increasing concern and focus across the Circumpolar North. With increasing stressors from multiple pathways, including climate change, resource extraction and development, and socio-economic shifts

(Cunsolo Willox et al. 2014; Ford et al. 2014), communities and regions across the North are developing strategies to support community wellness, enhance individual and community resilience, and preserve and promote cultural heritage, cultural knowledge, and cultural continuity from within Inuit culture itself. Programs such as *IlikKuset-Ilingannet* have the potential to assist both youth and adults in connecting together in positive environments dedicated to knowledge transmission and skills development, which, as this research indicates, can create strong cohesion among generations and develop and enhance relationships based on cultural pride.

Connected to culturally-based pride and relationships, this research indicates that this program also contributed to preserving and promoting Inuit cultural skills and knowledge, supporting the continuation of skills and knowledge in the region, enhancing health and wellness, and promoting cultural sustainability via an emphasis on cultural promotion and resurgence. These findings resonate with the increasing emphasis on community-based and community-led cultural reclamation programming in other Northern and Indigenous contexts in Canada. For example: the *Aullak, Sangilivallianginnatuk* (Going Off, Growing Strong), program, which provides Inuit youth with the opportunity to learn hunting skills and participate in food-sharing activities in Nain, Nunatsiavut; the *Piqqusilirivvik* Inuit Cultural School in Clyde River, Nunavut that provides opportunities for participants to learn various Inuit cultural skills through traditional teachings and practices; the Dechinta Bush University in Dene Territory in the Northwest Territories, a land-based educational program, premised on decolonizing approaches and aimed at creating learning environments to promote cultural resurgence (Freeland Ballantyne 2014); the Akwaesasne Cultural Restoration program in Québec that links Mohawk youth with mentors and knowledge-holders to learn land-based skills and revitalize cultural wisdom (Alfred 2014); and the Chisasibi land-based healing program in Cree Territory in Québec that fosters healing and wellness through the promotion of Cree traditions and cultural knowledge (Radu et al. 2014). All of these programs, including the *IlikKuset-Ilingannet* program, have, at their core, a commitment to ensuring that traditional and cultural knowledge, wisdom, and skills are not only passed on to younger generations, but are also reclaimed, revitalized, and renewed, thus supporting Indigenous cultural resurgence through intergenerational knowledge exchange (Stephenson et al. under review).

Another important aspect of the *IlikKuset-Ilingannet* program was providing the spaces and opportunities for youth and mentors to connect with culture through 'making and doing'. This approach to knowledge transmission resonates with Inuit culture, epistemologies, and ontologies (Healey and Tagak 2014; c.f. Stephenson et al. under review). Returning to the concept of *piliriqatigiinniq* ('working in a collaborative way for the common good') (Healey and Tagak 2014), programs such as *IlikKuset-Ilingannet* not only promote cultural knowledge and preservation, but they also promote Inuit ways of knowing and reflect Inuit values and socio-cultural organization. By bringing together different generations, and creating a collaborative learning environment, *IlikKuset-Ilingannet* highlighted the need for people to work collaboratively for the common good of preserving and promoting Labrador

Inuit culture and heritage and, proudly "wearing their teachings" (Simpson 2014: 11) or displaying their learnings for all to see, appreciate, and connect with.

This approach to cultural programming also connects to other epistemological concepts that Healey and Tagak (2014) outlined: *Inuuqatigiittiarniq* (being respectful of all people), which was fostered through the development of mutual respect between the youth and mentors as they learned together; *unikkaaqatigiinniq* (storytelling), which developed naturally as mentors shared stories from the past with youth—stories containing wisdom of survival, resilience, heritage, and adaptation; *pittiarniq* (being kind and good), which was encouraged through working together individually and in small groups for cultural knowledge sharing and with a desire and drive to preserve and promote Labrador Inuit culture; and *iqqaumaqatigiinniq* (all things coming into one), which was experienced by the youth and mentors as they worked together to discover and highlight the interconnected nature of Inuit knowledge, cultural skills, survival, and adaptation in the twenty-first century.

Through this research and the voices shared, we contend that issues and notions of sustainability across the North must encompass, incorporate, and integrate understandings of cultural sustainability. In the holistic Inuit worldview, where all things are tied together, connected, and relational, separating cultural sustainability from environmental or infrastructural sustainability would be artificial at best, and could serve to undermine other efforts for creating sustainable futures or for adaptation. Indeed, this program emerged from research that was initially examining climate change, mental health, and adaptive capacities in Nunatsiavut, and quickly grew beyond this focus—and, with requests and input from over 100 people in the region, a focus on (re)connecting with culture was considered not only desirable to support intergenerational knowledge sharing and celebrate Inuit heritage, but also essential for fostering resilience and supporting adaptation to multiple stressors and to any change by building from the strength, wisdom, and power of Inuit culture.

This research, then, underscores the importance not only of focusing on preserving, promoting, and supporting cultural knowledge and skills for wellbeing, but also creating stronger, more robust, more sustainable individuals and communities who and that are more resilient to any type of change. Indeed, revitalizing connections to culture and fostering cultural sustainability may be a key to adaptation across the North through this type of programming because it draws from the strengths and innovations of Inuit knowledge, wisdom, and culture. In so doing, *"the old ways are coming to the new generation, and maybe the younger generation can adapt somehow."*

Acknowledgements A huge thank you to all the youth and mentors involved in this program. Without you, nothing would have been possible! This program would also not have been possible without the support and participation of the communities of Postville, Makkovik, and Rigolet, in the Nunatsiavut region of Labrador and the Inuit Community Governments of Rigolet, Makkovik, and Postville. Special thanks to Charlotte Wolfrey, Herb Jacque, Diane Gear, Chris Brennen, and Melva Williams for all your project support, advice, and guidance. Thanks also to Jordan Konek/ Konek Productions for editing the final video and Joanna Petrasek MacDonald for video assistance. Many thanks to editors Gary Wilson and Gail Fondahl for all their editorial assistance, and to the anonymous reviewer for the helpful, supportive, and insightful comments. This research was supported through funding from Health Canada, through the First Nations and Inuit Health Branch *Climate Change and Health Adaptation in Northern First Nations and Inuit Communities* program. Complementary funding was received from the Nasivvik Centre for Inuit Health and Changing Environments.

References

Alfred, T. (2014). The Akwesasne cultural restoration program: A Mohawk approach to land-based education. *Decolonization: Indigeneity, Education, and Society, 3*(3), 134–144.

Allen, J., Hopper, K., Wexler, L., Kral, M., Rasmus, S., & Nystad, K. (2014). Mapping resilience pathways of Indigenous youth in five circumpolar communities. *Transcultural Psychiatry, 51*(5), 601–631.

Christensen, J. (2013). 'Our home, our way of life': Spiritual homelessness and the socio-cultural dimensions of Indigenous homelessness in the Northwest Territories (NWT), Canada. *Journal of Social and Cultural Geography, 14*(7), 804–828.

Collings, P. (2014). *Becoming Inummarik: Men's lives in an Inuit community*. Montreal: McGill-Queen's University Press.

Cunsolo Willox, A., Harper, S., Ford, J., Landman, K., Houle, K., Edge, V., & the Rigolet Inuit Community Government. (2012). "From this place and of this place:" Climate change, sense of place, and health in Nunatsiavut, Canada. *Social Science & Medicine, 75*(3), 538–547.

Cunsolo Willox, A., Harper, S., Ford, J., Edge, V., Landman, K., Houle, K., Blake, S., & Wolfrey, C. (2013a). Climate change and mental health: An exploratory case study from Rigolet, Nunatsiavut, Labrador. *Climatic Change, 121*(2), 255–270.

Cunsolo Willox, A., Harper, S., Edge, V., Landman, K., Houle, K., Ford, J., & the Rigolet Inuit Community Government. (2013b). 'The land enriches our soul:' On environmental change, affect, and emotional health and well-being in Nunatsiavut, Canada. *Emotion, Space, and Society, 6*, 14–24.

Cunsolo Willox, A., Stephenson, E., Allen, J., Bourque, F., Drossos, A., Elgarøy, S., Kral, M., Mauro, I., Moses, J., Pearce, T., Petrasek MacDonald, J., & Wexler, L. (2014). Examining relationships between climate change and mental health in the Circumpolar North. *Regional Environmental Change, 15*, 169. doi:10.1007/s10113-014-0630-z.

Dybbroe, S., Dahl, J., & Muller-Wille, L. (2010). Dynamics of Arctic urbanization. *Acta Borealia: A Nordic Journal of Circumpolar Societies, 27*(2), 120–124.

Ford, J. (2012). Indigenous health and climate change. *American Journal of Public Health, 102*(7), 1260–1266.

Ford, J., Berrang-Ford, L., King, M., & Furgal, C. (2010a). Vulnerability of Aboriginal health systems in Canada to climate change. *Global Environmental Change, 20*(4), 668–680.

Ford, J., Pearce, T., Prno, J., Duerden, F., Berrang Ford, L., Beaumier, M., & Smith, T. (2010b). Perceptions of climate change risks in primary resource use industries: A survey of the Canadian mining sector. *Regional Environmental Change, 10*(1), 65–81.

Ford, J., Cunsolo Willox, A., Chatwood, S., Furgal, C., Harper, S., Mauro, I., & Pearce, T. (2014). Adapting to the effects of climate change on Inuit health. *American Journal of Public Health, 104*, e1–e9.

Freeland Ballantyne, E. (2014). Dechinta Bush University: Mobilizing a knowledge economy of reciprocity, resurgence, and decolonization. *Decolonization: Indigeneity, Education, and Society, 3*(3), 67–85.

Healey, G., & Tagak, A. (2014). Piliriqatigiinniq "Working in a collaborative way for the common good": A perspective on the space where health research methodology and Inuit epistemology come together. *International Journal of Critical Indigenous Studies, 7*(1), 1–14.

IPCC. (2013). *Climate change 2013: The physical science basis. Contribution of Working Group 1 to the fifth assessment report of the Intergovernmental Panel on Climate Change*. Geneva: IPCC.

IPCC. (2014). *Climate change 2014: Impacts, adaptation, and vulnerability. Contribution of Working Group II to the fifth assessment report of the Intergovernmental Panel on Climate Change*. Geneva: IPCC.

Kirmayer, L., Tait, C., & Simpson, C. (2009). The mental health of aboriginal peoples in Canada: Transformations of identity and community. In L. Kirmayer & G. Valaskakis (Eds.), *Healing traditions: The mental health of Aboriginal peoples in Canada* (pp. 3–35). Vancouver: UBC Press.

Kirmayer, L., Dandeneau, S., Marshall, E., Phillips, M. K., & Williamson, K. J. (2011). Rethinking resilience from Indigenous perspectives. *Canadian Journal of Psychiatry, 56*(2), 84–91.

Kral, M. J., & Idlout, L. (2012). It's all in the family: Wellbeing among Inuit in Arctic Canada. In H. Selin & G. Davey (Eds.), *Happiness across cultures* (pp. 387–398). Dordrecht: Springer.

Kral, M. J., Idlout, L., Minore, J. B., Dyck, R. J., & Kirmayer, L. J. (2011). Unikkaartuit: Meanings of well-being, unhappiness, health, and community change among Inuit in Nunavut, Canada. *American Journal of Community Psychology, 48*(3–4), 426–438.

Kvale, S. (1996). *InterViews: An introduction to qualitative research interviewing*. Thousand Oaks: Sage.

Lehti, V., Niemelä, S., Hoven, C., Mandell, D., & Sourander, A. (2009). Mental health, substance use, and suicidal behaviour among young Indigenous people in the Arctic: A systematic review. *Social Sciences and Medicine, 69*, 1194–1203.

McDowell, G., & Ford, J. (2014). The socio-ecological dimensions of hydrocarbon development in the Disko Bay region of Greenland: Opportunities, risks, and tradeoffs. *Applied Geography, 47*, 98–110.

Miles, M., & Huberman, M. (1994). *Qualitative data analysis: An expanded sourcebook* (2nd ed.). Thousand Oaks: Sage.

Natcher, D. C., Felt, L., & Procter, A. (Eds.). (2012). *Settlement, subsistence, and change among the Labrador Inuit*. Winnipeg: University of Manitoba Press.

Pearce, T., Wright, H., Notaina, R., Kudlak, A., Smit, B., Ford, J., & Furgal, C. (2011). Transmission of environmental knowledge and land skills among Inuit men in Ulukaktok, Northwest Territories, Canada. *Human Ecology, 39*(3), 271–288.

Petrasek MacDonald, J., Ford, J., Cunsolo Willox, A., & Ross, N. (2013a). A review of protective factors and causal mechanisms that enhance the mental health of indigenous Circumpolar youth. *International Journal of Circumpolar Health, 72*(1), 1–18. Available at: http://www.circumpolarhealthjournal.net/index.php/ijch/article/view/21775

Petrasek MacDonald, J., Harper, S., Cunsolo Willox, A., Edge, V., & Rigolet Inuit Community Government. (2013b). A necessary voice: Climate change and lived experiences of youth in Rigolet, Nunatsiavut, Canada. *Global Environmental Change, 23*(1), 360–371.

Petrasek MacDonald, J., Cunsolo Willox, A., Ford, J., Shiwak, I., Wood, M., The IMHACC Team, & The Rigolet Inuit Community Government. (2015). Youth-identified protective factors in a changing climate: Perspectives from Inuit youth in Nunatsiavut, Labrador. *Social Science and Medicine, 141*, 133. doi:10.1016/j.socscimed.2015.07.017.

Prowse, T., Furgal, C., Chouinard, R., Melling, H., Milburn, D., & Smith, S. (2009). Implications of climate change for economic development in Northern Canada: Energy, resource, and transportation sectors. *Ambio, 38*(5), 2712–2781.

Radu, I., House, L., & Pashagumskum, E. (2014). Land, life, and knowledge in Chisasibi: Intergenerational healing in the bush. *Decolonization: Indigeneity, Education, and Society, 3*(3), 86–105.

Richmond, C. A. M., & Ross, N. A. (2009). The determinants of First Nation and Inuit health: A critical population health approach. *Health and Place, 15*, 403–411.

Simpson, L. (2014). Land as pedagogy: Nishnaabeg intelligence and rebellious transformation. *Decolonization: Indigeneity, Education & Society, 3*(3), 1-25.

Statistics Canada. (2014). 2011 Community profiles. Available at: http://www12.statcan.gc.ca/census-recensement/2011/dp-pd/prof/index.cfm?Lang=E

Stephenson, E., Cunsolo Willox, A., Pearce, T., Ford, J., Kaodloak, L., Klenkenberg, L., Brennen, R., Ulukhaktok Community Corporation, & The Postville Inuit Community Government. (under review). "I'm going to do my part to carry it on": Inuit cultural mentorship programs in Ulukhaktok, Inuvialuit Settlement Region, and Postville, Nunatsiavut

Way, R. G., & Viau, A. E. (2014). Natural and forced air temperature variability in the Labrador region of Canada during the past century. *Applied and Theoretical Climatology, 121*, 413. doi:10.1007/s00704-014-1248-2.

Wexler, L. (2009). The importance of identity, culture and history in the study of Indigenous youth wellness. *Journal of the History of Childhood and Youth, 2*(2), 267–278.

Wexler, L. (2014). Looking across three generation of Alaska Natives to explore how culture fosters indigenous resilience. *Transcultural Psychiatry, 51*(1), 73–92.

Wexler, L., Joule, L., Garoutte, J., Mazziotti, J., Baldwin, E., Griffin, M., Jernigan, K., & Hopper, K. (2014). Being responsible, respectful, trying to keep the tradition alive: Cultural resilience and growing up in an Alaska native community. *Transcultural Psychiatry, 51*(1), 73–92.

Chapter 22
Practicing Sustainable Art in the Arctic: Two Case Studies

Herminia Din

Abstract The environment is a global concern, evidenced both by global warming and pollution caused by human activity. Global warming has become increasingly evident, particularly in Alaska. Art can contribute to an increased awareness of these concerns. This paper makes a case for more creative programs related to conceptualizing and creating sustainable art activities and displays. It describes a collaborative learning model by focusing on two case studies: *Junk to Funk* and *Winter Design Project*. These projects benefitted from participation in the University of the Arctic's Arctic Sustainable Art and Design (ASAD) Thematic Network. They demonstrate how recycled materials can be used to create functional artwork and also examine "ice and snow" from a new perspective. The outcome of these efforts is to encourage further artistically inspired solutions using sustainable media.

Keywords Art • Art education • Sustainable art • Community engagement

22.1 Introduction

Global warming and pollution are caused by human activity and are environmental issues that are global concerns; these issues are especially important in the Arctic. Art can contribute to increased awareness of these concerns, especially given widespread interest in connections with the Arctic. Responding to the recycled-based art movement, *Junk to Funk* was created in 2008. The premise was that reducing consumption and reusing waste materials could be an essential link in the recycling effort to save and repurpose valuable resources, especially in the Arctic region. In 2013, the *Winter Design Project* was developed in Anchorage, Alaska to strengthen our collaboration with the University of the Arctic (UA) Thematic Network on

H. Din (✉)
Department of Art, University of Alaska Anchorage, Anchorage, AK, USA
e-mail: hdin@alaska.edu

© Springer International Publishing Switzerland 2017
G. Fondahl, G.N. Wilson (eds.), *Northern Sustainabilities: Understanding and Addressing Change in the Circumpolar World*, Springer Polar Sciences, DOI 10.1007/978-3-319-46150-2_22

305

Arctic Sustainable Arts & Design (ASAD).[1] This project provided University of Alaska Anchorage (UAA) faculty and students with an opportunity to explore and create an outdoor winter space, and to look at "ice and snow"—a truly sustainable medium—from a new perspective. These programs highlight the importance of projects conceptualizing and creating sustainable activities to inspire further artistic creation and/or creative solutions to global problems. The purpose of this paper is to describe the collaborative learning experiences of University of Alaska Anchorage (UAA) faculty and students from diverse disciplines in the context of practicing sustainable art.

22.2 Junk to Funk: An Overview

In Anchorage, community efforts for conservation traditionally have been directed toward the separation of recyclable materials. Few conservation opportunities and activities existed for the community to participate in ways that both "reduce and reuse" waste products. After a review of available recycling programs in the community, it was determined that there were no art classes or art programs for adults, families, or even college students that utilized waste materials or recyclables to support recycling efforts.

The recycled-based art movement offers a new perspective on looking at waste materials; this movement emphasizes that artistic elements can be applied to creative process for all ages while also raising awareness of human relationships to the environment. The art educator Jarvis Ulbricht (1998: 34) observes: "We cannot help but develop a socially responsible environmental art education curriculum in which values and aesthetic are combined in an instrumental manner for the benefit of all." Hicks and King (2007: 334) write: "Art education is well situated to address environmental problems that emerge at the point of contact between nature and social life." By following the current directions of eco-art, sustainable art, and environmental art, greenmuseum.org (2010) has summarized a clear paradigm of this art practice.[2] This paradigm:

- Informs and interprets nature and its processes, or educates about environmental problems;
- Is concerned with environmental forces and materials, creating artworks affected or powered by wind, water, lightning, and even earthquakes;

---

[1] More detailed information about the University of Arctic ASAD Thematic Network, http://www. uarctic.org/organization/thematic-networks/arctic-sustainable-arts-and-design/, and http://www. asadnetwork.org

[2] Read more about this topic in "Interview: Patricia Watts On the Eco-Art Movement" by Moe Beitiks, May 23, 2011, from Inhabitat.com, a weblog devoted to the future of design, tracking the innovations in technology, practices and materials that are pushing architecture and home design towards a smarter and more sustainable future. http://inhabitat.com/interview-a-cop15-arts-wrap-up-with-patricia-watts/

- Re-envisions our relationship to nature, proposing new ways for us to co-exist with our environment; and
- Reclaims and remediates damaged environments, restoring ecosystems in artistic and often aesthetic ways

It is within this context that *Junk to Funk* was created in 2008. It began as a community engagement project that used recycled materials to create beautiful and finished functional artworks. The UAA Center for Community Engagement & Learning (CCEL), and a UAA Faculty Development Grant supported this initiative.

The goals and objectives of the art education program at UAA are to help students develop an understanding of the principles and foundations of art education, particularly within Alaska. The *Junk to Funk* project was informed by Dewey's (1934) principle of hands-on experience-based learning and Petersen's (2008) advocacy on green curriculum.

Today's college graduates confront the first, truly worldwide environmental challenge of balancing the carbon budget — the stocks and flow of carbon through the biosphere — in order to ameliorate the negative consequences of global climate change. Colleges and universities have an obligation to ensure that they provide students with the knowledge and experience necessary to accomplish this challenging task.

In order to transform waste and/or recycled materials into functional artworks, the *Junk to Funk* art program used a hands-on, creative method to create awareness of the environment. Two approaches were applied. First, when working with art education students, art lessons were designed to develop the artistic skills and techniques necessary to create finished and quality products; no prior art skills and/or experiences were required from the participants. Each hands-on project utilized common household recyclables such as cereal boxes, plastic grocery bags, t-shirts, papers, notecards, catalogues, and magazines that participants could bring from home. The workshop utilized an open studio art environment to promote creativity and exploration, and participants were encouraged to experiment with different materials and ideas. Six *Junk to Funk* art projects were developed, including a "plarn" (plastic yarn) shopping tote, a magazine/wrapping paper bowl, an old magazine notecard, a recycled paper book using traditional sewing bookbinding techniques, a t-shirt scarf, and telephone wire and paper-bead eco-jewelry (Fig. 22.1). A transformation was observed as the participants began looking at "waste" materials differently. Perhaps the most rewarding aspect was the participants' awareness of the quality of art that could be created from recycled products.

The project also organized two major events during the academic year. In the fall semester, it worked with the UAA Bookstore to offer a *First Friday* fundraising event to raise money for Kids' Kitchen, a non-profit organization whose purpose is to provide nutritional meals for children at no cost. In the spring, an event and auction sale of the recycled products was organized in collaboration with the *Grassroots: A Fair Trade Store* in midtown Anchorage, and the UAA Office of Sustainability, as part of the "Earth Day" celebration.

Fig. 22.1 *Junk to Funk* Art Projects including magazine note card, recycled notebook, magazine bowl, "plarn" bag, telephone wire & paper bead eco-jewelry, and t-shirt scarf

Throughout the project, participating students were asked to keep a weekly journal to document their perceptions. One student quoted Suzi Gablik, an artist and art historian: "Art is an instrument. It can be used to make a difference to the welfare of communities, the welfare of societies, and to our relationship with nature." Another student reflected:

> *I both enjoyed and learned a great deal by taking this course, which incorporated learning, participation, community involvement, and practical experience. …. We learned different lessons from each experience about the various aspects and relationships between art education, community service, and event organization.*

> *What was surprising to me was the unfolding of a larger network of community connections and the involvement of people, which occurs as the result of an art event. It is remarkable how art can bring together and motivate so many. It was an intense experience in learning by doing and pushing your creativity when under the clock of the opening date of the show.*

Another wrote:

> *Art education has favorably progressed since Modernism's emphasis on self-sufficiency, autonomy, and 'art for art's sake'. We now see art integrated in the community as evidence of a growing sense of social responsibility. "Art for society's sake" has developed in many forms through the education of elementary and adolescent students.*

Community-based art education encourages the social responsibility of the artist and educator. When students learn how they can play a vital role in the health of their community through the arts, an integrated perspective is gained. Students find self-expression in relation to the world around them and the community is strengthened in the process.

The *Junk to Funk* project also gained broader attention within the Anchorage community. In the summer of 2008, a news article documenting the effort, *Recycling into Art – Green-minded Alaskans* by Dawnell Smith, was published in the *Anchorage Daily News*. In the article, Jerelyn Miyashiro, a former UAA art education student, was recognized for her contribution to the project. With this encouragement, further collaborations were undertaken working with other community agencies to promote the recycling effort. These included: the Homeward Bound/Rural Alaska Community Action Program that houses homeless people and works to find resources to help encourage self-sufficiency; the Alaska Youth for Environmental Action Group, a high school environmental education and leadership program of the National Wildlife Federation; and the Older Persons Action Group, that provides statewide advocacy of older persons' issues through community action programs. In March, 2014, three 90-min *Junk to Funk* art series workshops (family-friendly workshops teaching visitors how to make treasures from trash) were offered at the Anchorage Museum in conjunction with *Gyre: The Plastic Ocean* exhibit. By collaborating and interacting with other community agencies, *Junk to Funk* inspired new ideas and elicited creative action from people who participated in the workshops.

Secondly, to encourage non-art major college students to be creative and to look at recycled materials differently, the practice of making and creating sustainable artwork was incorporated into Art Appreciation, a General Education Requirement course, as a major course assignment. The project description follows as listed in the syllabus.

Junk to Funk Art Project (15/100 points)
Use library and Web resources for inspiration to help you create **an original artwork** that uses **recycled materials** only. To complete this project, you need to type or upload one-page profile of your artwork in the following sequence: (1) a title of your artwork, (2) a picture of your artwork, (3) a list of materials you used for this project, (4) a detailed description (200 ± words) of your artwork with **a strong emphasis of its functionality and purpose**, and (5) a short reflection (200 ± words) of the creative process. Remember, one piece of junk, with creativity and imagination, can turn into something fun, artistic, beautiful and <u>functional</u>.

During the semester, students explored ideas of using recycled materials to create functional artwork, and discussed artwork created by artists Richard Lang, Judith Selby Lang, Jean Shin, Derick Melander, Ron Van Der Ende, Steven McPherson, among others. The course also studied Fair Trade recycled artwork as an inspiration. The following are select comments reflecting creative outcomes (Figs. 22.2 and 22.3).

Materials: *2 VGA computer monitor cables, 1 PC power supply cable, 2 Coaxial Cables (varying length), ≈ 200 zip tie cables*

Fig. 22.2 *Computer Cable Basket* by a non-art major student in Art Appreciation, Spring 2014

Description: *This is a basket created using various found cables and zip ties. It stands about 5 inches from base to rim. The opening is about 8" in diameter, sloping down to an approx. 4" diameter at the base. The construction resulted in a surprisingly sturdy piece that should be serviceable for many years. It performs all the standard functions of a basket including: collecting, containing, and isolating varying objects of a small to medium size. I will probably use it to hold keys, or in what I find a humorous turn, small cables, adapters, and other small electronics. The basket is primarily black in color due to the cables from which it was created. The cable ends also contribute chrome, blue, and even purple. These colors come from the materials themselves and add a nice bit of contrast. These zip tie ends protrude around the outside of the basket and give it an interesting studded look. The cable ends also protrude from the basket at various points and are incorporated into the outside structure.*

Reflection: *When I saw this project in the syllabus I knew that I wanted to use some of the cables I have collected over the years. The idea of introducing order to the "snake nest" of random cables in my closet was an amusing thought to me. I wanted to use whole cables, not cutting, or modifying them in any way, I think part of the fun of this piece is seeing how the ends of the cables peeking out from the structure, and trying to see where one cable begins and another ends. I began assembly using an old yogurt container as a guide to ensure I got a nice circular shape for my first few rows. I was unsure how to proceed initially, however a pattern soon emerged and the work went more quickly than I had expected. I have to say this project came out much better than I had expected. I am quite proud of this piece. I took it to work and had it on my desk briefly; it greatly hampered my productivity as people kept asking about how I had made it. It was a real joy to transform everyday clutter into something I will appreciate and use for many years. What a fun project.*

Fig. 22.3 *Ski Stool* by a non-art major student in Art Appreciation, Spring 2014

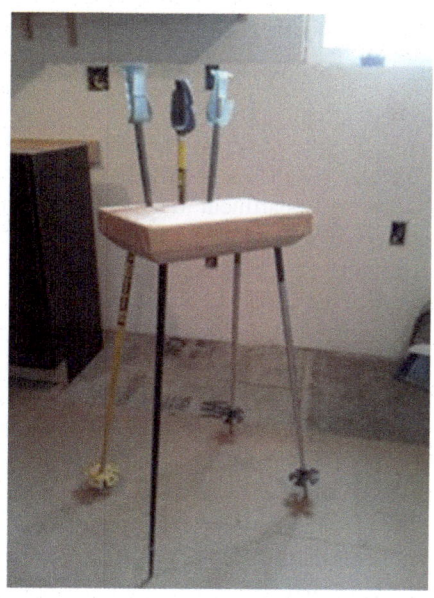

Reflection: *My* Junk to Funk *project started out by chance really. I took some old ski poles and cut them up and built a bar stool with an old piece of lumber I had laying around from my kitchen remodel project. It was a pretty easy project to complete but I would contribute that to already having all of the tools around. I have always been a fan of the repurposed outdoor gear based furniture and as this project came along I found myself with the opportunity to build my own. This stool has a bigger purpose though and I will not be keeping it. My neighbor who donated the ski poles will be getting the stool back as a gift. He can no longer ski and is battling cancer, I am hoping that he can use the stool while he enjoys his new hobbies and maybe it can remind him of some great memories.*

Construction of the stool was pretty simple. I started by beveling the bottom side of the seat just to add some character and line to the piece. I then sanded it until my heart was content, and then a little extra. All that was left was to bore some ski pole sized holes and then hammer the cut lengths of ski pole into the seat. I added a screw to each leg (pole) just to be sure they wouldn't slip out. I then added some of the remaining ski handles to create a back.

Junk to Funk has been implemented as an ongoing project in the UAA art education program. For the past 6 years, it has inspired over 1000 art majors and students who are not art majors to look at recycled materials differently and encourage artistic practices on daily basis using common materials. The goal is a continued awareness of waste consumption and our environment.

22.3   Winter Design Project: A Creative Practice of Sustainable Art

In the spring of 2011, Professor Timo Jokela from the University of Lapland and Professor Glen Coutts from the Institute for Northern Culture in Finland invited the UAA Art Department to become one of the founding members of University of the Arctic (UArctic) Thematic Network on Arctic Sustainable Arts & Design (ASAD). The goal of the thematic network is to promote cooperation and collaboration between universities, institutions, and communities focused on northern and arctic issues in the disciplines of arts, design, and visual culture. Its objectives are to identify and share contemporary and innovative practices in teaching, learning, research, and knowledge exchange in arts, design and visual culture education.

To continue the promotion of sustainable art on campus, planning for the *Winter Design Project* began in late 2013. It was a pilot project designed to transform a winter outdoor space through broad campus participation. To create a "winter design" project, UAA faculty members were invited to create a one-course assignment for their students within their discipline to be displayed in Cuddy Quad during the UAA Winterfest[3] celebration. Cuddy Quad is an open area with two main crosswise walking paths connecting classroom buildings from each side. Many student activities occur in the Quad during the spring, summer, and fall terms. However, it becomes very quiet during the winter months. Consequently, a goal of the project was to create a common space in the winter that could be a performance, an installation, or an interactive or participatory experience.

With administrative support from Facilities & Campus Services, Student Life and Leadership, Dining Services, University Advancement, the Office of International and Intercultural Affairs, the Office of Sustainability, and the Center for Community Engagement and Learning, along with help from individual faculty members, staff and students, the project successfully transformed the usually empty quad into an interactive winter outdoor playground using a sustainable medium.

There were over 55 staff and faculty and more than 250 students directly involved in the creation of the 2014 *Winter Design Project* (Fig. 22.4). A daily average of 100 people visited Cuddy Quad to experience the display, either by making a snowman, creating graffiti on snow, or taking photos of campus life. It was evident that the variety of these educational experiences enabled the faculty and students to participate in and transform a campus outdoor space (Gonzales 2014, UAA Green & Gold News 2014a, b). Many faculty members were able to design and incorporate a class assignment within their disciplines and, most importantly, to engage in a participatory experience in an under-utilized winter outdoor space. The *Winter Design Project* required much time and effort, but the collaborative energy produced a greater interest among faculty and students about core sustainability issues and allowed them to further explore how to create a more participatory experience on campus during wintertime.

---

[3] See a list of events and programs of the UAA Winterfest, http://www.uaa.alaska.edu/sll/cpb/winterfest.cfm

Fig. 22.4 Variety of ice and snow displays including a small colored igloo, a snow sculpture, and a group of dancing snowmen for the 2014 *Winter Design Project*

In addition, through the UArctic ASAD Thematic Network, international guests were invited from Nesna University College in Norway. Three Nesna faculty members in natural sciences and art, Mette Gårdvik, Wenche Sørmo and Karin Stoll, traveled to UAA with three students, specifically to be a part of the *Winter Design Project*. They presented a public lecture, and organized a snow sculpture workshop about the Norwegian education system, the environment, building snow sculptures, and student engagement. The students documented their experience though ALASKAPOST—Amazing stories being written every day![4]

The project has had a number of significant outcomes. On a professional level, it strengthened and expanded participation with the UArctic ASAD Thematic Network and created opportunities to collaborate with colleagues throughout UAA. On an educational level, it enabled UAA students to better understand sustainable, creative displays in the broader community. It provided a unique interdisciplinary and international learning opportunity for students and faculty from other fields. Technically, the project demonstrated the use of creative displays to encourage a deeper learning and understanding of our own outdoor winter environment. Finally, it provided UAA faculty members and students with an opportunity to explore and create an outdoor winter space, and to look at "ice and snow" from a new perspective. This has inspired further creative solutions using a truly sustainable medium.

22.4   Reflections on Personal Teaching

During the past 6 years, my personal teaching and learning have been impacted by my involvement in creating, developing, and implementing the *Junk to Funk* art education program, and the *Winter Design Project*. By submitting grant applications, working on new course development, project design and execution, and cultivating community partnerships, this process has enhanced my teaching paradigm beyond the traditional classroom setting. Most importantly, it has contributed to greater community involvement and environmental awareness.

Community engagement projects require time and commitment. Planning began at least one semester and in some cases a full year prior to the semester when the course was taught. Clear communication with students at the beginning of the

---

[4] Read blog entries posted by students from Nesna University College in Norway, http://alaska-posten.blogspot.com/

semester about the scope and expectations of the project was critical. In addition, the project needed to be aligned with semester scheduling and allow for flexibility, since many activities would take place outside regular class meeting times.

Multiple online resources were used for communication, sharing documents, and organizing weekly schedules. In 2008, we used *Epsilen*, a course learning management system, to document the learning experience. We built a public course site[5] that included an introduction to the project, and its activities and events. Each student was assigned one section of the site and they were individually responsible for reporting the event, including **creating** a photo journal. Students could use these websites for highlighting their academic accomplishments in the project. Recently, we began to use a Facebook page to disseminate information about a *Junk to Funk* fundraising event. Also, the *Junk to Funk Project* was featured in *Northern Light*, the UAA campus student newsletter (Mauigoa 2012).

In addition, the UAA Community Engaged Student Assistants (CESAs) program at the Center for Community Engagement & Learning awarded tuition credit waivers to assist faculty members involved in community-engaged projects, and community partnership development at the beginning of the semester. Chelsea Klusewitz, a CESA recipient, demonstrated her leadership skills and commitment to service learning on campus and in the community through the project. She presented her work in a poster presentation at the annual Community Engagement Forum in the spring semester of 2013. Other coverage of the 2014 *Winter Design Project* was posted on the UAA web site to promote the event (Gonzales 2014) (Fig. 22.5).

Fig. 22.5 *Faculty Spotlight* on UAA Green and Gold, Feb 26, 2014

---

[5] See the *Junk to Funk* course website, http://www.epsilen.com/crs/8673

22.5   Conclusion

Grounded in educational theory and practice, the *most* meaningful outcome of *Junk to Funk* and the *Winter Design Project* was engaging students in hands-on learning experiences focused on a theme of global significance and environmental conscience. The programs gave students a strong foundation of "best practice" in community or environmentally based teaching and learning, and cultivated their artistic practice for future endeavors. The projects also provided university faculty members with an opportunity to fully connect with the Anchorage community, the satisfaction of serving others through sustainable art, and opportunities to make connections with issues related to the North; they reinforced the benefits of collaborative effort directly related to artistic expression. Hopefully, these experiences will provide students with a lasting foundation in art education that will have a positive impact in their understanding of and shared responsibility for the environment.

References

Dewey, J. (1934). *Art as experience* (repr.). Carbondale: Southern Illinois University Press, 1987.

Gonzales, J. (2014, February 19). *Reimagining a winter campus during Winterfest 2014*. Retrieved from http://greenandgold.uaa.alaska.edu/blog/21837/

Greenmuseum. (2010). *What is environmental art?* Retrieved from http://greenmuseum.org/what_is_ea.php

Hicks, L. E., & King, R. J. H. (2007). Confronting environmental collapse: Visual culture, art education, and environmental responsibility. *Studies in Art Education, 48*(4), 332–335.

Mauigoa, N. (2012, November 19). Junk to Funk fundraiser offers treasure made from Trash, *UAA The Northern Light.* Retrieved from http://www.thenorthernlight.org/2012/11/19/junk-to-funk-fundraiser-offers-treasure-made-from-trash/

Petersen, J. (2008). A green curriculum involves everyone on the campus. *The Chronicle of Higher Education, 54*(41), A25.

Smith, D. (2008, August 3). *Recycling into art – Green-minded Alaskans revel in the reusable.* Anchorage Daily News (AK) (Final ed.), D1. Retrieved 15 April, 2013, from NewsBank online database (Access World News).

UAA Green and Gold News. (2014a, February 27). *Winter design project extends thanks to all contributors*. Retrieved from http://greenandgold.uaa.alaska.edu/blog/22286

UAA Green and Gold News. (2014b, February 26). *Faculty spotlight: Herminia Din*. Retrieved from http://greenandgold.uaa.alaska.edu/blog/22085/

UAA Junk to Funk Facebook Page. (2014). Retrieved from https://www.facebook.com/junkto-funk.uaa

Ulbricht, J. (1998). Changing concepts of environmental art education: Toward a broader definition. *Art Education, 51*(6), 22–24 + 33–34.

Chapter 23
Meaning and Means of "Sustainability": An Example from the Inuit Settlement Region of Nunatsiavut, Northern Labrador

Rudolf Riedlsperger, Christina Goldhar, Tom Sheldon, and Trevor Bell

Abstract A diverse body of literature discusses the importance and application of concepts related to sustainability in the Arctic and Subarctic, with a considerable portion of scholarship being developed outside of Northern regions. However, rather than applying external definitions of sustainability to the Arctic and Subarctic, it is important to recognize Northern Indigenous methodologies and epistemologies, including inherently sustainable worldviews or philosophies and locally grounded tools, processes, or strategies to address sustainability challenges. We present a case study that highlights the relevance of Inuit approaches to sustainability transformation. *SakKijânginnatuk Nunalik* (the Sustainable Communities initiative, or SCI) is located in the autonomous Inuit region of Nunatsiavut, Labrador. The SCI informs best practices and provides guidance for community sustainability in the coastal Subarctic under changing environmental, social, and economic conditions. Its overarching goal is to ensure individual and community well-being in climate adapted communities. We discuss the preliminary successes and challenges of the initiative and conclude with an outlook on how approaches to meet sustainability challenges in the Arctic and Subarctic can contribute to non-Northern sustainability research and concepts.

Keywords Inuit • Nunatsiavut • Sustainability challenges • Sustainability transformation • Co-creation of sustainability • Sustainability indicators

R. Riedlsperger (✉) • T. Bell
Department of Geography, Memorial University of Newfoundland, St. John's, NL, Canada
e-mail: r.riedlsperger@mun.ca

C. Goldhar
Nunatsiavut Secretariat, Nunatsiavut Government, Nunatsiavut, NL, Canada

T. Sheldon
Department of Lands and Natural Resources, Environment, Nunatsiavut Government, Nunatsiavut, NL, Canada

© Springer International Publishing Switzerland 2017
G. Fondahl, G.N. Wilson (eds.), *Northern Sustainabilities: Understanding and Addressing Change in the Circumpolar World*, Springer Polar Sciences, DOI 10.1007/978-3-319-46150-2_23

23.1   Introduction: Sustainability and Sustainable Development in the Canadian Subarctic

The *SakKijânginnatuk Nunalik* (Sustainable Communities) initiative (SCI) is a transdisciplinary and community focused initiative aiming to inform best practices and to provide guidance for enhancing community sustainability in Nunatsiavut, northern Labrador. The SCI aims to improve the quality of life for residents in Nunatsiavut through addressing current concerns around housing, food, energy, and community development and planning, among other areas. The initiative is grounded in indigenous methodologies and epistemologies that support community-based tools, processes and strategies to address contemporary sustainability challenges. It promotes a holistic approach that takes into account social, cultural, environmental, and economic aspects of sustainability.

This paper is written from the perspective of four researchers and administrators who are closely involved with the project. We discuss the beginnings of the SCI and the early implementation phases of sub-projects currently underway. We give an overview of the successes and challenges of the initiative, before concluding with an outlook on how the challenges may be overcome, and how the SCI may provide guidance for building sustainable communities elsewhere in the North.

Northern regions are subject to significant environmental, social, and cultural changes, which are in part are driven by processes and pressures originating elsewhere, including climate change, environmental pollution, global economic processes, and intense resource competition (Bock 2013). These changes cause various challenges. For example, climate change affects individual and community livelihoods through its impact on sanitation and water facilities, food security, transportation infrastructure, and the prevalence of infectious disease (Parkinson 2010). Environmental pollution causes contaminants to accumulate in the Arctic, threatening the safety of fauna and humans who depend on wildlife for food (Muir and deWit 2010). Resource developments cause environmental degradation and pollution and social stress within some Arctic and Subarctic communities (Parlee and Furgal 2012). Other detrimental effects of environmental change are more subtle, including social and cultural transformations that increase vulnerability to environmental and economic pressures. Examples include feeling a loss of control, the disruption of cultural continuity, a weakening of local knowledge systems, and a loss of the social capital necessary to thrive and survive within Arctic and Subarctic communities (Ford 2012; see also Crate's chapter in this volume). Furthermore, substance abuse, domestic violence, child abuse, suicide, and unintentional injury are associated with rapid cultural and social change (Parkinson 2010).

Most Northern regions are looking for long-term solutions that nurture the well-being of humans and their environments, while also placing an emphasis on sustained economic growth (Bock 2013). In sustainability science, a discipline that strives towards achieving sustainable societies, such processes pertaining to the transition of unsustainable to sustainable states and dynamics are referred to as "sustainability transformation" (Komiyama and Takeuchi 2006; Miller et al. 2014).

One approach for framing sustainability transformation is through the implementation of sustainable development, a concept made widely known by the World Commission on Environment and Development, also referred to as the "Brundtland Commission". Its report, "Our Common Future," published in 1987, conceptualized sustainable development as development that "seeks to meet the needs and aspirations of the present without compromising the ability to meet those of the future." (Brundtland et al. 1987: 51).

The Brundtland report provides a framework that brings together the natural or environmental limits of development and the potential for new directions in social development contained within those natural limits (Chance and Andreeva 1995). The incorporation of local knowledge, and the participation of local residents in the achievement of sustainable development are central tenets of the approach. However, at least on a global scale, action on sustainability has consistently lagged behind society's concern over sustainability (Nel and Ward 2015). There is substantial criticism on the implementation of sustainable development, in part alluding to a realization that local engagement or participation may be limited in practice, and thus far may have produced little in the way of real change. Instead of providing a means for the local pursuit of social and natural well-being, Graf (1992: 553) argued that sustainable development, as advocated by the Brundtland Commission, "vindicates the hegemony of the classes and interests, which are the present beneficiaries of the international economic order". In other words, those advocating sustainable development may not be primarily concerned with the well-being of humans and their environment, but with the prolongation of the status quo, which prioritizes economic performance and growth. Similarly, Crate (2006: 295) noted that the Brundtland report "confirms a dominant, western top-down economic worldview that bases ecosystem management on generalized prescription rather than specific contexts". Referring to the top-down problematic of implementing sustainable development strategies, Keith and Simon (1987: 209) warned early on that sustainable development would be rejected and fail in the Arctic if it allowed the "[e]xclusion of local peoples from the decision-making process for both development and conservation initiatives."

If an important aspect of sustainability pertains to how communities envision and pursue social and natural well-being (Miller et al. 2014), then a crucial factor in Northern sustainability is to recognize that its inhabitants have a right to drive their own sustainable futures. As Hugh Brody succinctly observed: "What must be defended is not the traditional as opposed to the modern but, rather, the right of a free indigenous people to choose the components of their lives" (cited in McCannon 2012: 256). Indeed, researchers have demonstrated that Northern sustainability may best be expressed through or begin with community goals and visions, and their importance for individual and community livelihoods (Kruse et al. 2004; Crate 2006). Among the enabling factors for accomplishing locally and regionally meaningful sustainability transformation is the increasing political self-organization of indigenous communities, which brings capacity and decision-making opportunities back to the North, and empowers communities and regions to address sustainability challenges first hand (Southcott 2009).

Importantly, however, Crate (2006) warns that without actions, visioning sustainable futures remains nothing more than a theoretical exercise. Our paper discusses how sustainability transformation may be fostered by gearing efforts to reflect the interests and perspectives of local residents. Using the example of the SCI, we reflect upon what is meant by "sustainability" in a particular Northern context, and illustrate how it might be conceived and acted on to achieve well-being. Among others, the paper may be of interest to researchers whose primary focus or audience relates to policy development and program delivery.

23.2 *SakKijânginnatuk Nunalik*: The Sustainable Communities Initiative

23.2.1 *Nunatsiavut: Homeland of the Labrador Inuit*

Located within the province of Newfoundland and Labrador, Nunatsiavut is part of *Inuit Nunangat* (Inuit homeland), representing one of the four autonomous Inuit regions of Canada. Prehistoric cultures occupied the region for thousands of years, while modern Inuit descended from Thule Inuit migrating from Alaska toward Greenland in the 1400s (Wenzel 2009). Settlers first arrived in Labrador in the early nineteenth century as professional trappers, fishers, and seal-hunters. They did not live in organized communities, but instead set up houses along the bays and inlets of the Labrador coast. Before then, in the late eighteenth century, the Moravian church began establishing missions in the region (Ben-Dor 1966). Prior to the establishment of Nunatsiavut, the two main governing bodies in the region were the Moravian church and the Governments of Newfoundland, which included the separate Dominion of Newfoundland until 1927, the British controlled commission government of Newfoundland from 1927 to 1949, and the Province of Newfoundland and Labrador from 1949 to 2005 (Anderson 2007). Nunatsiavut achieved the right to self-government through the Labrador Inuit Land Claims Agreement (LILCA) in 2005. Thirty years of negotiations between the Labrador Inuit Association and the federal and provincial governments preceded this outcome (Nunatsiavut Government 2012).

As a polity, Nunatsiavut operates on two interacting scales. The Nunatsiavut Government (NG) is the regional Inuit consensus-based democratic government. In addition, municipalities or Inuit Community Governments (ICGs) were established for each of the five Nunatsiavut communities: Nain (the administrative capital), Hopedale (the legislative capital), Postville, Makkovik, and Rigolet (Fig. 23.1). These communities have a total population of about 2500, 90 % of whom are beneficiaries to LILCA (Statistics Canada 2012a, b, c, d, e). Beneficiaries include all residents of Inuit descent and the *Kablunangajuit* of Nunatsiavut. The latter term is an Inuktitut word meaning "resembling a white person" and includes non-Inuit residents formerly referred to as settlers (Natcher et al. 2012). All *Kablunangajuit* who

Fig. 23.1 Map of Labrador including Nunatsiavut, which is indicated by the shaded regions on this map and comprises Labrador Inuit Settlement Areas (*medium gray*) and Labrador Inuit Lands (*dark gray*). Map produced by Charles Conway, Department of Geography, Memorial University, 2014

have lived in the region since before 1940, or who were born before 1990 and have ancestors who lived in the region since before 1940 can apply for beneficiary status. The term *Kablunangajuit* is commonly used to encompass all beneficiaries, including those who reside outside of Nunatsiavut. As of October 2014, the total number of beneficiaries was about 7200 (Nunatsiavut Government). They form the electorate of Nunatsiavut (Felt 2011).

23.2.2   Situating SakKijânginnatuk Nunalik *Within Provincial and Federal Legacies*

Before Nunatsiavut came into being in 2005, decision-making for the region largely took place outside of Labrador in the provincial and federal capitals of St. John's and Ottawa, respectively (Anderson 2007). As a result, Labrador Inuit had limited opportunity for meaningful participation within decision-making processes. Provincial and federal legacies also left a mark on Nunatsiavut, as illustrated here with three indicative examples related to military contamination, housing pressures, and food security.

The Second World War led to a growing militarization of the global Arctic and Subarctic, which in North America reached its full extent during the Cold War with the installation of the Distant Early Warning (DEW) Line and countless air force bases and military sites operated by the United States (McCannon 2012). The north coast of Labrador was of particular strategic military importance due to its proximity to Greenland and Europe. Negative environmental and social impacts of these sites are still felt today. In Hopedale, the operation of a US air force base (initially established in 1953 as part of the Pinetree Line and finally closed in 1968) is linked to areas of buried debris and soil contamination (ESG 2012). As a result, certain areas within the municipal boundaries remain closed for subsistence activities, such as hunting and berry picking, and expensive environmental monitoring and remediation projects are necessary to ensure the safety of residents (Aivek STANTEC 2014). The exclusion areas have also led to extreme pressure on available building land in the community.

Similar to indigenous groups in other parts of the Arctic and Subarctic, Labrador Inuit were resettled in the 1950s and 1960s "not exactly by force, but neither with their full agreement" (McCannon 2012: 257). The effects of these resettlements on Labrador Inuit have been documented by Ben-Dor (1966), Zimmerly (1975) and Kennedy (1982), among others. In some communities, such as Makkovik, populations doubled almost overnight, leading to immense housing, subsistence and social pressures. While perhaps not as pronounced today, residential housing shortages have remained an on-going concern in the region. Throughout Nunatsiavut, infrastructure deficits affect housing, as demonstrated by high levels of overcrowding, mould, and repair and plumbing problems (NG Regional Housing Needs Assessment 2012). Contributing variables include expensive yet inappropriate housing design that was intended for climates that are not subject to the same intense freeze-thaw cycles common in northern coastal environments (Goldhar and Sheldon 2014). Finding strategies to provide affordable, durable, and culturally appropriate housing is therefore an important contemporary challenge in Nunatsiavut.

Food security is an important component of socio-economic health. Throughout *Inuit Nunangat*, subsistence activities contribute significantly to the food security of individuals and communities (Ford 2009). Natcher et al. (2012) note that over 80 % of Nunatsiavut residents participate in subsistence activities, both directly (as hunters and gatherers) and indirectly (as recipients of country foods). Residents commonly cannot obtain the quantities of food necessary to sustain a household (Egeland 2010). An Inuit health survey conducted in 2008 and 2009 found that 44 % of homes in Nunatsiavut were food insecure. The survey defined food insecurity as an inability to access a sufficient amount of healthy calories. Egeland further found that 16 % of households were severely food insecure, as characterized by disrupted eating patterns and reduced food intake. This is twice the national average for Canada (8 %). Reasons cited by Egeland include households without hunters and the prohibitive costs of gasoline and ammunition that are necessary today to "go off" on the land and ice, the local term for act of engaging in subsistence activities.

Similar to Southcott's (2009) observations for other parts of Northern Canada, the establishment of the Nunatsiavut Government brought capacity, in the form of human and physical capital, and the opportunity for decision-making power in the region. The Sustainable Communities Initiative (SCI) is a concrete outcome of this capacity. Founded in 2012, the SCI is led by the Nunatsiavut Government and rooted in the communities. The Joint Management Committee of the Nunatsiavut Government (which includes the *AngajukKât* – community leaders or mayors from each of the communities – among other members) and Nunatsiavut's Executive Council guide the direction of the SCI, thereby fostering local and regional representation (Goldhar et al. 2013). At the same time, the SCI is what Trencher et al. (2013) identify as a co-creation of a sustainability initiative: a project that involves various actors ranging from university to government to the private sector, with the aim of transitioning society from unsustainable to sustainable conditions. The SCI incorporates all of these stakeholders. Its overarching goal is to inform appropriate practices in community planning and development while providing guidance for community sustainability under changing climatic, socioeconomic, and environmental conditions. To accomplish this, the SCI seeks to put into practice locally developed tools, guides, and strategies that are adapted to the changing climatic and socioeconomic realities of Nunatsiavut (Goldhar et al. 2013).

The SCI holds as its foundation the importance of processes that are locally appropriate and that reflect Inuit philosophies and epistemologies, which are similar to approaches to envisioning sustainability being applied in other parts of the North (Healey and Tagak 2014). Specifically, all SCI processes are based upon principles that include transparency, respect, accountability, collaboration, and holisticness. These guiding principles were decided upon by the Joint Management Committee of the Nunatsiavut Government (Table 23.1; Goldhar et al. 2012). Their application is also expected from initiative partners, including community planners and architects, and university researchers. The Nunatsiavut Government Research Advisory Committee, chaired by the regional Inuit research advisor, is in charge of evaluating their application (Nain Research Centre n.d.).

Table 23.1 Guiding principles of the Nunatsiavut Government's Sustainable Communities Initiative

Transparent to community leaders, decision-makers and community members
Respectful of Inuit values, individual thoughts, community contexts, and priorities
Accountable to all residents of Nunatsiavut communities
Collaborative with community members, regional decision-makers, governments, industry stakeholders and university partners.
Holistic in its sustainability approach to consider the impacts of today's decisions on future generations.

Adapted from Goldhar et al. (2012)

Table 23.2 Emerging sustainability challenges in Nunatsiavut

Sustainability challenge	Description
(1) Infrastructure, housing, development	To enhance design, durability, cultural appropriateness, environmental suitability, and life span of the built environment.
(2) Food security	To support healthy families through improved access to diverse country and market foods that are affordable and high in quality
(3) Energy security	To improve access to and reliability of energy (diesel, oil, and wood supply) and support alternative/renewable energy and energy efficiency
(4) Transportation and emergency services	To improve critical transportation and emergency infrastructure, including airports and wharfs; establish public transportation in larger communities
(5) Safe communities	To advance human health and support a healthy environment by addressing concerns related to water, dust, contaminated sites, diesel generators, quarries and garbage dumps in and around communities.
(6) Valued spaces and places	To protect natural spaces, important buildings and landmarks, trails and roads (both traditional and modern), native vegetation and water bodies.

23.2.3   Understanding Sustainability Challenges of Nunatsiavut

The existence of various sustainability challenges, notably those listed in the previous section, are no secret to *Nunatsiavummiut*, the people of Nunatsiavut. To better understand the scope and interdependencies of these challenges, however, a series of workshops titled "Learning from the Coast" (LfC) invited community members to talk about community priorities, challenges, opportunities, and visions for the future (Goldhar et al. 2012). The workshops were organized by the *AngajukKât* and facilitated by principal research partner Trevor Bell. Approximately fifty community members attended a total of five workshops, from which six themes on sustainability challenges in the region emerged (Table 23.2).

Infrastructure, housing, and community development were among the priority concerns for Nunatsiavut communities. In addition to the housing problems described above, the lack of culturally appropriate housing design, including the layout of houses and the size and location of building lots, inhibits well-being and quality of life in Nunatsiavut (Goldhar et al. 2012). For example, inadequate design limits the ability of Labrador Inuit to prepare, cook and store country foods (for similar challenges in Nunavut see Tester 2009).

Nunatsiavummiut also expressed concern about the availability of both country and store-bought foods, and promoted the development of community gardens, community freezer programs, and local food cooperatives to help support those in need and to improve food quality and diversity (Tester 2009).

Energy insecurity was identified as an additional source of stress. Limited energy capacity in all Nunatsiavut communities was a significant topic for workshop participants (Goldhar et al. 2012). Wood heat is the preferred source in the region (NG Regional Housing Needs Assessment 2012) and when access to firewood is compromised residents struggle to heat their homes. This is especially the case in mild winters when poor snow and ice conditions limit travel (Riedlsperger 2013). Firewood can be accessed relatively easily in the southern communities (Rigolet, Makkovik, and Postville), but residents in northern communities, where trees become less abundant, have to travel further to access wood (Riedlsperger 2013). While the use of woodstoves is cause for political debate in some Arctic and subarctic communities and towns due to air pollution (e.g., Fairbanks, Alaska; see Murphey 2013), such issues have not yet been documented in Nunatsiavut. This may be due to low population numbers and the coastal location of the communities. However, in the recent past individuals who preferred electric heat sources encountered real and perceived barriers because of a lack of electrical capacity in the communities (NG Regional Housing Needs Assessment 2012).

Challenges surrounding firewood illustrate transportation and emergency services as priority needs, including physical infrastructure such as trails, roads, wharfs, and airstrips, but also access to medical transport and medical personnel in the communities. Transportation infrastructure is linked to energy security through winter trail access to firewood sites (see above), and to food security through trail access to subsistence areas (Goldhar et al. 2012).

Safe communities were envisioned as built spaces that support human health and foster a healthy environment by addressing inadequacies related to water, dust, contaminated sites, diesel generators, quarries and garbage dumps in and around communities. Recreational facilities such as skating rinks, ballparks and playgrounds are also included as they offer youth safe spaces to play and grow.

To address some of the sustainability challenges, participants from all communities wanted to become more involved in community planning to ensure that local goals and values were respected and considered during decision-making processes. Specifically, residents wanted to preserve valued places, spaces and activities that comprised an integral component of quality of life in their communities and that in large part depends on outdoor and subsistence activities.

Table 23.3 Examples of ongoing SCI projects in Nunatsiavut

Project	Description
InosiKatigeKagiamik Illumi: Healthy homes in Nunatsiavut	Providing appropriate models for sustainable housing in Nunatsiavut
DISC: Digital Information System for Communities	Community mapping to aid municipal planning and craft development regulations
Adaptation planning for food security and climate change	Building resilience to ensure food availability, accessibility, and quality to positively influence the health of Nunatsiavummiut
Aullak, sangilivallianginnatuk: Going off, growing strong	Teaching subsistence skills and knowledge to youth
Regional energy strategy	Developing a sustainable energy strategy that meets changing demands for communities
SmartICE: Sea ice monitoring and real time information for coastal environments.	Generating observations of changing sea ice conditions combined with user-based satellite image classification of sea ice conditions
Sustainable water and wastewater management	Developing tools and community based strategies for water and wastewater management

23.2.4   Towards Sustainable Communities in Nunatsiavut

To address these sustainability challenges, the SCI is implementing various projects that explore solutions and establish best practices to foster resilient and sustainable coastal Nunatsiavut communities (Table 23.3). A key step in addressing these sustainability challenges was to recognize the complexity and interconnectedness of many of the community needs and hence the need to adopt a holistic approach in designing multiple projects. For example, the lack of available building land, in combination with inadequate housing design, results in overcrowding, which in turn may trigger physical and mental health problems (Ford 2012; Bourque and Cunsolo Willox 2014). A holistic approach means that sustainability challenges and solutions are evaluated for their environmental, cultural, social, and economic dimensions. An important early impact of the SCI is the adaptation of a framework by the Nunatsiavut Government Executive Council that allows government departments to effectively support cross-departmental programs and initiatives. The following sections highlight some of the most prolific projects at the moment, which address adaptation planning for food security and climate change, climate change hazard assessment and building constraint mapping, the development of an Inuit-based sea-ice classification system for on-ice travel, and the development of tools and strategies for sustainable water and wastewater management (Goldhar and Sheldon 2014).

23.2.4.1 InosiKatigeKagiamik Illumi: Healthy Homes in Nunatsiavut

Long-term solutions to current housing challenges must be approached through a holistic lens in order to be effective, as community development stresses are inherently complex. Examples of current housing problems in the region include poor construction, infrastructure failure, increased costs, overcrowding, and negative effects on physical and mental health. These challenges have in some cases led to outmigration, both somewhat voluntarily (if better housing options arise) and through the lack of alternatives (as is the case with senior citizens in need of assistive care). To address the precarious housing situations in the communities, a program called *InosiKatigeKagiamik Illumi* (Healthy homes in Nunatsiavut) aims to construct culturally relevant, affordable, climate adapted housing, thereby contributing positively to the mental and physical health and well-being of residents (Goldhar and Sheldon 2014). This program is directly contributing to, and a part of, the development of a broader Nunatsiavut housing strategy that aims to provide affordable and appropriate housing to all residents of the region.

23.2.4.2 DISC: Digital Information System for Communities

Housing and building land concerns are closely related, as functional houses depend on physically and culturally suitable areas of land for construction. The DISC (Digital Information System for Communities) component of the SCI addresses questions of appropriate building land in communities. At the core of DISC is the production of planning constraint maps that identify available, suitable areas for residential or commercial development across a range of land uses under current and projected future climate states. Each community database compiles digital information on community infrastructure and resources, landscape characteristics and hazards, regulated land areas, protected and valued spaces and places, climate scenarios and environmental modeling (Lee et al. 2014, 2015, also see Baikie et al. 2013). The planning constraint maps combine existing community information with Inuit knowledge gathered through participatory mapping based on map-biography methods (see Tobias 2009) and new geoscientific data in a georeferenced information database to support community infrastructure planning and development decisions. DISC is providing databases accessible to Inuit Community Governments to allow staff to access and update spatial information relevant to their communities.

23.2.4.3 Adaptation Planning for Food Security and Climate Change

Adaptation planning to ensure food security includes: the creation and support of community and regional networks of food security actors; the collection, collation, and synthesis of existing information on food security in Nunatsiavut; and the documentation and review of existing food supporting landscapes and seascapes in Nunatsiavut (Furgal 2014). Subsistence is about both obtaining country foods and maintaining social relationships (Wenzel 2013). SCI has developed a program that addresses the socioeconomic as well as the sociocultural aspects of food security: the *Aullak, sangilivallianginnatuk* (Going off, growing strong) program pairs at-risk youth with experienced hunters and elders from the communities to learn about land skills and Inuit knowledge. Youth spend time traveling, fishing, hunting, trapping, learning about cultural traditions and customs, building *komatiks* (sleds) and smokehouses, and delivering and sharing country foods with elders. There are early indications that the program is indeed improving the resilience of youth participants (Hirsch et al. 2014).

23.2.4.4 SmartICE: Sea Ice Monitoring and Real Time Information for Coastal Environments

Sea ice, an integral part of Inuit livelihoods, is fundamental in Nunatsiavut for subsistence, recreational, and commercial or industrial purposes. SmartICE (Sea Ice Monitoring And Real Time Information for Coastal Environments) addresses pressures associated with deteriorating snow and ice conditions that affect safe travelling and navigation on the ice through integrating adapted technology, remote sensing, and Inuit knowledge to promote safe travel in the region. The three main elements of SmartICE include *in situ* sensors to measure sea ice conditions, satellite imagery to map sea-ice conditions, and information technology to integrate information and generate outputs useful for various user groups, ranging from ice navigation managers to Inuit ice experts and recreational ice users. SmartICE, represents a Nunatsiavut Government-community-university-industry collaboration with a vision to provide a sea ice information service for Inuit by Inuit (Briggs et al. 2014; Bell et al. 2014).

23.2.4.5 Regional Energy Strategy and Sustainable Water and Wastewater Management

Two important aspects of safe communities relate to energy security and water safety. SCI is currently in the planning stages of an energy security strategy that seeks to ensure all communities have access to a sufficient, affordable source of sustainable energy to meet residential and economic development needs. The strategy seeks to: help the region to better project the short and medium term energy demand trends and requirements; document the impacts of current energy sources and constraints on social, environmental, and economic conditions; and identify options for energy demand reductions (Goldhar and Sheldon 2014). To address water safety, SCI is undertaking a project to develop tools and methodologies for the planning and design of sustainable water and wastewater management systems in Nunatsiavut. The underlying rationale is that sustainable community development in Nunatsiavut depends on the provision of safe drinking water. In Rigolet the cost of servicing a single lot for residential housing has risen to over $200,000

(2013 values) due mainly to the requirements for water distribution and wastewater collection and disposal. High costs for water and wastewater infrastructure development limit the ability of the local and regional governments to adequately provide for other needs, such as appropriate wastewater management practices.

The novelty in the approach undertaken by the SCI to address these problems lies in the proactive inclusion of community members in the development of adaptation strategies. This is achieved through participatory activities that, for example, let community members rank and identify locally appropriate infrastructure design criteria. The overall aim is to develop and validate approaches, methodologies and tools that can be applied to the planning and design of sustainable water and wastewater infrastructure in all Nunatsiavut communities (Gordon and Farahbakhsh 2014). Specific objectives include the evaluation of current opportunities and risks, the identification of best management practices in northern communities, the development of water management and decision making tools, and the development of adaptation strategies and scenarios (Gordon and Farahbakhsh 2014).

23.3 Discussion and Outlook

SakKijânginnatuk Nunalik strives to be an example of sustainability transformation, where the notion of "sustainability" is inextricably linked to current local priorities, individual and community visions for the future, and Inuit philosophies and worldviews. The SCI shows that meanings of sustainability must not necessarily be derived from southern ideas and/or governments that are distant in terms of geography and mindset. Instead, the meanings of sustainability as expressed by *SakKijânginnatuk Nunalik* ought to stem from individuals and communities and be directly applicable to their immediate environments. Similarly, the means to accomplish sustainability transformation should be rooted in local knowledge and local skills that result in local strategies with the aim of yielding immediate, applicable, and noticeable results.

Within academic circles, the SCI has already received recognition. Plenary and individual presentations and panel discussions at arctic conferences including the ArcticNet annual science meeting (Vancouver 2012, Halifax 2013, Ottawa 2014) and the International Congress of Arctic Social Sciences (Prince George 2014) were well received. For its knowledge mobilization and action plan, the SCI was a recipient of the 2013 Arctic Inspiration Prize (NLEC 2013). The prize is awarded for

multidisciplinary teams who have made a substantial, demonstrated and distinguished contribution to the gathering of Arctic knowledge and who have provided a concrete plan and commitment to implement their knowledge into real world application for the benefit of the Canadian Arctic, its Peoples and therefore Canada as a whole. (Arctic Inspiration Prize n.d.)

Such recognition is important as it provides a boost in morale for those involved with the SCI, and as the Arctic Inspiration Prize involves a sizeable cash award that supports their action plan on healthy homes in Nunatsiavut.

Despite its successes on the ground and in more distant academic circles, the SCI currently lacks awareness at the community level, as was apparent during a recent visit (August 2014) to two Nunatsiavut communities. While turnout and enthusiasm for local community mapping sessions contributing to the DISC project were good, mapping participants were not necessarily aware of the existence of the SCI or its overall goals. Neither were community-based researchers directly involved with specific sub-projects aware of the larger context of their important work. To move forward with and implement the projects listed above, the SCI currently strongly relies on a small group of key leaders from the Inuit Community Governments and the regional Nunatsiavut Government. For an initiative that is rooted in local communities this is a serious problem. While the SCI employs a mixed top-down/bottom-up approach, it may be useful to contemplate shifting more strongly towards the latter, to ensure the sustainability of the initiative. Scholars in other parts of the world have noted that sustainability initiatives do rely strongly on popular support or buy-in to ensure their prolonged effectiveness (Fraser et al. 2006).

There are two main additional reasons why it is important to create on-the ground awareness of the SCI in Nunatsiavut. First, even residents who may not be actively involved with or aware of the SCI at the moment may benefit from the overall initiative or a specific subproject in the future. It is important that residents who experience sustainability related problems know that an initiative exists that they can consult. Second, the guiding principles of the SCI (Table 23.1) imply the obligation of the initiative to proactively reach out to the people of Nunatsiavut. In order to remain accountable, the SCI must hear about concerns, including those of residents who may not be supportive of the initiative or specific subprojects. This can only be accomplished through visibility and dialogue. With respect to its guiding principles, the integration of Inuit values is indeed crucial for the SCI. However, there is a need to gain a better understanding as to how the realities of the SCI reflect local indigenous knowledge in action. While some projects, such as "Going off – growing strong", provide positive examples, concerns related to limited engagement with community members may indicate that, going forward, local indigenous knowledge needs to be more carefully considered and operationalized. Some Inuit regions, Nunavut in particular, have captured the importance of specific indigenous philosophies through *Inuit Qaujimajatuqangit (IQ)*. *IQ* can be understood as a reflection of Inuit cultural expectations arising after the Nunavut Land Claims Agreement, taking the form of guiding principles for the Government of Nunavut (Wenzel 2004). *IQ* is a concept in progress. It emerged as an attempt to move beyond the narrower concept of traditional ecological knowledge. Tester and Irniq argue that

> IQ can bring together generations of Inuit in a common challenge. That challenge is to hold in check relations that seriously threaten Inuit culture and, in so doing, put before us relationships between and among people, animals, and landscapes relevant to all of us that might otherwise be absorbed by a very different, totalizing logic. (2008: 59)

Similar to approaches taken in other parts of *Inuit Nunangat*, Nunatsiavut and the SCI may benefit from making their conceptualization of Inuit values more explicit. It is important to acknowledge that, despite its intentions, the research methods the SCI employs have their inherent biases and worldviews, which may not always be in line with Inuit values the region may wish to promote. To prevent research myopia, the actual levels of engagement with community members and the integration of Inuit values need to be evaluated on an ongoing basis.

These challenges prompt several key questions for the SCI to address going forward. How can we assess or track the direction of the SCI and its sub-projects, and evaluate the rate of change and success? How can we ensure responsiveness to the priorities, needs, visions, concerns, and values of community members? And how can we continue to explore complex, interrelated questions in an integrative, interactive manner?

One approach to address current challenges is through the implementation of a sustainability indicator framework. Such frameworks could help to measure phenomena important to the SCI and its sub-projects (for example, the number of single unit dwellings; proficiency in Inuktitut; subsistence harvesting characteristics), thereby gathering valuable data (Larsen et al. 2010). Indicators also aid in tracking the direction and rate of change (for example: how many youth engage in subsistence activities in any given year?), and they can become tools in informing policy and guiding decision-making (Fraser et al. 2006).

Sustainability indicators have been developed in local contexts, albeit at this point mostly outside of the Arctic (for example see Holden 2006). Exceptions include the Arctic Social Indicators project (ASI), which aimed at crafting sound indicators that allowed the tracking of trends in human development (Larsen et al. 2010). The ASI made use of domains such as health and population, cultural well-being and cultural vitality, contact with nature, and fate control, among others. These domains were populated with indicators in order to detect trends of development. As Rydin et al. (2003) noted, creating successful indicators relies on their integration into processes of governance. In other words, indicators have to be relevant for policy making. Hence, the intent of a second ASI project is to allow for the integration or implementation of arctic social indicators in community based monitoring programs (Larsen et al. 2015).

Sustainability indicator programs can enhance our overall understanding of environmental and societal problems in Nunatsiavut and can facilitate capacity building through the concerted engagement of community members with community challenges that goes beyond the collection of anecdotal evidence. These programs would benefit the SCI greatly in terms of staying on track and being able to feel the pulse of local residents. In other words, using sustainability indicator programs would benefit the balance between community level and regional governmental level steering of the initiative, which in turn may translate into policy and the development of purposeful projects. Not least, a sustainability indicator framework may open up the SCI for a more formal or rigorous comparison with sustainability transformation projects in other regions. While the specifics of such a program as part of

the SCI are being worked out, initial feedback from the Inuit Community
Governments is positive.

23.4 Concluding Remarks

From a global perspective, "sustainability" may be described primarily as a reaction
to the negative effects of western industrialized capitalism, going back as far as the
Industrial Revolutions of Western Europe and North America in the late eighteenth
and nineteenth centuries respectively. We note that this period was characterized by
the reorganization of manufacturing and labour to allow for the mass production
and shipment of goods, and accompanied by population increases and the growth of
towns. The era also observed public reactions, spurred by intellectuals of the time,
to the deterioration of living conditions, the destruction of the environment, and
social ills detrimental to individual health and well-being, including increasing
alcoholism and the suppression of leisure (de Vries 1994). The inherent tensions
between "economic efficiency criteria, social justice and the equitable distribution
of wealth, and environmental conservation in perpetuity" were well identified then
(Lumley and Armstrong 2004: 71). "Sustainable development" as proposed by the
Brundtland Commission in the late 1980s and promoted by the "Earth Summit" in
Rio in the early 1990s continues to grapple with these constant tensions. As outlined
above, some question the effectiveness of a philosophy rooted in the same processes
that cause the problems it is trying to solve. The sustainability initiative discussed in
this paper offers a fresh perspective on sustainability and sustainable development
in a specific northern setting. We hope that those working towards sustainability
transformation outside of the Arctic and Subarctic may benefit from lessons learned
in northern communities through initiatives such as *SakKijânginnatuk Nunalik,*
where local approaches are being used to address problems that extend far beyond
their locality.

Acknowledgements The authors would like to acknowledge the *AngajukKât* of Nunatsiavut
Tony Andersen (Nain), Wayne Piercey (Hopedale), Diane Gear (Postville), Herb Jaque (Makkovik)
and the former *AngajuKak* of Rigolet, Charlotte Wolfrey. Dr. Chris Furgal and Team (Trent
University) and Dr. Khosrow Farahbakhsh and Team (University of Guelph) provided invaluable
contributions to the SCI. The SCI would not be possible without support from the Aboriginal
Affairs and Northern Development Canada (AANDC) Climate Change Adaptation Program, from
Health Canada's Climate Change and Health Adaptation Program and from the Government of
Newfoundland and Labrador. The authors would further like to thank Arctic-FROST (PLR #
1338850) for funding meetings and workshops that proved highly valuable for writing this paper.

References

Aivek STANTEC. (2014). *Additional delineation and updated remedial action plan, former U.S. military site, Hopedale, Labrador*. Prepared for the Newfoundland and Labrador Department of Environment and Conservation. Retrieved from http://www.env.gov.nl.ca/env/env_protection/ics/hopedale/hopedale_updated_rap_1.pdf

Anderson, K. (2007). Influences receding Nunatsiavut self-determination: Historical, political and educational influences on the people of Northern Labrador, Canada. *Australian Journal of Indigenous Education, 36*, 101–110.

Arctic Inspiration Prize. (n.d.). About the prize. Available at: http://www.arcticinspirationprize.ca/prize/about.php.

Baikie, C., Riedlsperger, R., Hatcher, S., Bell T. & Lee, C. (2013, December 9–13). *Digital information systems for community planning in Nunatsiavut*. Poster presented at the ArcticNet annual scientific meeting, Halifax, Nova Scotia.

Bell, T., Briggs, R., Bachmayer, R. & Li, S. (2014, September 14–19). *Augmenting Inuit knowledge for safe sea-ice travel: The SmartICE information system*. Presentation for the Oceans '14 MTS/IEEE Conference, St. John's, Newfoundland.

Ben-Dor, S. (1966). *Makkovik: Eskimos and settlers in a Labrador community: A contrastive study in adaptation*. St. John's: Institute of Social and Economic Research, Memorial University of Newfoundland.

Bock, N. (2013). Sustainable development considerations in the Arctic. In P. Berkman & A. N. Vylegzhanin (Eds.), *Environmental security in the Arctic Ocean* (pp. 37–57). Dordrecht: Springer.

Bourque, F., & Cunsolo Willox, A. (2014). Climate change: The next challenge for public mental health? *International Review of Psychiatry, 26*(4), 415–422.

Briggs, R., Laing, R., Angnatok, J., & Bell, T. (2014). Development of an Inuit-based sea-ice classification for on-ice travel. In C. Goldhar & T. Sheldon (Eds.), *SakKijânginnatuk Nunalik: Understanding the risks and developing best practices for sustainable communities in Nunatsiavut, Phase 2 Report* (pp. 50–61). Nain: Nunatsiavut Government.

Brundtland, G. H., Khalid, M., Agnelli, S., et al. (1987). *Report of the World Commission on Environment and Development: Our common future*. New York: United Nations.

Chance, N. A., & Andreeva, E. N. (1995). Sustainability, equity, and natural resource development in northwest Siberia and Arctic Alaska. *Human Ecology, 23*(2), 217–240.

Crate, S. A. (2006). Investigating local definitions of sustainability in the Arctic: Insights from post-Soviet Sakha villages. *Arctic, 59*(3), 294–310.

De Vries, J. (1994). The industrial revolution and the industrious revolution. *The Journal of Economic History, 54*(02), 249–270.

Egeland, G. (2010). Inuit health survey 2007–2008 Nunatsiavut. International Polar Year Inuit Health Survey: Health in transition and resiliency.

Environmental Sciences Group. (2012). *Buried debris assessment, Hopedale, Labrador*. Kingston: Royal Military College.

Felt, L. (2011). *Land claims agreements and Aboriginal governance issues in Labrador: The Nunatsiavut experience*. Northern Policy Papers. Published by Action Canada.

Ford, J. D. (2009). Dangerous climate change and the importance of adaptation for the Arctic's Inuit population. *Environmental Research Letters, 4*(2), 1–9.

Ford, J. D. (2012). Indigenous health and climate change. *American Journal of Public Health, 102*(7), 1260–1266.

Fraser, E. D., Dougill, A. J., Mabee, W. E., Reed, M., & McAlpine, P. (2006). Bottom up and top down: Analysis of participatory processes for sustainability indicator identification as a pathway to community empowerment and sustainable environmental management. *Journal of Environmental Management, 78*(2), 114–127.

Furgal, C. (2014). Adaptation planning for food security and climate change in Nunatsiavut. In C. Goldhar & T. Sheldon (Eds.), *SakKijânginnatuk Nunalik: Understanding the risks and developing best practices for sustainable communities in Nunatsiavut, Phase 2 Report* (pp. 6–15). Nain: Nunatsiavut Government.

Goldhar, C., & Sheldon, T. (Eds.). (2014). *SakKijânginnatuk Nunalik: Understanding the risks and developing best practices for sustainable communities in Nunatsiavut, Phase 2 Report.* Nain: Nunatsiavut Government.

Goldhar, C., Bell, T., Sheldon, T., Andersen, T., Piercy, W., Gear, D., Wolfrey, C., Jacque, H., Furgal, C., Knight, J., Kouril, D., Riedlsperger, R., & Allice, I. (2012). *SakKijânginnatuk Nunalik: Understanding opportunities and challenges for sustainable communities in Nunatsiavut, Learning from the coast.* Nain: Nunatsiavut Government.

Goldhar, C., Bell, T., & Sheldon, T. (2013). *Learning from others: Recommendations for best practices in adaptation of the built environment to changing climate and environment in Nunatsiavut.* Nain: Nunatsiavut Government.

Gordon, A., & Farahbakhsh, K. (2014). Development of tools and strategies for sustainable water and wastewater management in Nunatsiavut. In C. Goldhar & T. Sheldon (Eds.), *SakKijânginnatuk Nunalik: Understanding the risks and developing best practices for sustainable communities in Nunatsiavut, Phase 2 Report.* Nain: Nunatsiavut Government.

Graf, W. D. (1992). Sustainable ideologies and interests: Beyond Brundtland. *Third World Quarterly, 13*(3), 553–559.

Healey, G., & Tagak, A., Sr. (2014). Piliriqatigiinniq 'Working in a collaborative way for the common good': A perspective on the space where health research methodology and Inuit epistemology come together. *International Journal of Critical Indigenous Studies, 7*(1), 1–14.

Hirsch, R., Angnatok, D., Winters, K., Pamak, C., Sheldon, T., Furgal, C., & Bell, T. (2014). *Going off to grow strong? Towards a framework for understanding and improving community resilience in Nain, Nunatsiavut.* Research snapshot. Retrieved at http://www.ideas-idees.ca/sites/default/files/bell_en.pdf

Holden, M. (2006). Revisiting the local impact of community indicators projects: Sustainable Seattle as prophet in its own land. *Applied Research in Quality of Life, 1*(3–4), 253–277.

Keith, R. F., & Simon, M. (1987). Conservation with equity: Strategies for sustainable development. In *Sustainable development in the northern circumpolar world* (pp. 209–225). Gland: IUCN.

Kennedy, J. (1982). *Holding the line: Ethnic boundaries in a Northern Labrador Community* (Social and economic studies No. 27). Institute of Social and Economic Research. Memorial University of Newfoundland.

Komiyama, H., & Takeuchi, K. (2006). Sustainability science: Building a new discipline. *Sustainability Science, 1*(1), 1–6.

Kruse, J., White, R., Epstein, H., Archie, B., Berman, M., Braund, S. R., et al. (2004). Modeling sustainability of arctic communities: an interdisciplinary collaboration of researchers and local knowledge holders. *Ecosystems, 7*(8), 815–828.

Larsen, J., Schweitzer, P., & Fondahl, G. (Eds.). (2010). *The Arctic Social Indicators report.* Copenhagen: Nordic Council of Ministers.

Larsen, J., Schweitzer, P., & Petrov, A. (Eds.). (2015). *The Arctic Social Indicators report II – Implementation.* Copenhagen: Nordic Council of Ministers.

Lee, C., Bell, T., & Riedlsperger, R. (2014). Climate change hazard assessment and building constraint maps: Digital information systems for community planning in Nunatsiavut. In C. Goldhar & T. Sheldon (Eds.), *SakKijânginnatuk Nunalik: Understanding the risks and developing best practices for sustainable communities in Nunatsiavut, Phase 2 Report* (pp. 16–49). Nain: Nunatsiavut Government.

Lee, C., Bell, T. & Riedlsperger, R. (2015). *Digital information systems for community planning, Part Two: Nain, Rigolet, Makkovik, and Postville.* Report prepared for Aboriginal Affairs and Northern Development Canada.

Lumley, S., & Armstrong, P. (2004). Some of the nineteenth century origins of the sustainability concept. *Environment, Development and Sustainability, 6*(3), 367–378.

McCannon, J. (2012). *A history of the Arctic: Nature, exploration and exploitation.* London: Reaktion Books.

Miller, T. R., Wiek, A., Sarewitz, D., Robinson, J., Olsson, L., Kriebel, D., & Loorbach, D. (2014). The future of sustainability science: a solutions-oriented research agenda. *Sustainability Science, 9*(2), 239–246.

Muir, D. C., & de Wit, C. A. (2010). Trends of legacy and new persistent organic pollutants in the circumpolar arctic: Overview, conclusions, and recommendations. *Science of the Total Environment, 408*(15), 3044–3051.

Murphey, K. (2013, February 16). Fairbanks area, trying to stay warm, chokes on wood stove pollution. *Los Angeles Times.* Retrieved from: http://articles.latimes.com/2013/feb/16/nation/la-na-fairbanks-air-pollution-20130217

Nain Research Centre. (n.d.). Research process. Retrieved from http://nainresearchcentre.com/research-process

Natcher, D. C., Felt, L., McDonald, J., & Ford, R. (2012). The social organization of wildfood production in Postville, Nunatsiavut. In D. Natcher, L. Felt, & A. Procter (Eds.), *Settlement, subsistence, and change among the Labrador Inuit: The Nunatsiavummiut experience* (pp. 171–188). Winnipeg: University of Manitoba Pres.

Nel, W. P., & Ward, J. D. (2015). Towards a rational sustainability framework. *Sustainability Science, 10*, 1–6.

NG Regional Housing Needs Assessment. (2012). Nunatsiavut Government, Nain, Newfoundland and Labrador. Retrieved from http://www.nunatsiavut.com/wp-content/uploads/2014/03/Nunatsiavut-Indicators-Results-Overview.pdf

NLEC (Newfoundland and Labrador Department of Environment and Conservation). (2013). Provincial government supports sustainable communities initiative. Press release. Retrieved from http://www.releases.gov.nl.ca/releases/2013/env/1230n06.htm

Nunatsiavut Government. (2012). *Nunatsiavut Strategic Plan 2012–2015.* Nain: Nunatsiavut Government.

Parkinson, A. J. (2010). Sustainable development, climate change and human health in the Arctic. *International Journal of Circumpolar Health, 69*(1), 99–105.

Parlee, B., & Furgal, C. (2012). Well-being and environmental change in the Arctic: A synthesis of selected research from Canada's International Polar Year program. *Climatic Change, 115*(1), 13–34.

Riedlsperger, R. (2013). *Vulnerability to changes in winter trails and travelling: A case study from Nunatsiavut.* MA thesis, Memorial University of Newfoundland, Department of Geography.

Rydin, Y., Holman, N., & Wolff, E. (2003). Local sustainability indicators. *Local Environment, 8*(6), 581–589.

Southcott, C. (2009). The socio-economic impacts of climate change and sustainable development in Canada's northern communities. In *Climate change and Arctic sustainable development: Scientific, social, cultural and educational challenges* (pp. 174–185). Paris: UNESCO.

Statistics Canada. (2012a). Makkovik, Newfoundland and Labrador. Census profile. 2011 Census. Statistics Canada Catalogue no. 98–316-XWE. Ottawa. Released February 8, 2012.

Statistics Canada. (2012b). Postville, Newfoundland and Labrador. Census profile. 2011 Census. Statistics Canada Catalogue no. 98–316-XWE. Ottawa. Released February 8, 2012.

Statistics Canada. (2012c). Nain, Newfoundland and Labrador and Newfoundland and Labrador. Census profile. 2011 Census. Statistics Canada Catalogue no. 98–316-XWE. Ottawa. Released October 24, 2012.

Statistics Canada. (2012d). Rigolet, Newfoundland and Labrador. Census profile. 2011 Census. Statistics Canada Catalogue no. 98–316-XWE. Ottawa. Released October 24, 2012.

Statistics Canada. (2012e). Hopedale, Newfoundland and Labrador. Census profile. 2011 Census. Statistics Canada Catalogue no. 98–316-XWE. Ottawa. Released October 24, 2012.

Tester, F. J. (2009). Iglutaasaavut (our new homes): Neither "new" nor "ours" - housing challenges of the Nunavut Territorial Government. *Journal of Canadian Studies/Revue d'études canadiennes, 43*(2), 137–158.

Tester, F. J., & Irniq, P. (2008). Inuit Qaujimajatuqangit: Social history, politics and the practice of resistance. *Arctic, 61*, 48–61.

Tobias, T. (2009). *Living proof*. Vancouver: Ecotrust Canada.

Trencher, G. P., Yarime, M., & Kharrazi, A. (2013). Co-creating sustainability: cross-sector university collaborations for driving sustainable urban transformations. *Journal of Cleaner Production, 50*, 40–55.

Wenzel, G. W. (2004). From TEK to IQ: Inuit Qaujimajatuqangit and Inuit cultural ecology. *Arctic Anthropology, 41*(2), 238–250.

Wenzel, G. W. (2009). Canadian Inuit subsistence and ecological instability – If the climate changes, must the Inuit? *Polar Research, 28*, 89–99.

Wenzel, G. W. (2013). Inuit and modern hunter-gatherer subsistence. *Études/Inuit/Studies, 37*(2), 181–200.

Zimmerly, D. (1975). Cain's land revisited: Culture change in Central Labrador, 1775–1972. Social and economic studies No. 16. Institute of Social and Economic Research. Memorial University of Newfoundland.

Index

© Springer International Publishing Switzerland 2017
G. Fondahl, G.N. Wilson (eds.), *Northern Sustainabilities: Understanding and Addressing Change in the Circumpolar World*, Springer Polar Sciences, DOI 10.1007/978-3-319-46150-2

Printed by Printforce, the Netherlands